ELEMENTARY STATISTICS: AN APPLIED APPROACH

ELEMENTARY STATISTICS

AN APPLIED APPROACH

Neil R. Ullman
COUNTY COLLEGE OF MORRIS

JOHN WILEY & SONS, New York • Chichester • Brisbane • Toronto

Library of Congress Cataloging in Publication Data:

Ullman, Neil R

Elementary statistics.

Bibliography: p.
Includes index.
1. Statistics. I. Title.
HA29.U37 519.5 77-10828
ISBN 0-471-02105-9

Printed in the United States of America

10 9 8 7 6 5

PREFACE

I have pursued this book with one major goal in mind: to present and clarify the purposes and concepts of statistics to people who have little mathematical background.

In my teaching and dealings with students and workers in diverse fields I have observed a widespread awe and fear of statistics. The first encounter with statistical reasoning is often traumatic, even for those with fairly good mathematical preparation. However, a statistical awareness and understanding is becoming more and more necessary to a well-rounded education.

This textbook should be able to be used with a minimum of mathematics. The statistical work will mainly involve arithmetic. Some algebra is used to simplify the presentations. A separate chapter on probability is also provided, but the main thrust of that chapter is the relationship of probability to statistical reasoning.

I have observed that a major problem expressed by students in a first statistics course is their realization that it is not an exact study. They anticipate a mathematics course that is precise; they end up with a course in "uncertainty," with an emphasis on reasoning and decision making. I have presented these ideas of statistics in a way that the average nonmathematical student can understand. I have found that this manuscript and the textbook from which it was largely derived, *Statistics: An Applied Approach*, have been easily understood and used by a wide range of people, including students in areas from mathematics to law enforcement and humanities, faculty members studying for graduate degrees, and secretaries.

The organization of the textbook is designed to be flexible. Although the first three chapters should be covered in that order, most of the remaining chapters can be studied without other previous reading. Thus, Chapters Four, Five, Six, Eight, Eleven, Thirteen, and parts of Chapters Seven and Twelve can be read in any sequence. Although I have placed the discussion of the normal distribution (Chapter Five) before the binomial distribution (Chapter Six), there is no need to follow this order (except to defer the normal approximation until completing the study of the normal distribution or to rely on the empirical rule). The only other major sequential chapters are that Chapter Ten (tests involving the binomial distribution) assumes completion of Chapter Six, and that Chapters Five, Seven, and Eight, should be completed before Chapter Nine (on tests involving the normal distribution) is attempted.

Most of the problems require calculations, and I have provided a sufficient number of problems in all chapters that can be performed without calculators. Additional problems are also provided with more realistic numbers for which a calculator becomes important. Several special topics and some more difficult problems are also presented and noted with an asterisk (*).

I thank the many students, faculty members, and reviewers who have contributed numerous corrections, suggestions, and ideas for the organization and details of this textbook.

I am grateful to the Literary Executor of the late Sir Ronald A. Fisher, F.R.S., to Dr. Frank Yates, F.R.S. and to Longman Group Ltd., London, for permission to reprint Tables III and VII from their book *Statistical Tables for Biological, Agricultural and Medical Research* (sixth edition, 1974).

Neil R. Ullman

CONTENTS

INTRODUCTION

Copyright © 1974 United Feature Syndicate, Inc.

The use of statistics has become so common today that even poor Charlie Brown cannot escape from them. Look how just one single sentence can summarize so much so simply. A whole summer of baseball is lumped into one number, and yet all of us can feel the terrible frustrations that number represents.

And so it often is that a single "statistic" can tell us so much.

But Linus still has more disturbing information to report.

Copyright © 1974 United Feature Syndicate, Inc.

Statisticians like to think that they can assist others in finding solutions to questions and problems. Apparently Lucy isn't interested. How easy it is to find excuses or to read the results of a study the way you want it in the first place. Scientists, researchers, all inquiring people, must keep their minds open.

You must have a willingness and a desire to understand and to use the tools available to you. Statistical principles and methods will be one such set of valuable and useful tools. Use them wisely and they will serve you. Ignore them and you will miss opportunities and waste time and effort. Unfortunately, you must also be conscious of how not to use them poorly—because then you may come to wrong decisions.

Copyright © 1974 United Feature Syndicate, Inc.

Obviously, we can devise situations for all types of questions. But we must be careful that our information is accurate and complete. If Linus had overlooked the bubble gum he might have missed the most important item that Snoopy has been consuming.

I already said that the ultimate goal of all the work of our statistician is usually to help us in arriving at decisions. Charlie Brown is attempting to understand his problem. Unfortunately, he is pretty much alone.

Copyright © 1974 United Feature Syndicate, Inc.

I sometimes wish it were so easy. Just remember, though, that changing statisticians doesn't eliminate the problem of the team.

Most people are engaged in serious work. One part of many of their jobs is the gathering of data, or at least the studying of some data. But we do not gather data for fun. We hope to gain as much meaningful information as possible from our work.

As I have already said, the use of statistical methods is one important way that we can hopefully gain such meaningful information. I hope that you will embark on this subject with enthusiasm and interest.

CHAPTER ONE
AN OVERVIEW

Pick up a copy of a large daily newspaper and carefully look through it. You will usually find a wide array of articles that present the results of different studies. A typical assortment might include:

A survey that reports 60% of the population of the United States is dissatisfied with the way the government is handling the economy.

An advertiser claiming miles per gallon increased because of a new additive in the gasoline.

One article presenting arguments that "increases in a liver disease may be related to marijuana smoking," while another report describes the results of birth control efforts in a developing country.

We also find a continual deluge of statistics about unemployment, consumer prices, athletics, and so on. Very little of our everyday lives is not touched on, at least partly, or examined somehow with statistics.

Even more important is the fact that a large part of somebody's job is to gather information or data, evaluate or interpret it, and then present it for others to read and use. More and more people in virtually every field are involved in these activities. Even if you will never prepare such material, you will inevitably need to read, understand, and evaluate some amount of statistical information. Therefore, you should try to gain a clear understanding of the *concepts* of statistics.

1.1 NON-MATHEMATICAL DESCRIPTIONS

All too often the first question that comes to mind as you open this textbook and begin to study statistics is, "why bother?" This chapter provides some initial understanding of when we need statistical methods.

Every one of us has a need to provide descriptions. For example, I ask you to describe an egg. Let us try by using words, comparisons, and pictures.

Words

You might start by saying that an egg is a smooth, oval, or spheroidal-shaped object. Here, we are describing the object with a collection of words that are merely symbolic representations or condensations of the actual object. *Provided you and I both understand the words in the same way*, we can convey to each other an accurate image of the real thing. The set of words thus serves as a *substitute* for the object. The substitute never fully replaces the object, but often it is the only adequate way of transmitting information regarding the object.

You may claim that the above description is not complete, and you would be correct. Words alone can never fully *replace* the object. We can expand the definition from four words to 40 or even 400 words by enlarging on details and by including descriptions of additional aspects of an egg. Perhaps we would introduce the fact that it consists of an outer layer or shell and that inside this shell is contained a fluid of varying composition. Of course, this description could still be elaborated on and it would still be inadequate as a *complete* description. We will, however, have a very useful substitute, one that can be carried around completely independent of the real object, one that will not break and make a mess, one that can fit in my head as well as in a book.

Although words are useful tools, the immediate problem with them is in the area of interpretation of the words. Different people derive different meaning from the same words. We must be aware not only of the incompleteness of the description, but also of the *different ways in which different people interpret individual words*. The complete set of words can combine to give quite a diverse set of images different for each person. Remember, you just have the description "a smooth, oval, or spheroidal-shaped object, having an outer layer or shell and containing a fluid of varying composition." From this description you have been asked to construct an image of an object. Will it be an egg? Try to improve the description.

Comparisons

From the preceding we see that the direct use of words to replace an object is limited. Try to keep these limitations in mind as you read. They will recur later in a different setting—with numbers.

We can add another dimension to our description if we establish some standards, some item that we will mutually agree on as a *basis for comparison*. We can then compare our object to the chosen standard. For instance, we may select as standards two objects with which both of us are equally familiar. The two objects selected may be a marble and a 100-watt light bulb. We could further state that we will select a marble of a particular size and a light bulb of a particular brand name and catalog number, thereby establishing more exact standards.

In our future work we will find that we will need standards for comparison and that they will serve as important aids in describing many problems. We will also encounter the notion of exactness in setting these standards. We will state at times that the standard is a theoretical value.

We may say, for example, that the marble is much smaller than the egg. The light bulb is much larger. We observe that a marble is a perfect sphere. What about the egg? It is generally longer in one direction than in another, so it is not nearly as spherical as the marble. The light bulb is even more oddly shaped, with a concave section.

Let us stop and look at what I have attempted to do. I have attempted to make some direct comparison of the shape of several objects by using words only. It has been a rather poor attempt. I ask you to improve on it. Remember that the person you are talking to does not know what an egg looks like. I think you will experience the same kind of difficulty that I have encountered. Once again, we have found words limited in their ability to perform certain operations. We have observed that word descriptions may not be sufficient for complete comparisons.

Pictures There is an old adage that you have undoubtedly heard that goes, "A picture is worth a thousand words." So now let us resort to a picture to gain some further insight or description of our egg.

Figure 1.1*a* is a drawing of an egg; with no additional information the general shape of the egg is fairly obvious. The shape is more easily conveyed with a picture than with the use of the words. The picture, however, is pretty much limited to showing us the general shape or outline and outer surface characteristics. It does not tell us much about the structure or the variation of composition that we know exists.

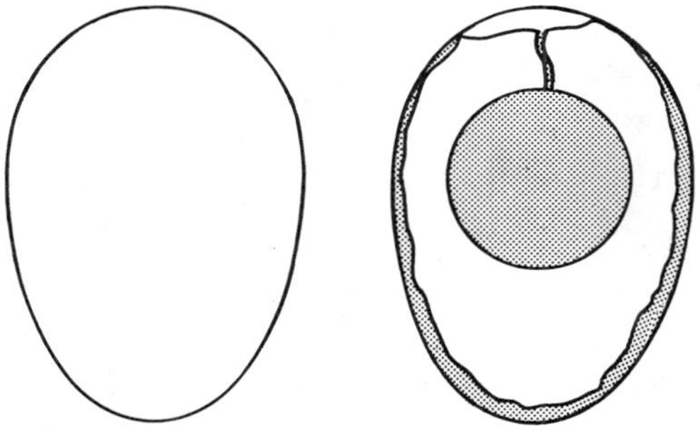

(*a*) (*b*)

FIGURE 1.1 (*a*) A picture of an egg. (*b*) A cross section of an egg showing the shell and some internal structure.

If we refine our drawing or change the type of drawing, we may find that we can describe even more than we did before. If we make a cross-sectional drawing, we can illustrate the shell thickness and some of the internal structure. In Figure 1.1*b*, we have a better image of how the various compositions of liquid are distributed inside the egg, but we have lost some of the information about surface.

So we now see that *no picture is perfect* and *different pictures can be made for the same object*. The different pictures may give different kinds of insight, answer different questions, or provide different descriptions for the object. Thus, we might want a whole collection of pictures, each of which may serve a different purpose or elaborate on a different facet of the object.

What about comparisons? Can pictures be used to make comparisons? Of course. In fact some of the difficulty we ran into before with an egg and a light bulb is almost nonexistent. In Figure 1.2 we have a picture of a marble, an egg, and a light bulb. With this picture the general size and shape relations are easily communicated. You can quickly see certain similarities and differences among these three objects. But again, the picture is limited both in the total scope of information it can possibly convey and in the nature of the questions it can answer.

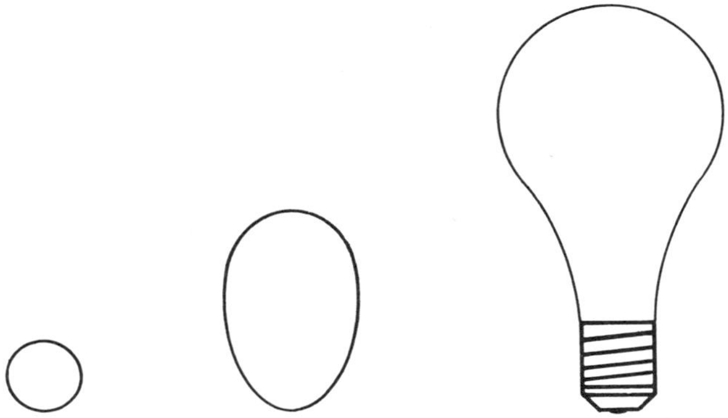

(*a*) (*b*) (*c*)

FIGURE 1.2 Pictorial comparison of (*a*) a marble, (*b*) an egg, and (*c*) a light bulb (100 watt).

AN OBSERVATION

At this point we may be able to stop. We have been asked to provide a description of an object. An attempt was made to explain the characteristics of that object by providing first a verbal or word description and, second, a pictorial description. It *is* possible that these methods of communication are sufficient. If so, then the subject of statistics is not applicable to the problem and we need not go any further. The key word is *possible*. A certain amount of description, comparison, and even inference can be made through the use of the technique we have been discussing. However, our capacity for comprehensive examination of even just one egg is extremely limited. The more detailed and more exact descriptions involved in most of today's world generally requires measurements of some type. We use numerical and mathematical quantities as better ways of characterizing or describing an object.

1.2 DATA

What if I wanted further information and more detailed descriptions about some particular characteristics of the egg. A chemist could provide the chemical composition or the amount of cholesterol in it. A dietician can tell me its protein value. An economist will tell me what it costs. A packaging engineer could tell me its breaking strength. And a market researcher might explain how consumers respond to the quality of eggs they buy, the quantities they use, and the way they use them.

All of these characteristics, and many more, can generally be described through the use of numerical values of some type. Thus a particular egg might be found to be of grade A, medium size, contain 200 milligrams of cholesterol and 14% protein, and cost 5.5 cents. A particular person may have consumed four eggs during the past week and regarded them as very good.

Thus we obtain some form of numbers, which will be called data. Data will be regarded as any set of numerical information or individual recordings. Hopefully, the purpose of having these data is for future evaluation and analysis.

Data generally are gathered in one of two primary ways: *counting* and *measuring*.

Counting takes place, for example, in performing surveys, taking a census, or taking an inventory.

We measure all types of things, from the heights, weights, intelligences, and incomes of people to the mass of an atom, the temperature of the sun, and the distance to another galaxy.

EXAMPLE 1.1

Over the course of several weeks the number of speeding tickets issued at radar trap locations was recorded. A total of 112 separate quantities was listed and presented in the form of Table 1.1. We could describe this set of numbers as raw data.

TABLE 1.1 Traffic Tickets Issued

14	17	18	14	12	15	12	10	11	14	11	9	14	13
11	3	12	8	10	13	11	8	6	9	3	7	12	11
13	15	11	11	14	13	19	10	11	9	10	12	13	11
9	10	11	9	9	11	9	6	9	1	2	8	9	2
15	15	9	10	10	12	14	10	12	10	8	13	14	12
4	11	5	8	10	13	2	7	9	9	9	12	10	10
11	8	12	10	9	15	13	9	12	12	10	12	11	17
8	10	8	10	8	9	6	3	6	7	8	9	4	7

Comment on Example 1.1 Because of the way that these values are presented, we are unable to say very much about the data except for very gross approximations. We can observe that all the numbers are not the same. By this statement we recognize that a distribution, a collection of different values, exists. If we are very diligent and carefully examine the data, we might also make a reasonable guess about the nature of the distribution. Later we will discuss some formal ways of reducing this mass of data into some small set of easy-to-use numbers and some meaningful graphs.

Of course, the purpose of this or similar information might be to aid in determining:

1. Efficiency of personnel who issue traffic tickets.
2. Proper timing in the use of a radar trap.
3. Change in location of radar trap or deployment of personnel.
4. An estimate of revenues to be obtained through tickets.
5. Number of speeders on different roads.

EXAMPLE 1.2
A debate arose as to the time it takes to get to school. A route was chosen and one trial was made. The value was 16.5 minutes. What can you conclude? Additional trials were made and all were recorded in Table 1.2.

TABLE 1.2 Time (in Minutes) to Get to School

16.5	16.5	15.1	16.4	15.8
16.1	14.1	15.7	14.3	15.1
13.6	15.0	15.9	16.4	15.6
16.0	16.7	15.2	14.5	15.5

Comment on Example 1.2 What should you have replied to the question presented above? From one piece of data, one individual recording, you should realize that not much can be stated about the next trial run or about the typical length of time it should take. You do know a lot more than you did before you made that reading. However, after you obtained one value, you still do not know how much confidence to place in it. Depending on the car, the weather conditions, the driver, the driving conditions (city or highway), and so forth, we can end up with all types of different values on subsequent tests. Even under identical conditions, *random variations* will disperse any additional values.

When we look at the collection of data that was accumulated, we should try to observe patterns and a distribution. Hence, collections of data convey information; individual observations just give a point of departure.

You might also note that this data could represent diverse items such as: time (in seconds) for a rat to run a maze in a psychological study; number of hours per week of television watching by a group of children; and number of days for the germination of corn seeds. If the numbers did not have to be time, they could represent breaking strength of plastic rods in thousands of kilograms; carbon monoxide levels in blood samples in parts per million; or rate of increase in price of some consumer products.

EXAMPLE 1.3
A survey was conducted to determine certain characteristics and attitudes of college students. The responses of six students to several questions were as follows:

QUESTION	STUDENT					
1. Sex (M, F)	M	F	M	M	F	M
2. Hours worked last week	15	12	1	24	6	21
3. Rating of cafeteria food (10 = excellent; 0 = worst)	5	4	5	7	3	6
4. Where do you eat? (cafeteria, home, lounge, none)	C	L	H	C	N	L
5. On a scale of 0–10 do you approve of the legalization of marijuana? (0 = No; 10 = Yes)	6	8	7	6	2	10
6. If you were to vote would you approve legalizing marijuana? (Yes) (No)	Yes	Yes	Yes	No	No	Yes

These questions illustrate a number of different types of scales with which you may need to work.

In several of the questions the answer that a student gives is not a number but, instead, fits into a *category* or is classified according to an *attribute*. Thus, the first student is male, answers yes to the last question, and eats in the cafeteria. At best, we can identify the number of students who gave each answer: there are four males and two females; four yes and two no; and two cafeteria, two lounge, one home, one none. The raw data are the responses; these summaries then provide numbers. But there is no true scale. We call data of this type *categorical* or *attribute* data, and the scale is generally referred to as a *nominal* scale. Several chapters will deal with data of this type.

Questions 3 and 5 require answers consisting of numbers that fit on a scale. But what kind of scale? Think carefully about those scales. How do you distinguish between a 3 and a 4 or a 9 and a 10 as responses to these questions? We do not usually consider these differences as the same. The first difference is often a toss-up (between a 3 and a 4). But you will usually think hard before going from a 9 to a 10. Scales like these have a definite pattern of numbers that either increase or decrease, and the numbers are related to the quantities or qualities being measured. There is a definite order to the numbers, and the scales are called *ordinal* scales. In this textbook we will not devote much time to data measured on ordinal scales. Almost all of our work with numerical scales will require an additional assumption that will be discussed below. You should, however, be aware of the fact that other scales exist, and you should be prepared to research other techniques in those cases.

Most of the numerical data we will deal with will use scales where all the intervals have the same meaning. The number of hours between 1 and 2 hours, a 1-hour difference, means the same as a difference between 24 and 25 hours. Similarly, when we measure lengths an interval of 1 centimeter is the same whether it is between 4 and 5 centimeters or between 64 and 65 centimeters. A temperature difference of 1 degree is the same from 12 degrees to 13 degrees as from 112 degrees to 113 degrees. This type of scale is called an *interval* scale.

An even "better" scale is one that preserves ratios. For many types of measurements some form of an absolute zero exists, and we are able to say that on scales that have an absolute zero, 100 is twice as big as 50, 10 is twice as big as 5, and 1000 is 10 times as big as 100. For example, there are temperature scales called Rankine and Kelvin scales, where 100° is twice as hot as 50°, and 1000° is 10 times as hot as 100°. However, these ratios are not true for temperature measured on the Fahrenheit or the Celsius scales (100°F is not twice as hot as 50°F). Similarly, we can measure how many inches taller you are than my son, who is 48 inches tall. Thus you might be 30 inches taller than him (or 78 inches) and someone else could be 15 inches taller (or 63 inches). But the first person is *not* twice as tall as the second person. On a ratio scale, however, if we would measure one person as 30 inches (a small child) and another as 15 inches (an infant), the ratio of 2 to 1 does hold true.

Although we have looked at several types of measurements scales, the two kinds we will be working with will be categorical or attribute and interval.

1.3 ERROR

Now we come to one of the main problems with gathering data; this problem is why we need to understand statistics. As long as there is some sort of instrument to measure a quality or quantity, (1) the possibility of error in measurement, and (2) the presence of variation in measurement exist.

These concepts are generally easiest to recognize with instruments that measure physical quantities such as length and mass. For example, consider the measurement of the "length" or longest dimension of an egg. Pick one from your refrigerator. No matter what instrument you will use to measure the egg, you must be aware of its limitations. Perhaps the egg has a length (unknown to you) of about 1.864 inches. If you use a ruler that measures to the nearest 1/16 inch there is no way you can measure a length of 1.864 inches, since the nearest values are 1 13/16 (1.812) and 1 7/8 (1.875) inches. Thus, when I say the egg is a little less than 1 7/8 inches, you have no way of knowing exactly how much less. We will be making an error of at least 0.011 inches. If we use a more precise instrument, one with a finer scale, we could obtain a reading of perhaps 1 55/64 (1.859) inches. But this is still not an exact value. Similarly, if we use a vernier caliper, a micrometer, or any other measuring instrument, we would still have the impossible task of measuring the true absolute size. The true size (which will always remain unknown) may be 1.864379185..... No known instrument can measure this precisely.

To see what I am describing, try measuring the line below. First use markings on a ruler of 1/8, then 1/16, 1/32, and 1/64 inch (or divisions of

1 centimeter, 5, 2, and 1 millimeters). Observe how the reading is different, depending on the scale you use. Which is the best or closest reading you found? How close do you think you are to the true value on each of the different measurements?

A second cause of discrepancy in reading, which might also be called error, I classify as variability. We will encounter variability in several different ways and most of the ways are present with our one egg.

If we remeasure we may have variability from reading to reading.

Why?

If I look at the egg a second time with the ruler, my eyes will probably *not line up exactly the same way as the first time*. Look at Figure 1.3. The first arrow on the right shows the direction I looked the first time. The next time I made an observation my eye was toward the side and the reading was different.

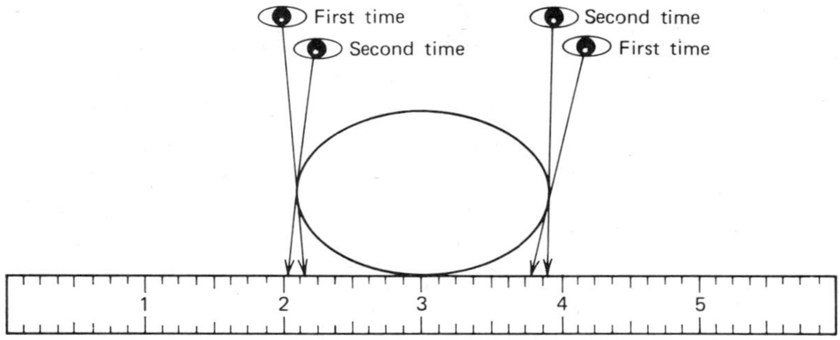

FIGURE 1.3 Trying to measure an egg by eye with a ruler. Two times that you look will probably give two different readings.

With either instrument (ruler or vernier caliper), if another person made the next measurement, it is quite likely to have been done with a slightly different way of seeing and of interpreting the values. The vernier caliper requires a certain amount of feel or touch and judgment in closing against the egg. Even if I took the second reading, I might press a little harder and compress the egg a little, thus getting a smaller reading. Or I might not press as much and get a larger reading. I might read the instrument a little differently and be off by a small amount. Thus, there would be a *variation* from measurement to measurement.

Another, perhaps more obvious notion, is that the egg might not be perfectly uniform. If we chose two slightly different points on the egg as our points to measure between, we quite probably would have different actual, true values. If the instrument is precise enough, it would be affected by this variation. Then repeated measurements would be different, partly because of the *variation in the object* we are measuring.

The problems of error caused by the precision of the instrument and the variability in readings can also often apply to categorical data. This is especially true with questions on a poll or a survey.

Consider, for instance, the question measuring attitude toward cafeteria food (in Example 1.3). If only two possible answers (good, bad) are provided, you can see that this is very imprecise. You have no way of giving a "no opinion" answer, or of saying that the food is perhaps average. If five answers are not permitted (excellent, good, average, poor, terrible), we have a more precise instrument. However, there is still no way of saying "slightly below average." The 10-point scale also given in the example improves on this.

We also can have variation in the response of a single person. You may feel that the answer really belongs somewhere between average and poor, but there is no place to indicate this. One time you might check off average, and another time you would state poor. Or different people having the same attitude would tend to choose different answers. The net effect is that you must expect variations and error in *all* types of data.

There is still an additional cause of variation that is perhaps the most important one: in almost all situations we are dealing with *samples*. It is doubtful that we would ask everyone in the college about the cafeteria food. Ratings of TV shows are conducted every day by keeping track of the programs watched by about 1200 "representative" families. Polls on the popularity, and the like, of the president usually involves less than 2000 people who are considered representative of the attitudes of over 200 million people.

I may check the size of one egg, or a dozen eggs, but surely not 1000 eggs. Even with the first 1000 eggs from a particular chicken or farm, I will not really be able to tell what the next egg will be like. I cannot see into the future and obtain units from the future.

Some types of tests, such as tests involving breaking things, wearing out things, or using things, require us to destroy and use up the item being tested. Obviously, you cannot test all the items, since there would be none left to sell or use in the future. In these cases we must take samples, because it is impractical to test an entire population or collection of individuals. In other situations, such as blood tests, it is unrealistic to examine all of the material. If you don't believe in sampling, we must examine all of your blood instead of a small portion.

The main point is that every time you gather, read about, look at, or evaluate data, you must keep in mind the presence of error and variation. If neither of these were present, then one of the main reasons for the existence of statistics would be eliminated. But as long as they are present, we generally need help in working around the difficulty presented by these uncertainties. Statistics are really just a set of tools that can assist us in making decisions or answering questions. They take into account that there is in fact a certain degree of uncertainty associated with our information.

1.4 STATISTICS

So now we've got some numbers, some data. It may be about the egg, the traffic tickets, the time to get to school, or the questionnaire. *What are you going to do with these numbers*?

This question is critical, since there is no reason to obtain data if we have no idea of what to do with it. To begin, I offer two basic possibilities.

1. You will attempt to *describe* the egg or eggs, the traffic tickets, . . .
2. You will attempt to *infer* something about the egg or eggs, the traffic tickets, . . .

At first you may say that these two statements seem to be the same. They are not. Since the rest of the text will deal with these two concepts, let us briefly examine them.

Descriptive statistics

In the examples in this chapter you have seen some sets of raw data. In order to convey more easily the information contained in the data, we employ a pair of methods to condense or simplify the data. These methods are ways of describing the data itself which, in turn, are ways of describing the things we measured. They fall into two categories of descriptive methods that we previously discussed: *words* and *pictures*. The difference, however, is that the words are now *mathematical* values (numbers or equations), and the pictures are *graphs*.

In Chapter Two we will examine graphs called *bar charts* and *histograms* or *frequency* diagrams. These pictures will provide a way of visualizing where the concentration of measurements lie and to what extent they are distributed, spread around, or differ from one another.

The mathematical "words" are used to *summarize* the data. We calculate quantities, such as various averages, which also help us to describe the data. These quantities are called statistics because they are *derived from the original data*. They are calculated with the use of some or all of the individual values but they are not the actual values. They are called *descriptive statistics* because they are used to describe the set of numbers.

First, they tell us approximately where the *center of the original numbers are*.
Second, they tell us to what extent the numbers are *distributed or spread out around that center*.

These two characteristics are the primary ones we consider important about a collection of data.

The descriptive facet of statistics is very important. Many of the statistics we read about in newspapers, and most of the statistics included in sources such as the federal government's various volumes of statistics and statistical abstracts fall under the category of descriptive statistics. They are merely mathematical summaries of multitudes of numbers. However, most of the other statistical methods rely on such an initial condensation of the data.

Inferential statistics

A second and perhaps more important type of statistical study involves the making of *inferences*. Here we begin to extend our work into more general, broader statements, judgments, or decisions.

In order to understand this aspect of statistics we must begin by recalling our brief discussion of samples. All of the data we have been examining consist of samples. But we must ask very pertinent, practical questions. What *population* are you really interested in? Are you only interested in that one egg? Only that collection of traffic tickets? Just the 16 trips? Or the opinion of only six students?

If you answer yes to any of these above questions, we can stop and deal only with measurement error. But this is not usually the case. The more realistic situation involves a desire to extend your information. You generally will want to answer questions more like: What should I expect eggs as a rule to measure? How many tickets should we expect on the average? How long should trips take in general? What does the entire student body think?

Now you must consider this egg as merely a typical one out of a collection of many eggs. Thus, although you measure this egg and arrive at a value for one of its characteristics, you really want to be making statements—inferences—about an entire population of eggs. Think about the situation. You measured one egg. Is it typical of all eggs? How reliable an estimate do you think you can make of the size of another egg?

Before you really can proceed, you will need to know something about the variability of eggs. You will need to look at a sample of eggs. But there still will remain an amount of uncertainty unless you examine the entire population of eggs. What about this population? I ask again, "What population are you interested in?" You might be concerned with:

The dozen eggs in the box you bought today at the local supermarket.
The two cases of eggs delivered today to the local supermarket.
The eggs delivered by a particular farm.
The eggs laid today by the hen Sally.
The eggs laid this month by the hen Sally.
The eggs laid today by the hens in coop A.
The eggs laid this season by the hens in coop A.
The eggs laid today by a particular variety of hen.
The eggs laid by any type of hen.

Similarly, we must state what population we are interested in when we examine our other samples of data.

Assume that we have a sample from a population that we desire to study. What can we determine? We should first think a little more about the population. Every population has characteristics that we have been discussing. There is variation or a distribution of values within the population. Thus there is a distribution of all sizes of eggs that will be laid by a particular hen. Furthermore, *if* we ever had the complete set of values for the population, we would be able to calculate certain mathematical quantities, such as certain averages, for the entire population. These mathematical quantities are called *parameters* because they describe the true population.

Unfortunately, we almost never have the ability to determine these parameters. But we try to make educated guesses. So we calculate certain types of statistics, or quantities that use our data, to assist in our guesses. We make *estimates* of the true population parameters (the average size for example). After estimating the exact average size of eggs, we recognize that we have only used a sample from the total population. We may have missed large ones or small ones. If we were to grab another egg or another sample of eggs, the new sample undoubtedly would be different. If the sample were different, our estimate of the population would be different. So, we extend our approach to estimating by determining an *interval* within which we believe the actual population value, the particular average, lies. This interval would state, for example, that we believe the population or true size of eggs is between perhaps 2.10 and 2.25 inches. Of course, we would have to calculate these values in some particular way using the sample values. We will discuss the technique and certain important assumptions throughout the textbook. Furthermore, we cannot be absolutely certain that the interval does contain the true value. The statistical methods that will provide the *limit* or ends of the interval will also provide us with a statement about the chance that we are wrong—that the interval does not include the true value. This chance will be expressed in terms of the *probability* of making an error.

The idea of making inferences from our data is really where our main study of *statistical analysis* will begin. The descriptive part of statistics is necessary because in many cases we actually work with the condensed values it provides. We will often manipulate these condensed values instead of the actual data. Thus, we will arrive at newly computed quantities that we also call statistics. Based on these new statistics, we will make statements about the original data and, perhaps even more important, about the *source* of the data.

Quite often it is not the estimate of the population value that is most important. Instead, we may ask if a *change* has occurred in a particular population, or if a difference exists between two populations. We may ask a question of whether feeding a hen different food *lowers* the cholesterol in the egg, if the temperature of the room will cause a significant *difference* in the number of eggs to be laid, or if lowering the price *increases* the sales of eggs. All of these questions and many more are asked in the form of a *hypothesis*. We then obtain data from our samples and try to determine how correct the hypothesis might be. This concept and procedure can be applied to many different types of questions and problems, and we will devote several chapters to these concepts.

1.5 GATHERING DATA

Obviously, there should be a reason for you to gather data, although the determination of just what kind of data and how to proceed may not always be apparent. Most of the considerations of what questions to ask on a survey, what measurements to take on an automobile emission test, and what chemicals to measure in the blood, as well as how to perform these measurements, are up to the researcher. However, the number of tests and as much of the "design" or arrangement of the experiment as possible should be determined with the statistical analysis in mind.

Often we have data that could be called *accumulated data*, because it has been gathered with essentially no control. The traffic ticket data are just such an example. We can do some planning, such as determining where the traps will be located and perhaps deciding on the hours to check, but we have no control over the weather, the number of cars coming past, or if an accident occurs that might affect the traffic pattern. And even more often there is no control or information on the manner of gathering the data—we are often at the mercy of the whimsy of the people recording it.

Hopefully, we will have *designed the experiment*. Thus, when we take the trial runs to determine the time to get to school, we choose the number of trips. Then we select the times to start *randomly* from among the possible times we might have started. If I might normally start at any minute in the 10 minutes from 8 to 8:09 o'clock, I will use a table of *random numbers* to pick out the time to start for the experiment.

In appendix Table A.7 you have what is called a set of random numbers.* These consist of digits, the numbers 0 to 9, that are arranged in such a way that no matter where you start, the number appearing could be any of the digits and has the same chance of appearing as any of the other 9 digits. The next digit, no matter in what direction you choose to go, could also be any digit and with an equal chance of appearing. So, if we start on the top of page 333 and look at the first set of digits on the left, we find the numbers 40603 16152 83235. . . .

The first digit is a four, so the first day I would start at 8:04 o'clock (the digit 4 is being used as the number of minutes past 8 o'clock, in this case 4 minutes past 8 o'clock). The second digit in the list is a 0, and this represents the number of minutes past 8 o'clock on the second day, or 8:00. The third day I would start at 8:06, and so on. If we follow this procedure we will be assured that the times to start will have no bias or influence by me in their selection. They will have been chosen in a completely *random* manner, and randomness in choosing our samples is vital.

Of course, we will be able to enlarge the use of the random number tables to include larger sets of digits. For example, we could assign a number from 000 to 999 to each and every one of the 1000 students in our population of students in Example 1.3. Then when we pick our random numbers we would instead take groups of 3 digits to serve as the number for the student to be chosen. Thus, if we looked at the second row in the same set of digits on page 333, we would find the numbers 409. Thus the first student to be

* These values came from a larger source of random numbers called *One Million Random Digits with 100,000 Normal Deviates*, by the Rand Corporation. The method used to produce truly random numbers is quite difficult and special techniques are used. If you just pick numbers in your head you will tend to emphasize certain ones and not include others enough. See Problem 1.15.

polled would be student number 409. The next student would be 415, the third 358, the fourth 569, and the next 958. In this way we will select our students completely at random.

Returning to our designed experiments, you should be aware that we could have much more complicated problems. For example, a biologist may study the blood sugar level of male and female rats of different ages, on different diets, and at different times of the day, but using a single set of rats to answer all of these questions at the same time. Designing such an experiment and analyzing the data are extremely involved and beyond a first exposure to elementary statistics. But you should be aware of the scope of the situations that can be best analyzed with modern statistics.

Finally, we have survey-type data. Here, among the many special concerns that are important, we must consider the fact that the questionnaire must be properly prepared and pretested, and that the individuals must be suitably and randomly chosen. Ultimately, we hope that the surveyed information will portray the population we are interested in understanding; therefore, extensive planning is necessary.

1.6 CALCULATIONS

Whenever most people think about statistics, they picture mountains of numbers and tons of calculations. However, two major developments have occurred during the last several years that have helped to relieve much of the burden of the calculations. They are (1) the availability of very inexpensive calculators, and (2) the use of time-sharing computer programs.

Calculators are now available for as little as $4 that can do addition, subtraction, multiplication, and divisions. You can do all your statistical work with no greater investment. However, one very valuable additional capability is what is called a true memory. Most statistical problems involve adding up a large set of products of various quantities. The true working memory feature allows you to multiply two numbers and add these products or accumulate them in the memory portion of the calculator. Thus you calculate a product, place it in the memory, perform another product, and add it to the previous amount in the memory. For example, $2 \times 3 + 3 \times 4$ is calculated first by multiplying $2 \times 3 = 6$ and entering the 6 into the memory. Next you multiply $3 \times 4 = 12$ and, when you add this 12 to the memory, you have $6 + 12 = 18$ stored in the memory, and you can either have that value recalled (displayed and read) or you can continue to add to it. Calculators with this feature are being sold for as little as $10.

The only additional features you might find helpful in elementary statistical calculations are a square root key and the ability to do squaring relatively easily.

The major problem that arises through using a calculator is the need to be realistic in reporting your calculated values. You should make sure to round off your results.

Time sharing is a technique whereby many people at many locations can share the same large computer. Each person works with a teletype unit that resembles a typewriter and couples together with the telephone. You just dial in to the computer and start computing. Problems may be quickly solved, modified, corrected, and rerun in short order. Perhaps even more important is that the computer has stored internally for your immediate call a large library of previously written programs. For many problems you don't have to write the difficult program; you just need the instructions on how to use the ones that are already there. There are drawbacks though. Not all problems may have programs available in every time-sharing system. Some of these programs may be on only a very few, but many of the basic ones are widely available. The other possible disadvantage is that for some problems time sharing is slow, and time is equal to money here. For most basic problems, however, this will not be the case, and we find time sharing quite satisfactory for our problems.

There sometimes are difficulties in using library programs, especially for sophisticated or complex types of statistical analyses. Before using any computer to work out statistical methods that you are unfamiliar with, see an expert. The often-quoted computer expression GIGO (garbage in, garbage out) is definitely true in this case.

REVIEW

Data • Count and Measurement

Error • Variation

Sample • Population

Statistic • Descriptive Statistic • Inference Statistic • Parameter • Estimate

Random Numbers

Kinds of Measurements—Categorical or Attribute • Ordinal • Interval

PROJECTS

Several small "projects" are to be introduced here and expanded on as we continue through the text. Choose at least one of these as your own to follow. You may find that more than one will interest you, and they should not be so difficult that you cannot do several. Conscientiously gather the data and work out the problems as we progress. The realities of using live data can be intriguing and will give meaning to the course.

Two questions should be answered for each problem at this point.

How are you going to assure that your sample is random?
What might be the purpose of the experiment?

1. Put a thermometer in your refrigerator (you can purchase an inexpensive one for about $1 or else try to scrounge up something—perhaps from a neighbor or through a lab). The thermometer should be capable of reading within 1 or 2 degrees. Take a collection of readings at different

times and carefully record the values. Do you have any observations to make? Are all the readings the same? What do you think might influence the values or cause them to vary?

2. Record your gas mileage over a period of time. What can influence the measurements?

3. Obtain a collection of sugar packages. You should be able to find them in the cafeteria or a local restaurant. Weigh the sugar in each packet. Try to use an accurate scale such as would be available in the chemistry lab. Also try weighing them on a scale such as a postage scale. How does the different scale affect the values you obtain? Are all the values the same? If not, why might this be expected?

4. Go to a local hospital and find out the birth weights (or lengths) of some babies born at that hospital. Are they all the same? Why not?

5. Record the number of songs per hour on a particular radio station.

6. Record the number of commercials per hour on a TV channel.

7. Drop a tennis ball about 15 times. Measure the height it bounces to each time. Are the values all the same? How easy was it to measure the height? What do you think would happen if you used several different tennis balls?

8. Using a spring scale (if available), measure the tear strength of cotton thread. Try a number of different pieces.

9. Measure your foot about 10 times with a 12-inch ruler. Does your foot vary?

10. As an alternative to Project 3, scoop out a teaspoon full of sugar and weigh it. Do this for a number of spoonfuls.

11. Measure your pulse rate 10 times.

12. Prepare several survey questions to investigate an important social question or local problem. Write some questions to obtain Yes or No answers and write others to have a scale (i.e., 0 to 6 or 0 to 10). Define the populations that you will survey.

PROBLEMS

SECTION 1.2

1.1 Try to describe what is good coffee. If you were going to try to determine what you should do to make good perked coffee, what are some of the different things that you might consider (e.g., different brands)? How might you go about devising an experiment to investigate this question?

1.2 Do the same as in Problem 1.1, but consider the quality of a razor blade, the selection of a brand of razor blade (or an electric razor) to use, and so forth.

1.3 For each of the following items describe several different *types* of measurements that can be taken. What kinds of instruments are used to perform the measurement and who might perform each of the kinds of measurements?

a. Reading (5th graders).
b. Eyeglasses.
c. Crime.
d. Ice cream (chocolate chip).
e. Stamps.

1.4 A bug spray is to be tested. What might be some ways of measuring its effectiveness? If you sprayed several sets of bugs why might you get different results?

1.5 Light bulbs are labeled according to life, wattage, voltage, and amount of light. Under what circumstances would you be interested in different qualities (don't forget about street lights or bulbs in out-of-the-way places)? How would your interest lead to different types of measurements?

1.6 Explain how the following items are measured (e.g., eggs are classed as small, medium, large, extra large, and jumbo). What types of scales are they?

a. Olives.
b. Toothpaste.
c. Cars.
d. Soap.
e. Shoes (both length and width).
f. Nails.
g. TV sets.

SECTION 1.3

1.7 Summarize the different kinds of error or sources of variation that can be present in measurements.

1.8 One reason that we must sample is because we destroy the item being measured or tested. Give some examples of items and the kind of measurements to be taken that would destroy or otherwise change the item so it cannot be used in the original way or remeasured.

SECTION 1.4

1.9 Look up the definition of the word "statistics" in several unabridged dictionaries. How do the definitions differ from the idea of inference statistics discussed in this chapter?

1.10 The ability to recognize the population you are dealing with or are interested in is extremely important. Provide a list of at least five populations from which each of the following could be samples. In each case also show how these populations might then be included in other larger populations.

a. College Board scores.
b. Number of traffic tickets.
c. Number of divorces or ages of persons getting divorces.
d. Mail.
e. Bug spray (or the results of using the bug spray).

1.11 One way of showing the relationship between populations that fall within each other is called a tree diagram. With this we can show how subpopulations branch out of larger populations. For example,

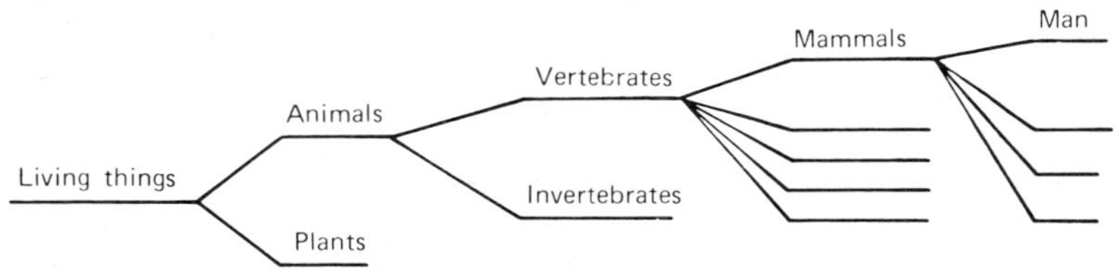

Using the same technique, show how different populations branch out of the following types of overall populations.

a. The people who live in the United States (consider various types of regions or ethnic groups).
b. Blood (include the question of blood samples).
c. Crimes.
d. Automobile repairs.
e. Light bulbs.

1.12 Explain in your own words the meaning of an inference about a population. Why does an inference about a population always have a chance of error associated with it (the inference)?

SECTION 1.5

1.13 Statistics deals with all fields of study that somehow accumulate or use data. A collection of problems follow that individuals working in different fields might pursue. For each statement discuss the following:

1. Who might be concerned and why?
2. What type of data and/or measurements might be gathered? (Specify if they are categorical, ordered, etc.)
3. What would be the source or population associated with the data?
4. What difficulties might distort the data, cause it to be in error, or lead to misinterpretation?
5. What could be the effect of an error in interpreting the data?
6. What problems are there in assuring that the following studies are random?

   a. The effect of a stimulant and a depressant on learning is studied.
   b. The noise level around a rock group is suspected of leading to deafness.
   c. Thickness of tire tread seems to be related to accidents and to the time that a person will purchase a new tire.
   d. A psychologist has determined how many of the college students in a particular college are well adjusted, satisfactorily adjusted, or poorly adjusted.

e. A new car dealer has been keeping a record of the kinds of defects that customers reported after picking up their new car.
f. A survey is taken to determine how many people watched Channel 2, 3, 4, and so forth.
g. The number of strawberries in a pint container is counted for several different containers.
h. The number of tomatoes per tomato plant is recorded.

1.14 Locate in journals or newspapers some examples of data and apply questions 1 to 6 of Problem 1.13 to them.

1.15 Pick a digit at random (0 to 9) and write it down. Do the same thing three more times, each time forgetting or ignoring all previous digits. Thus each digit is a random digit and you should have a set of 4 random digits. Repeat the process 25 times. You will then have a total of 100 digits in groups of 4.

a. Count the number of times you wrote down each digit. You should have about 10 of each digit. How close were you?
b. Examine each of the 25 groups of 4. Determine which groups have 2 (or more) of the same digits (e.g., 8, 6, 1, 8 has two 8's). What percentage of your groups have repeated digits?
c. Now examine the random number table. Look at 25 sets of 4 digits. What percentage of these sets have repeated digits? Comment.

1.16 You are to randomly choose 4 rats out of a group of 10 laboratory rats for an experiment. Describe how you would proceed to make sure the selection was random. Use the random number table to select the rats.

1.17 I have 27 students in my class in 3 rows of 9 students. Each period I select 3 students at random. How should I select the students? Which students should I select during the first 10 periods (identify the exact way you select them). Is anyone chosen more than once?

1.18 You are to pick a name and telephone number at random from a phone book. What procedure would you follow (you have different pages, different columns, and different individuals in columns). Use the random number table to pick 20 names.

1.19 Provide an alternate way of randomly picking telephone numbers. This time use the random number table to select the actual phone numbers. Pick 20 numbers. How does this problem differ from Problem 1.18? What if you also wanted the names of the people to be called?

CHAPTER TWO
PICTURES OF A DISTRIBUTION: HISTOGRAMS

You have been introduced to some sets of data. Our first task is to try to make some sense out of the mass, or rather mess, before us. Somehow there must be ways of reducing it to some reasonable collection of descriptive terms.

In Chapter One we discussed how words and pictures helped to convey a description of an object. We then observed that words and pictures can be used similarly to describe a set of data. In this chapter we will begin the task of formally organizing our data and drawing pictures that should aid us in gaining a clearer understanding of the nature of our data. In the next chapter we will deal with mathematical descriptions.

2.1 GRAPHICAL TECHNIQUES

There are several methods of displaying a collection of data. The ones we will be concerned with at this time are those that deal only with distributions of a *single set of data*. Later, in Chapter Thirteen, we will look at another type of pictorial or graphical technique.

Two general approaches will be dealt with in detail.

1. Those involving frequencies or numbers of times that particular individual values appear—histograms, bar charts, or frequency polygons.
2. Those involving accumulated frequencies or the total number of individuals less than or equal to some particular value—cumulative frequency diagrams.

In addition, we may speak of:

1. Frequencies—actual number of times different individual values appear.
2. Relative frequencies—the ratio or proportion of times a particular value occurs out of all the occurrences.

We are going to start by considering graphs. You should first recall that a graph has two axes that are usually labeled the X-axis (the horizontal or abscissa) and the Y-axis (the vertical or ordinate), as shown in Figure 2.1. There are scales along these two axes. A point that is above the X-axis is said to have an X-value associated with a location on the X-axis directly

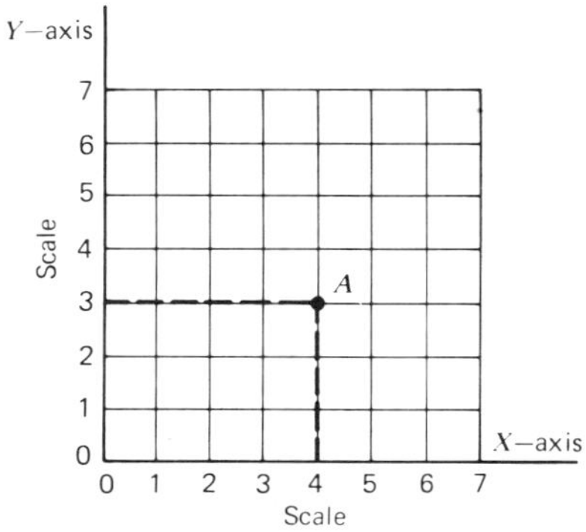

FIGURE 2.1 A graph with an X-axis and a Y-axis. A point on the graph has an X-value and a Y-value found by following a straight line to the axis. Point A has an X-value of 4 and a Y-value of 3.

below it. So point A has an X-value of 4. The same point has a Y-value associated with it, in this case 3.

2.2 BAR CHARTS

Whenever we encounter a mass of data, it is important to begin to organize it. The simplest initial technique is the use of a *tally sheet*. First you must provide a location to keep track of the times that each category occurs. Then you place a slash line next to the value or category each time that you observe a particular occurrence that fits in that category (using cross slashes to collect sets of five for convenience).

For example, when Linus kept score of a baseball game, his categories were teams, and each time a run was made he put a slash mark down next to the team that got the run (Figure 2.2). I think that little picture should convey a lot of meaning. Dwell a little more on the data presented by those slash marks. This is attribute or categorical data since, for each category, we only have a measure of the times that something occurred in that category (the number of runs in the particular inning by a team).

To provide a better picture, however, we will draw a graph. When we deal with *categorical* data, such as the "runs by home team" and "runs by visitors," we will draw a *bar chart*. The bar chart resembles the tally sheet. Alongside each category the number of times that the particular value occurred is recorded. This time, however, instead of making slashes, we introduce a formal second axis and a second scale. This second axis is the frequency axis. The height or length along the second axis indicates the

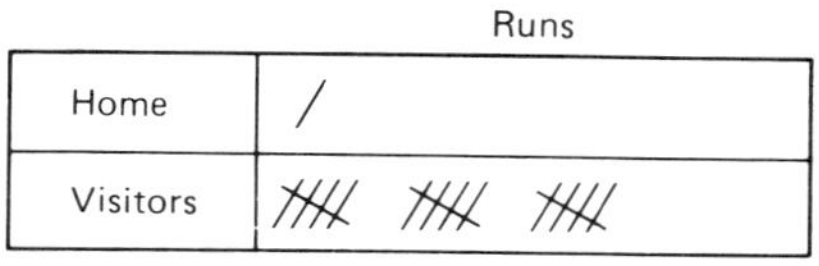

FIGURE 2.2 A tally sheet for a depressing baseball game.

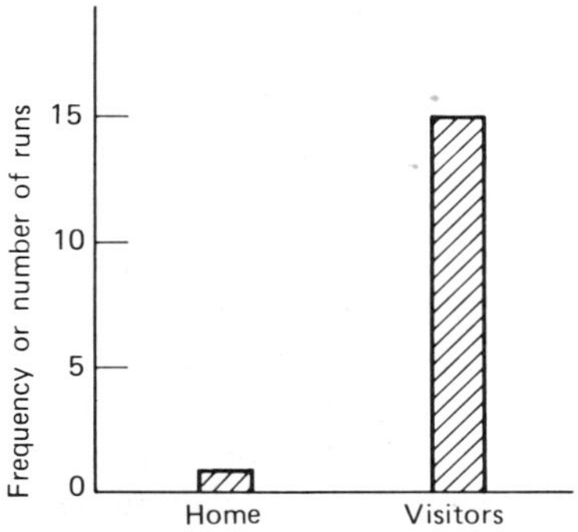

FIGURE 2.3 A bar chart for the number of runs for two baseball teams.

number of times the associated value occurred. The scale, which we will call the frequency scale, generally has equal subdivisions. It is a linear scale.

Figure 2.3 shows a bar chart for the runs by the two teams. As you can see, the construction is very simple.

2.3 HISTOGRAMS

Probably most of the data that are encountered involve actual measurements.

Thus, we have a collection of number of traffic tickets issued, minutes to travel to school, or hours worked by students. What we would like to have is a convenient recording device to simplify the set of data we have already gathered. In each case we have measured some values that we are keeping track of. These values will be associated with the scale along the X-axis. It is the measurement scale. Next we will look for the total number of times each measurement is encountered. This total becomes our frequencies. The scale for frequency makes up the second axis in the same way as on our bar chart. Let us refer to Example 1.1. The original data was presented in Table 1.1 on page 7.

First we should prepare a tally sheet, which would resemble Figure 2.4. Observe that I have provided a place for every possible value of number of tickets (including the value of 16 tickets that never appeared). For example, we see that on 3 different days only 2 tickets were issued (the 3 slashes above the number 2), on 17 different days 9 tickets were issued, and 15 tickets were issued on 5 different days.

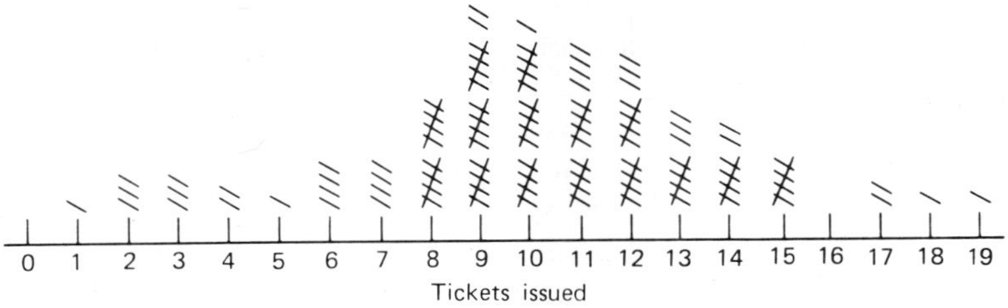

FIGURE 2.4 Tally sheet for Example 1.1—tickets issued per day.

The tally sheet is useful for organizing the data but, as with the bar chart, a graph will provide an even better picture. This time we have measurements that have a numerical value and we can therefore draw a scale along the X-axis that will be for measurements. In this case the measurements are numbers of tickets issued, and the values go from 0 to 20. They are the same values we used in the tally sheet. The second axis will be frequency, in this case "on how many different days the particular number of tickets were issued." We then draw bars above each X-value (number of tickets) as high as the frequency of those values (number of days). When these bars extend the full width of the particular category or interval along the X-axis and therefore touch each other, we call this type of chart a *histogram*. Figure 2.5 shows the histogram for traffic tickets issued.

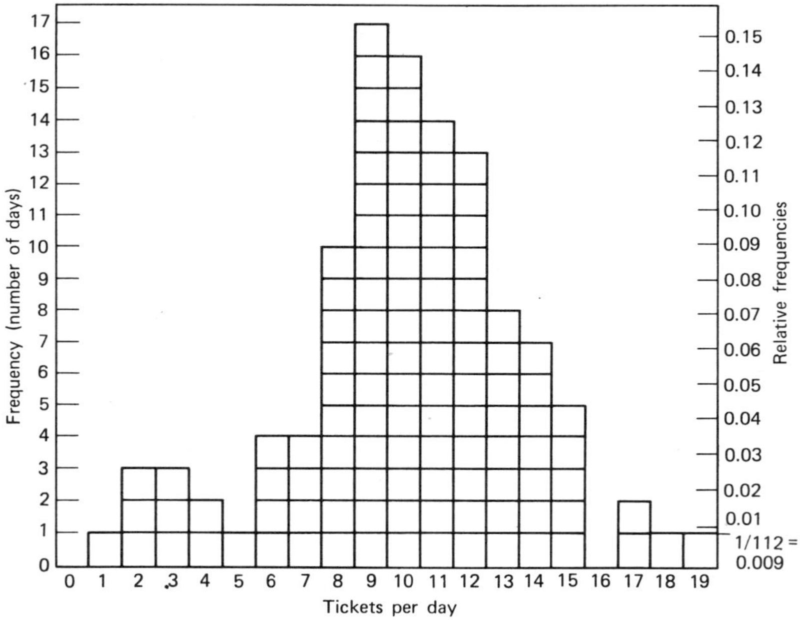

FIGURE 2.5 Histogram for number of times various numbers of traffic tickets were issued.

Sometimes you will encounter other similar graphs. In this case you could also draw a bar chart. Occasionally a graph is constructed by placing dots at the tops of the values on the bars. When the dots are connected, as in Figure 2.6, we call the graph a frequency polygram.

Even though all of these different techniques are used, I will mainly employ the histogram. It is the most commonly used method and will help in our future understanding of distributions. The primary reason that I will devote so much time to the histogram is that the histogram may be considered about the best picture of a distribution. If you can begin to

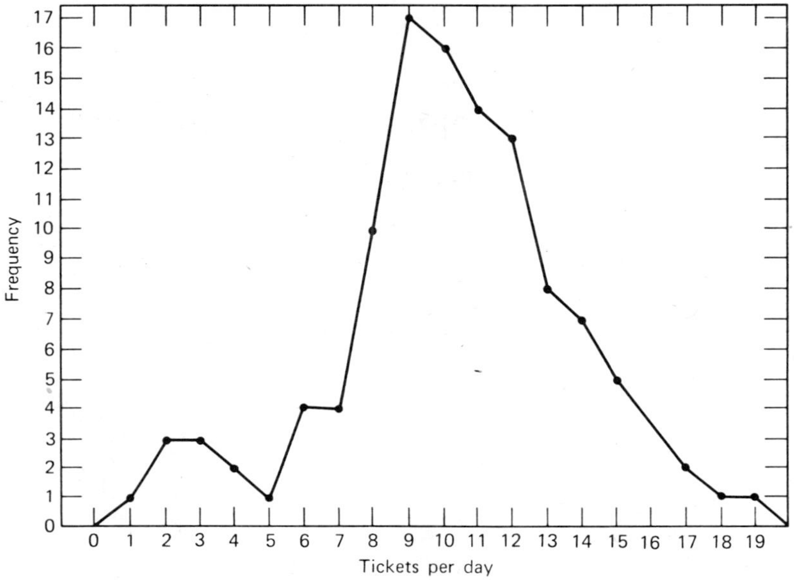

FIGURE 2.6 A frequency polygon for the number of tickets issued per day. Compare this to Figure 2.5, the histogram for the same data.

internalize a feeling of the relationship between this picture and the real distributions, much of the future work will be substantially easier. You might picture in your mind a day taken from a calendar and then placed in each of the little boxes of the histogram. After all, each box supposedly represents a day during which the particular number of tickets were issued.

Until now we have been recording actual frequencies. A very common alternate to using actual frequencies is the use of *relative frequencies*. For this type of scale we divide each of the actual frequencies by the total number of individuals included in the collection. In this example, the number of observations is the total number of individual days and is 112 (you will see this expressed as $n = 112$). The relative frequency for a particular frequency value is simply

$$\text{Relative frequency} = \frac{\text{Actual frequency}}{\text{Total number of observations}} \tag{2.1}$$

For our example, the relative frequency of 8 tickets being issued is

$$\text{Relative frequency} = \frac{10}{112} = 0.089$$

Since there were 10 days when 8 tickets were issued, what is the relative frequency for 14 tickets per day?*

* $7/112 = 0.062$.

We can also easily create a relative frequency scale on the histogram. Look back to Figure 2.5 and observe that there are two scales for the Y-axis, or frequency axis. The left scale has already been discussed and is for recording actual frequency. The right scale is a *relative frequency scale*. The first point on that scale corresponds to an actual frequency of 1 and is therefore a relative frequency of 1/112 or about 0.009.

In a similar manner other points can be located to establish the entire scale. Note, however, that the numbers on the scale do not necessarily line up with the ends of any particular bars on the graph (since they do not necessarily round off perfectly).

Two important observations should be made about relative frequencies. First, they are easy to convert to percents. *Multiply the relative frequency by 100 and you have percent of occurrences.* So our statistician can state that eight tickets are issued 8.9% of the time, since

$$\text{Relative frequency} \times 100 = \text{percent of occurrences}$$

$$\text{or } 0.089 \times 100 = 8.9\%.$$

TABLE 2.1 Frequencies and relative frequencies for number of tickets issued (Example 1.1)

NUMBER OF TICKETS X-VALUE	FREQUENCY	RELATIVE FREQUENCY
20	0	0.000
19	1	0.009
18	1	0.009
17	2	0.018
16	0	0.000
15	5	0.045
14	7	0.062
13	8	0.071
12	13	0.116
11	14	0.125
10	16	0.143
9	17	0.152
8	10	0.089
7	4	0.036
6	4	0.036
5	1	0.009
4	2	0.018
3	3	0.027
2	3	0.027
1	1	0.009
0	0	0.000
Sum	112	1.001

The second observation is that no relative frequency can be greater than 1.00, and the sum of all the relative frequencies must be 1.00 on 100%. This is a useful check on your computation.

In Table 2.1 I have presented a table with the X-values, and both the frequency and relative frequency for each X-value. Note that the sum is not exactly 1.00. This happens because 0.089 and most of the other values are rounded off.

If we carried out the calculations to more decimal places the sum would be closer to 1.00, but this value is still very close to 1.00 (it is within 0.001).

2.4 AREAS

When it comes to drawing the histogram, it is best if the intervals that we use are equal in size. If this is done, and if the frequency scale is also linear, then any histogram consists of collections of little rectangles. For a histogram with an "actual frequency" scale we can think of the little rectangles as being one unit along the frequency scale and one category wide (see Figure 2.5). If relative frequencies are used, then the height of the little rectangle would be $1/n$ units. In either case, areas of the histogram can be equated to the proportion of the time that the corresponding values occur. It is in this way that the picture presented by a histogram is most important.

If you think of the little boxes as weights, then the more frequently a value occurs the more weight it has. Another way of thinking about this is the higher the column for a value, the more weight or area the particular value has. The more area associated with the X-value, the more important that value is out of the whole collection of values or the entire distribution.

2.5 GROUPING AND MEASUREMENT DATA

Quite often we find it desirable to compress a large number of intervals into a smaller set of categories. For instance, it might be desirable for presentation purposes to use only seven or eight categories for our present example. In cases where we are dealing with true measurements, not just integer values, it is usually necessary to work with categories. Consider the type of numbers you would obtain if everyone in your class were weighed to the nearest gram (about 1/454th of a pound or 0.002 lb) or your height measured to the nearest 1/64th of an inch. (or nearest millimeter). How big would the histogram have to be to record all of the values? Obviously, we must simplify the graph. The procedure is called grouping.

To perform grouping, we are merely going to use a different set of categories for the same set of data. You should realize that each category has a *midpoint* and a *pair of limits*. If you look at Figure 2.7 you will find the first four of the categories of the complete histogram of traffic tickets (Figure 2.5).

The categories are the sections separated by vertical lines. These lines constitute our *interval limits* and a category (or class, as it is sometimes referred to) is considered as encompassing the complete set of possible values between these limits. A midpoint is the center of an interval or category and is the value you consider that category to represent. So the first interval is from 0.5 to 1.5 and the midpoint is 1.0. You therefore consider

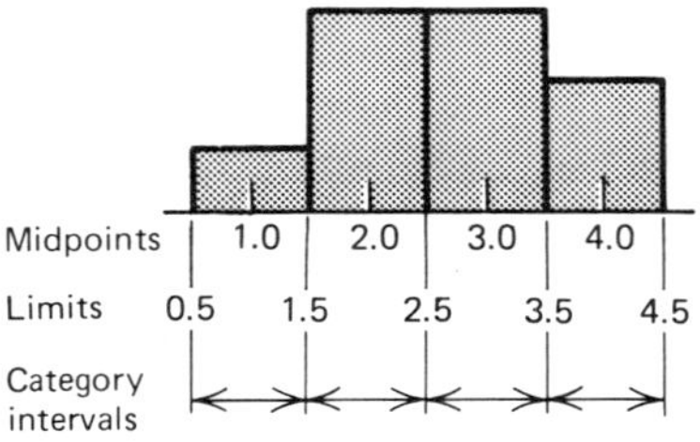

FIGURE 2.7 Midpoints, limits, and intervals for several categories of Figure 2.5.

any value in that category as though it were a 1.0. This will become obvious as we continue. In grouping we must choose a new set of class or category intervals or a new set of central points. If we decide on a set of intervals (defined by the limits of the intervals), then we automatically establish the centers. On the other hand, if we provide centers, the ends or limits are then established.

RULE 1 *A location must be provided for each and every possible value or measurement.* Sometimes this will appear to be unwieldy. For example, maybe on one day 30 tickets were issued. We would need 10 more positions on our scale just to satisfy the one additional value. Some type of simple adjustment should be available for this situation. We can provide a single additional category of perhaps "20 to 30" or "20 to 40." *But*, there must be extreme care in interpreting this category, especially since you can see that I could make this last category as large as I want to.*

RULE 2 *An individual value must fit into one and only one interval.* This means that we must provide definite *limits* to the interval that any location or category on the measurement axis represents. Here the values are discrete; there is no such thing as half a ticket. However, you will probably encounter more situations that involve continuous measurement scales. In those cases there is always an intermediate, inbetween possible value and we must be aware of the number of decimal places to which we will be measuring.

The limits are to be determined or stated so that there is no confusion as to which category a particular value belongs in.

RULE 3 *Each interval should have the same width (the same number of possible values).* This may seem obvious to you. I already indicated a possible exception while discussing Rule 1 above, but I also caution you on making interpretations from histograms that have unequal-sized intervals. It should be noted that commonly 2, 5, 10, 25, 50, or 100 is used. These usually provide convenient interval sizes to work with.

* For a discussion of how to properly draw such categories see, for example, Neil Ullman *Statistics: An Applied Approach*, Xerox Corp., 1972, pages 63–64.

RULE 4 *There should be at least six and no more than about 25 intervals.* This is the most flexible rule and is truly one of common sense. Too small a number of categories leads to a complete loss of information. Too many categories (particularly with measurement data) can lead to an enormously large histogram with many gaps. We then tend to lose our ability to summarize and interpret the data adequately.

Note that all of these rules deal with the measurement or X-axis. For our example, let us pick a set of intervals that appears reasonable. I have chosen as a first possibility the following set of intervals (remember—the rules for selecting categories still hold).

SET 2.1

0.5– 3.5	12.5–15.5
3.5– 6.5	15.5–18.5
6.5– 9.5	18.5–21.5
9.5–12.5	

The histogram for the interval Set 2.1 is shown in Figure 2.8.

Each category contains three integer values, since the only kinds of values in this problem are integers. As long as each interval contains the same number of integers, there is no confusion. Note that the centers of the

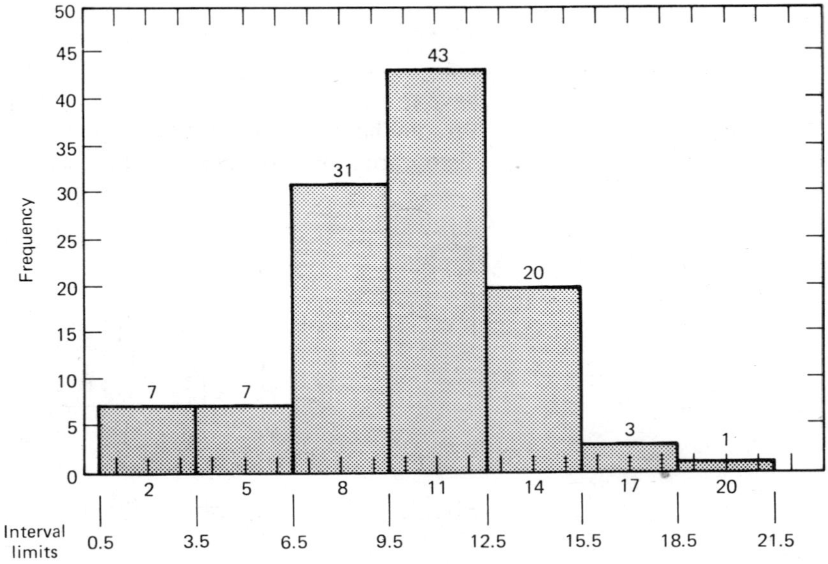

FIGURE 2.8 Grouped data for Example 1.1 using intervals of

0.5– 3.5	12.5–15.5
3.5– 6.5	15.5–18.5
6.5– 9.5	18.5–31.5
9.5–12.5	

intervals have also been established. The first one is 2. What are the remaining midpoints?* Also note that I have not rounded off the limits to 1–3, 4–6, 7–9, and so on. It is preferred not to leave any space between the end points of adjoining intervals.

Observe that our condensed version of the distribution, shown in Figure 2.8, is very similar in shape to the original histogram we drew (Figure 2.5). I will admit that our new histogram is not as accurate as the original, since much of the variation has been eliminated. *Whenever we group, we lose accuracy.* However, in most cases the loss is not great enough to be concerned with, provided we do not reduce the histogram to only a couple of categories. Common sense is the best rule here. Obviously, if you see just two categories, *almost nothing* can be learned from the data. In other cases, if, say, thousands of intervals are possible, then by size alone the histogram is also meaningless.

You might select a different set of midpoints and group into different intervals. The new histogram would probably look different, although the general shape would be similar. After all, we have been approximating and condensing data, and each approximation is somewhat different. In Problem 2.13 you can have the opportunity to see the effect of changing these intervals.

We should next consider graphing data that involve measurement-type scales.

EXAMPLE 1.2 (Cont.)

Let us draw a histogram for the data in Example 1.2 (page 8). Now we have to be more careful when considering the rules for intervals. If I can assume that the time was read to a definite one tenth of a minute, then the histogram in Figure 2.9 would be the proper one for showing all the data.

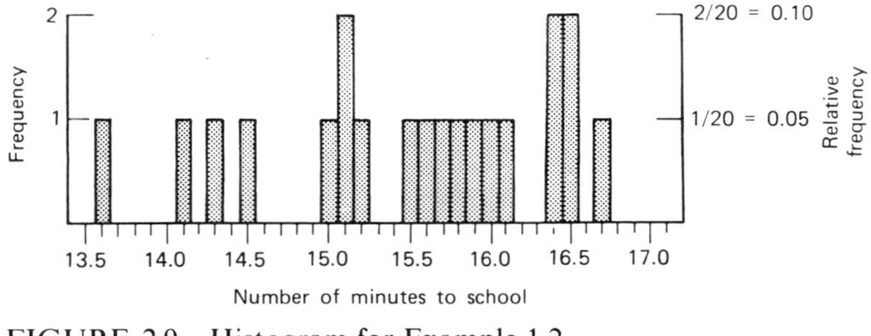

FIGURE 2.9 Histogram for Example 1.2.

The main reason I asked about the assumption is because with measurements we are always approximating. The approximation is usually established by the instrument used. With a more accurate device we could always measure to 1 or 2 more decimal places. Thus, 16.5 is an approximation and

* *Answer*. 5, 8, 11, 14, 17, 20.

does not necessarily mean 16.50 or 16.500 or 16.5000 but, instead, includes a range of possible values, one of which may be 16.50. We may consider that we are grouping all of the 10 values from 16.45 to 16.54, or all the 100 values from 16.450 to 16.549, and so forth, into one point that we identify as 16.5. We are rounding off in this case and obviously we already lost some accuracy. In most cases the loss is not enough to cause concern, but there is always some loss. The important thing to keep in mind is that the value 16.5 is really a *grouping of many different possible values* and that every location on our measurement axis is a grouping of finer, smaller points.

Note that if we were able to measure to 2 decimal places, 16.45 minutes is a possible value. It must, therefore, be placed in a category and there must be no question about what category it belongs to. This problem involves the general concept of grouping data. As a measurement, 16.45 must fit into a category. Perhaps the category is 16.5. How many other 2-decimal values would fit into the category 16.5?

For the values we have worked with up to now, we have already stated that the point 16.5 was really the center of an interval that extends from 16.45 to 16.54 (the interval depending on the number of decimal values you are able to work with). If, in addition to 16.45, there are the 9 values from 16.46 to 16.54, then 10 different possible 2-decimal cases fit in the class interval or the category we called 16.5. If we continued along the X-axis, our next midpoint would be 16.6 and would also contain 10 different 2-decimal individuals, 16.55 to 16.64. All other values would follow similarly. Figure 2.10 illustrates the interval from 16.45 to 16.54. Note how the actual center is at 16.495 but can essentially be treated as 16.50 with little additional loss in accuracy.

I would now like to group this data into, say, seven or eight categories. I will select a set of midpoints. One reasonable set of midpoints would be

13.5, 14.0, 14.5, 15.0, 15.5, 16.0, 16.5

They are equally spaced and include five values per category. Now, what

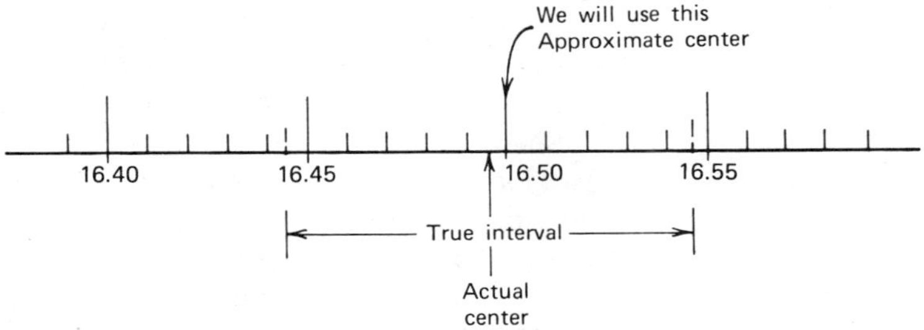

FIGURE 2.10 Line showing interval subdivisions. If the interval units to be used are 16.45–16.54, the actual or true center is 16.495, not 16.50. However, using 16.50 will not create much additional loss in accuracy.

would be the interval limits? First let us look at a magnified portion of the X-axis scale.

As you can see, the value 14.0 contains all of the 1-decimal values from 13.8 to 14.2. However, to provide a more continuous line, we should designate the interval as 13.75 to 14.25. Thus the intervals would be

SET 2.2

13.25–13.75	15.25–15.75
13.75–14.25	15.75–16.25
14.25–14.75	16.25–16.75
14.75–15.25	

A histogram for this set of points has also been drawn in Figure 2.11.

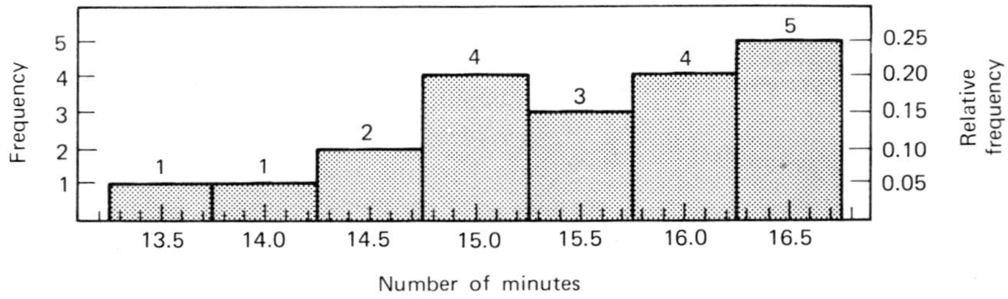

FIGURE 2.11 Grouped histogram for Example 1.2 using centers.

Have you made an important observation? The *intervals have been given to 1 decimal place more than to what the measurements are recorded*. This has been done so that there is no question of where to place any quantity. It also provides assurance that we will have a closer approximation to a continuous line for our measurement axis (the X-scale). If the possibility exists that a value would be recorded to more than 1 decimal place, you must carry the interval limits 1 decimal place more.

So far in this problem we have been dealing with actual frequencies. If you look at the right-hand columns of Figure 2.11 you will see that I have also included *relative frequencies*. These indicate (in Figure 2.11) that out of the total of 20 observations 4/20 or 1/5 of them had an approximate value of 15.0 (fell within the limits of 14.75 ot 15.25). We can also state it as 20% of the data had an approximate value of 15.0. What percent (or relative proportion) of the original histogram (Figure 2.9) had a value of 15.5?*

* *Answer:* 5%, since 1 out of the 20 had an "exact" value of 15.0.

Relative frequencies will generally be more useful and more informative for all types of data. This is especially true for large collections of data. In addition, when we encounter theoretical and infinitely large distributions, relative frequencies will be our only way of identifying frequencies and for drawing histograms.

2.6 CUMULATIVE FREQUENCY DIAGRAMS

Up to now we have dealt with graphs of frequencies for individual values. Another technique that conveys a considerable amount of meaning involves adding up all the numbers of times values less than or equal to a designated value occur. This is called a *cumulative frequency diagram.*

For instance, we may find it necessary to talk about the proportion or the probability of a bread weighing *less than* 20.00 ounces. We might want to know what our chance is of taking *more than* 18 minutes to get to school. Our ball player wants to know how frequently his team gets *more than* two runs. All these involve cumulative frequencies as opposed to frequencies or occurrences of exact, individual, or particular values.

In addition, some types of distributions will only be discussed from a standpoint of cumulative frequencies. When we encounter certain theoretical distributions, the only way to consider the histogram will be by using areas, which in turn will involve the use of cumulative frequencies.

EXAMPLE 1.1 (Cont.)

Now let us determine the cumulative frequencies and draw a cumulative frequency diagram for the number of tickets issued. In Table 2.1 we had listed the actual and relative frequencies for the number of tickets. These values are again repeated in the first set of columns of Table 2.2.

We should start with the value of 0. We find no times (0 times) that the quantity 0 (or less than 0) occurred.

Therefore, the cumulative frequency for the number 0 is 0. The smallest number of tickets issued was 1 and it occurred with a frequency of 1. So the cumulative frequency for 1 ticket (less than or equal to 1 ticket) is 1.

For 2 tickets issued we have a frequency of 3. The sum of all previous frequencies plus this case is 4, $(1 + 3)$. So the cumulative frequency for 2 tickets is 4. By the time we reach 5 tickets, we have accumulated a total of $0 + 1 + 3 + 3 + 2 + 1 = 10$. The cumulative frequency for 5 (up to and including 5 tickets) is therefore 10.

A complete table showing the frequencies and cumulative frequencies for each number of tickets issued are shown in Table 2.2.

Notice that the cumulative frequencies are found by simply adding the actual frequency of the particular value to the previous cumulative frequency. This is the simplest way to create the table of values. It also works in reverse (by subtracting) if you want to determine the individual frequencies from the cumulative frequencies—try it.

I have drawn the complete cumulative frequency diagram for Example 1.1 in Figure 2.12. Two scales have been provided along the frequency axis.

TABLE 2.2 Frequency Measures for Tickets Issued

NUMBER OF TICKETS	FREQUENCY	RELATIVE FREQUENCY	CUMULATIVE FREQUENCY	RELATIVE CUMULATIVE FREQUENCY	CUMULATIVE PERCENT
20	0	0.000	112	1.000	100
19	1	0.009	112	1.000	100
18	1	0.009	111	0.991	99.1
17	2	0.018	110	0.982	98.2
16	0	0.000	108	0.964	96.4
15	5	0.045	108	0.964	96.4
14	7	0.062	103	0.920	92.0
13	8	0.071	96	0.857	85.7
12	13	0.116	88	0.786	78.6
11	14	0.125	75	0.670	67.0
10	16	0.143	61	0.545	54.5
9	17	0.152	45	0.402	40.2
8	10	0.089	28	0.250	25.0
7	4	0.036	18	0.161	16.1
6	4	0.036	14	0.125	12.5
5	1	0.009	10	0.089	8.9
4	2	0.018	9	0.080	8.0
3	3	0.027	7	0.062	6.2
2	3	0.027	4	0.036	3.6
1	1	0.009	1	0.009	0.9
0	0	0.000	0	0.000	0.0
Sum	112	1.001			

One is actual cumulative frequencies and the other is relative cumulative frequencies. In the previous section you should have noted that relative frequencies or percents are easier to make interpretations with. From the relative frequency you can immediately determine percentages. With cumulative diagrams this is even more valuable. For example, you can tell where the bottom 50% of the numbers are by finding the value corresponding to a *relative cumulative frequency* of 0.50. For this problem it would be 10 (see the dotted lines).

One note on construction and interpretation. The diagram I presented contains abrupt jumps as each new value is accumulated (see the insert of Figure 2.12). You should interpret the cumulative frequency for a particular value as the upper level at that point. Thus for 8, use 0.250, not 0.161, as the relative cumulative frequency, since the term cumulative implies up to *and* including the frequencies of the particular value.

Sometimes the centers of the horizontal lines are connected, to give an approximation of a smooth curve (Figure 2.13).

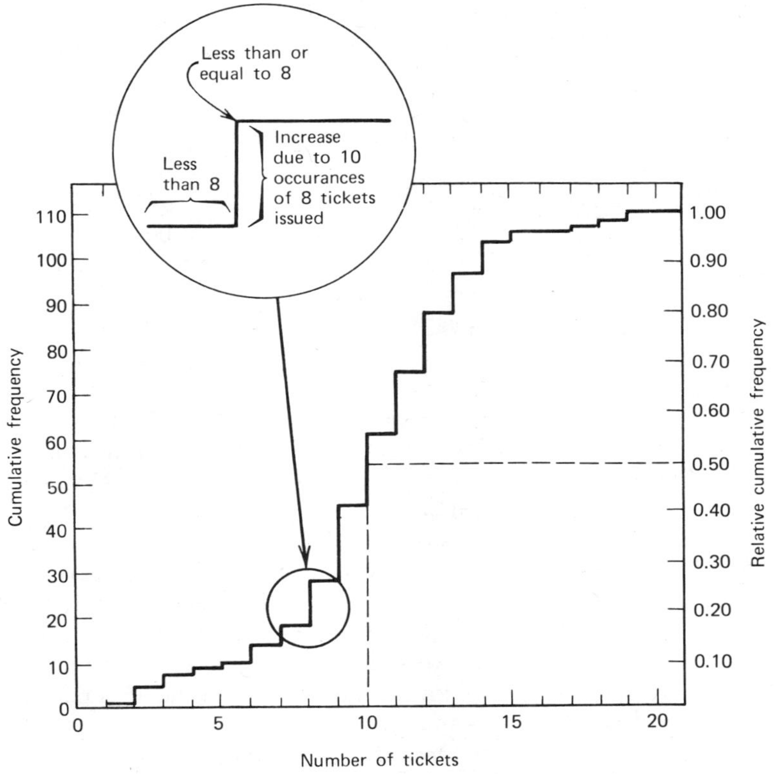

FIGURE 2.12 Cumulative frequency and relative cumulative frequency diagram for Example 1.1.

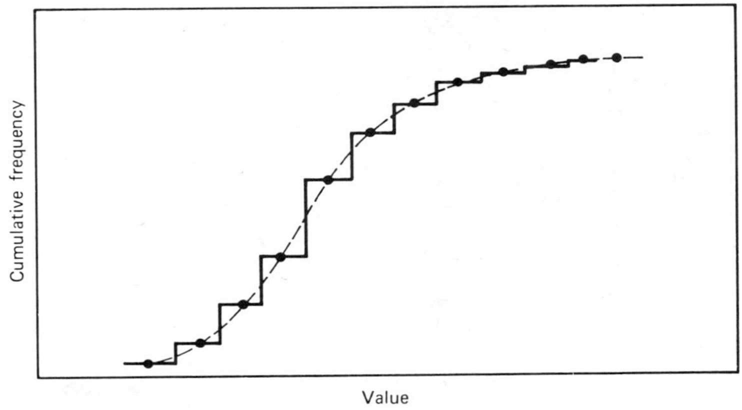

FIGURE 2.13 Cumulative frequency diagram drawn as an approximate smooth curve.

We can also determine *cumulative percents* by multiplying our *relative cumulative frequency* by 100. Thus, for example, we find, for 8 tickets per day

Relative cumulative frequency × 100 = Cumulative percent

$$\text{or } 0.250 \times 100 = 25.0\%.$$

Quite often it is useful to be able to list the X-values that correspond to particular cumulative percentages. We then also call these X-values as percentiles. Thus, the X-value for which 25% of the distribution fall below is called the 25th percentile (in the above example it is the value 8). The 67th percentile in this example is 11, and the 92nd percentile is 14 (refer to Table 2.2).

Frequently we write the percentile value as $P_{25} = 8$, where the P stands for percentile, the 25 represents the cumulative percent we are referring to, and the 8 is the corresponding X-value (and this is read as P sub 25). Similarly, $P_{67} = 11$ and $P_{92} = 14$.

Sometimes when we deal with very large sets of data it may be more useful to provide a list of percentiles or a conversion from the X-values to percentiles. This listing provides a better method of interpreting the relative position of a particular value. For example, perhaps you scored 65 correct answers on a particular reading comprehension test. Is that great? Well, if we know that this is also the 85th percentile score, we can say that your comprehension is as good or better than 85% of the population taking the test ($P_{85} = 65$). However, on a spelling test, perhaps you score 127 correct answers. Is this also good? This might only be the 40th percentile value, therefore you are only better than 40% of the population in this subject ($P_{40} = 127$).

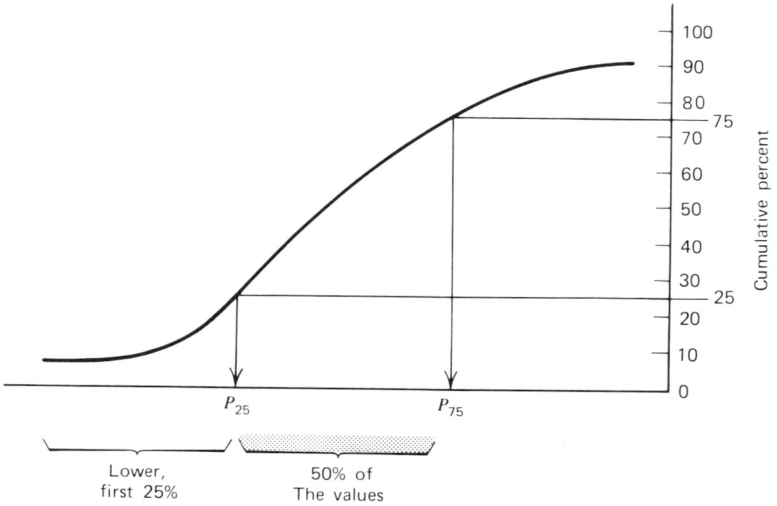

FIGURE 2.14 How to find the middle 50%.

Finally, cumulative percents or the corresponding percentiles can be used to determine particular areas of a histogram or portions of a distribution. To find the middle 50% of a collection of values, we need only find the 75th percentile and the 25th percentile, and all X-values between these two limits fall in the middle 50% (see Figure 2.14).*

2.7 SHAPE OF THE DISTRIBUTION

The histogram is a picture. It is a way of showing *how* our values are distributed and it has a very important property: shape. We have looked at a couple of distributions and you should have observed, first, how irregular they were and, second, the fact that the shape may be affected by slight changes in the method of grouping. In addition, as you look at a sample or the histogram of a sample, you should have more on your mind than just that set of values. You should be thinking about the population from which that sample came. The sample is a means to a more significant or general end—an understanding of the population. As a consequence, we should feel free to approximate a shape and smooth out the histogram of a sample.

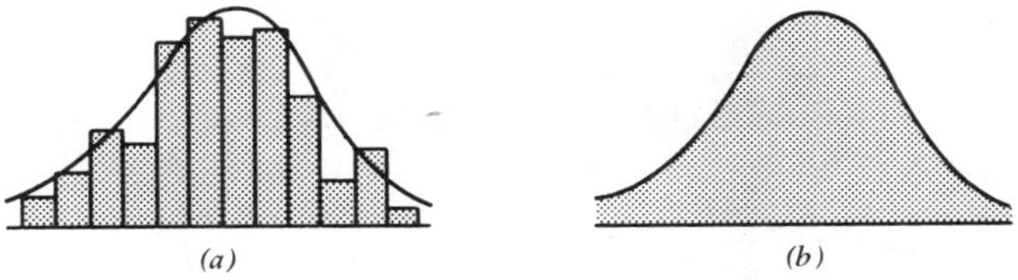

FIGURE 2.15 (*a*) A smooth curve drawn over a histogram. The curve is only an approximation of what we believe the actual population is like. (*b*) The smooth curve by itself.

You will be looking at the jagged peaks and jumps of your histogram, but I hope you will be seeing the overall picture and the smooth curve. We do this by drawing a continuous, smooth line through the peak areas and over the gaps, as in Figure 2.15*a*. After we obtain the smooth curve, we can somewhat erase from our minds the sample distribution and strictly work with the smooth curve (Figure 2.15*b*).

Next let me turn the situation around. Quite frequently I will talk about "a distribution" and I will show a smooth curve to represent that distribution. That smooth curve, whatever its shape, is an *ideal* curve. It is not the curve of a particular sample but, instead, an estimation of what the histogram for the infinite population (with intervals of essentially 0 width) might be. So, when drawing the smooth curve, remember that an actual set of data would not be exactly like it.

* In our example the 75th percentile is somewhere between 11 and 12. If, however, we approximate the 75th percentile as 12, then the middle 50% is between 8 and 12. This type of difficulty will often occur when the number of values used in calculating percentiles is not sufficient to have separate scores with each possible percentile or, as is true in this case, the scale used in measuring is not fine enough.

2.8 DESCRIBING A DISTRIBUTION

There are two ways that distributions are distinguished. The first involves a truly descriptive approach for which I will discuss three characteristic shapes a distribution may possess: symmetry, skewness, and bimodality. The second approach involves naming a hypothetical model or distribution that approximates the shape of the distribution (or histogram) we are evaluating. We will consider several types of models in later chapters.

Characteristic shapes

A curve might be *symmetrical*. A symmetrical distribution is one in which each half of the distribution is a mirror image of the other half (Figure 2.16*a*). In a symmetrical distribution one side, say the left one, could be flipped around the center and matched up with the other side, the right side (Figure 2.16*b*).

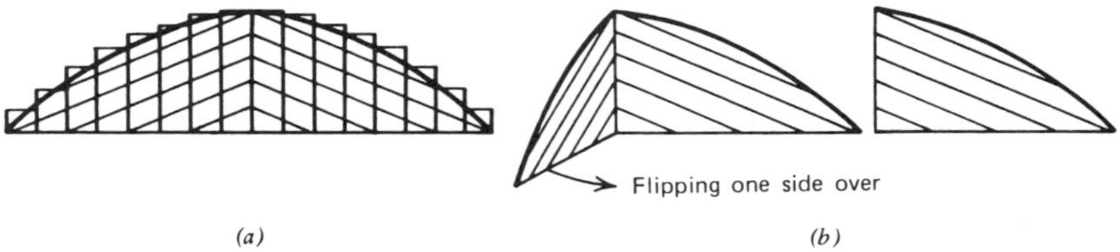

FIGURE 2.16 A symmetrical distribution.

Most distributions that are not symmetrical (nonsymmetrical) can be described as *skewed*. A skewed distribution has a tail on one side and a concentration on the other. If this tail is on the right, it is called positively skewed (Figure 2.17); if the tail is on the left, it is a negatively skewed distribution.

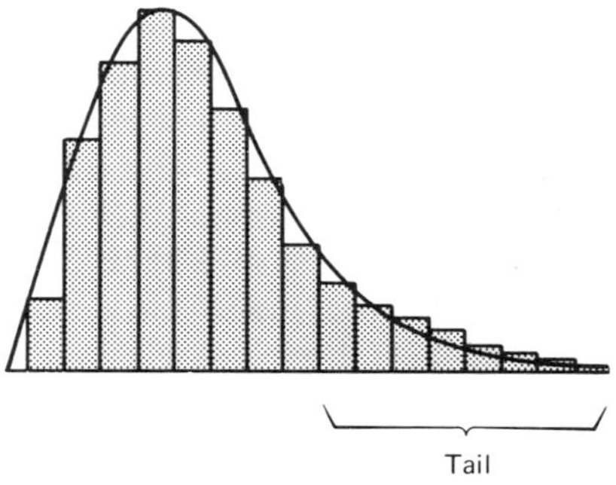

FIGURE 2.17 A positively skewed distribution. The tail or extended section is toward the right.

Skewness is very important in many of the types of statistical tests we will perform. If a distribution we will be working with is very skewed, certain techniques should not be used. Unfortunately, however, there is no very good measure of skewness that can be used for reliable comparisons of distributions.

In a continuous smooth distribution we can usually identify a peak. The X-value corresponding to the location of this peak is called a *mode* (see Figure 2.18*a*). In some cases a noticeable second peak and mode are present (Figure 2.18*b*). We call such a distribution a bimodal distribution. This may be a situation to be very concerned about. Many statistical techniques will be invalid if the distribution is bimodal. This is often a more serious condition than skewness.

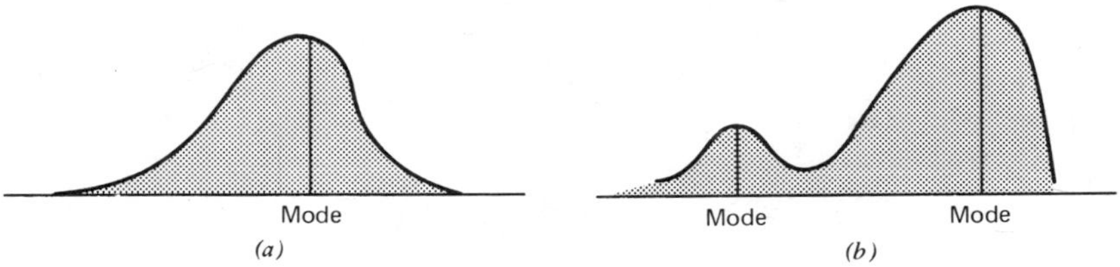

FIGURE 2.18 Distributions showing (*a*) one mode, and (*b*) two modes or bimodal.

When we observe a bimodal distribution, one of the primary reasons for the concern I have expressed is that *you probably do not have a single distribution.* You may not have a single source of the data. Other factors appear to be contributing to the overall distribution. Often we have two completely independent distributions that, when we combine them, merge to create one that is bimodal. When we look at Figure 2.18*b*, we should be wary. It is very infrequent that a single source will yield this shape.

Consider the following examples.

EXAMPLE 2.1

A ball is tossed 10 times to a person and the number of times he or she catches it with the right hand is recorded. Figure 2.19 shows a histogram for a sample of 124 different persons to whom a ball was tossed (10 times).

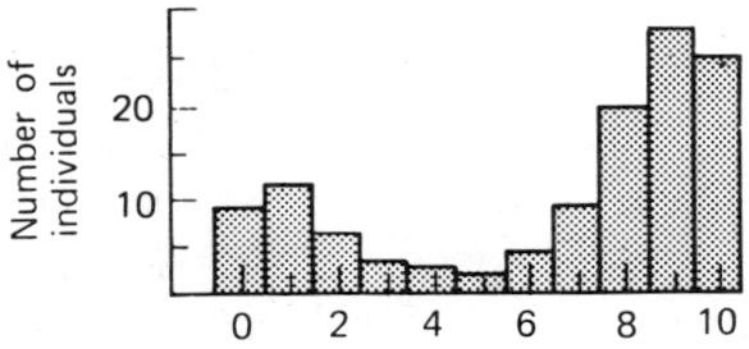

FIGURE 2.19 Number of catches with right hand out of 10 tosses for 124 different people. A bimodal distribution.

Here we have two different groups of individuals, right-handed and left-handed. The distribution is strongly influenced by the duality of categories present. We are able to attribute the shape to an expected dual cause system. However, if we did not know about right- and left-handedness, we would indeed have an interesting problem.

Often two distributions will have different locations or spreads and a combination of them will not be bimodal. The combined distribution will just be spread out much more than either of the individual distributions. This is illustrated in the next example.

EXAMPLE 2.2
The distribution of light intensity (in lumens) of two different styles of 60-watt light bulbs is shown in Figure 2.20*a* and 2.20*b*. If the bulbs are all lumped together and not identified as to types of light bulb, then the resulting distribution is the much broader, more spread out one, as in Figure 2.20*c*. However, by just looking at the overall distribution, we

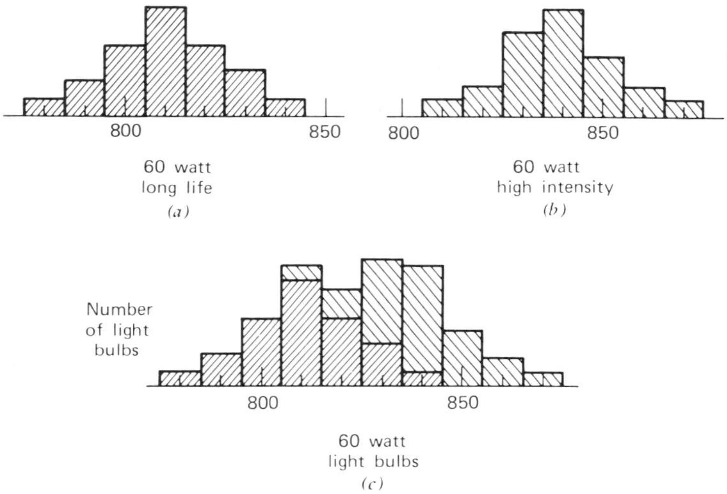

FIGURE 2.20 Intensity of light bulbs: (*a*) and (*b*) have different arithmetic averages; (*c*) disregards different types of light bulbs.

would not have reason to question the width of the overall distribution. We must also have the data broken down into subdivisions. If we never subdivide the data according to any additional classifications, then we never can determine if we have more than one factor at work. We cannot discover other important influences. In most real situations as many as 5 to 50 major factors might be considered. The intensity of the light bulb might depend on things such as the manufacturer, the actual voltage at the wall, the age of the bulb, the instrument used to measure the bulb, the color of the wall, the angle at which the instrument is held, or the temperature of the room.

In most practical situations many different factors will influence the final distribution. Some of these factors will create additional modes, others will just spread out or distort the distribution. The net result is that one rarely encounters simple distributions. Even less frequently will you find one that would also be theoretically expected.

REVIEW

Graphs • Talley Sheets • Bar Charts • Histograms
Frequency • Relative Frequency • Percent • Area
Category • Grouping • Midpoint • Intervals and Limits
Cumulative Frequency • Relative Cumulative Frequency • Percentile
Shape of a Distribution—Symmetrical • Skewed • Mode

PROBLEMS

SECTION 2.2 In each of the following problems you should also state which are X-values and which are frequencies.

2.1 Draw a bar chart for the number of different errors that Lucy made (see Introduction, page 1).

2.2 Draw a bar chart for the amount of food consumed by Snoopy during the past season (see page 2).

2.3 A collection of people were stopped in the hall and asked what meals they ate the previous day. B stands for breakfast, L for lunch, and D for dinner. A person who ate just breakfast and dinner are recorded BD. The following is the resulting data.

BLD, LD, BLD, BD, BD, LD, BL, D, BLD, BD, BD, BLD, B, LD, BLD, BLD, BD, BL, BLD, LD, BD, L, BLD, BLD,

Draw a bar chart showing how many people had which meals. What kind of data is this?

2.4 Referring to the data in Problem 2.3, draw a bar chart showing the number of people who ate breakfast, lunch, and dinner. What are the relative frequencies?

2.5 A poll of students found that 25% of the students favored the food in the cafeteria, 40% would never eat it, 15% tolerated it, 10% did not know there was a cafeteria, and 10% brought their own lunch.

a. What type of data do we have?
b. What are the categories?
c. What goes on the frequency axis?
d. Draw the bar chart

SECTION 2.3

2.6 The number of customers in a store on each day of the week for one week were found to be

127, 56, 32, 41, 37, 67, 167.

What are the X-values? Draw a histogram.

2.7 The number of hours of overtime worked during one week by the 125 employees of a company were as follows.

NUMBER OF HOURS	NUMBER OF EMPLOYEES	NUMBER OF HOURS	NUMBER OF EMPLOYEES
0	39	7	3
1	3	8	6
2	3	9	10
3	9	10	12
4	10	15	3
5	15	17	1
6	10	20	1

Draw a histogram to show the frequencies (numbers of employees) that the particular number of overtime hours were worked.

2.8 The number of absences per day in my last semester statistics class are recorded below. There were 25 students originally registered in the course and a total of 30 class periods to the semester.

0	1	2	1	0
1	2	0	1	0
0	0	0	2	1
1	1	1	2	2
3	2	4	4	2
6	3	2	2	2

Draw a histogram showing the number of days each number of absences was recorded. Provide a scale of actual frequency and relative frequency.

SECTION 2.5

2.9 I will measure the heights of the students in my class of 30. Discuss the following possible sets of category interval sizes. Which is most appropriate? Why are some very inappropriate?

a. To the nearest inch.
b. To the nearest foot or 0.25 meter.
c. To the nearest 0.5 inch or centimeter.
d. To the nearest 0.01 inch or millimeter.

2.10 Do the same as in Problem 2.9 if we are to record the total assets (in dollars) of 50 banks in New York City:

a. To the nearest billion dollars.
b. To the nearest million dollars.
c. To the nearest thousand dollars.
d. To the nearest cent.

2.11 Discuss when and why it is often necessary to group.

2.12 For the following situations provide a reasonable set of intervals and midpoints for preparing a histogram.

a. Sales of fresh doughnuts during a 3 month period have varied from 20 to 147.
b. Body temperature of 50 individuals is recorded to nearest 0.1°F (0.05°C) and ranges from 94.5°F to 102.1°F (34.7°C to 38.9°C).
c. Math college board scores of 100 freshmen in a community college varied from 290 to 674.
d. The lengths of 8-foot boards (or 2.438 meters) measured to the nearest 1/32 inch (or 1 millimeter) vary up to 2 inches larger or smaller (about 50 millimeters larger or smaller).

2.13 On page 31, I have grouped the data for Example 1.1 with midpoints of 2, 5, 8, and so forth. Group the data now using midpoints of 1, 4, 7, 10, and so on. List the interval limits. Draw a histogram and compare the resulting shape to that shown in Figure 2.8.

2.14 Repeat Problem 2.13, but use only three groups with midpoints 0, 10, 20 or 5, 15, 25. Now what happens when you try to interpret the histogram?

2.15 Loaves of white bread are baked in a factory. A hundred of these loaves were carefully measured to determine their actual weights. The data are presented below.

20.055	20.040	20.019	20.025	20.018
20.052	20.024	19.993	19.984	19.982
20.046	20.036	19.968	20.013	19.982
19.994	20.008	19.987	19.984	20.006
20.033	20.041	19.997	20.003	20.037
19.999	19.975	19.985	19.968	20.022
20.022	20.011	20.002	19.975	19.988
20.002	20.026	20.009	19.960	19.997
19.976	19.976	20.000	19.988	20.006
19.994	19.989	19.970	19.958	20.019
19.981	19.987	20.013	20.018	19.994
19.990	19.968	20.013	20.009	19.986
19.965	19.970	20.053	20.026	20.018
19.977	19.975	20.005	20.024	20.002
19.986	19.969	20.004	20.003	20.024
20.052	20.017	19.986	19.972	20.006
19.982	20.017	20.004	20.030	19.983
20.032	19.955	20.019	20.013	20.001
20.007	20.011	19.983	19.962	20.035
20.025	19.999	20.010	20.000	19.993

First select an appropriate set of intervals and midpoints to be used for categories in summarizing this data. Tally the weights (save the table, since you can use it for further work in this chapter and the next chapter) and then prepare a histogram.

Comment on how the histogram would probably have looked if you did *not* group.

2.16 A collection of 40 fine sutures used in surgery was tested to determine the number of grams they could stand before breaking. The results were

33.5, 27.1, 29.7, 35.3, 41.7, 22.5, 31.3, 29.6, 41.4, 51.3
41.4, 35.9, 28.5, 23.4, 31.3, 40.4, 30.1, 37.3, 26.6, 28.4
39.1, 44.1, 29.6, 24.1, 27.4, 29.1, 32.5, 27.8, 32.3, 39.0
31.1, 29.8, 34.6, 30.7, 40.2, 30.0, 29.9, 26.1, 25.5, 27.7

a. In looking at the results of this data, why might you say that it was really unnecessary for the researcher to have bothered recording the data to three significant figures?
b. You are to prepare a tally sheet and a histogram with about 12 to 14 categories. Determine the midpoints and intervals for each category. Provide the tally sheet and then draw the histogram.
c. Calculate relative frequencies.

SECTION 2.6

2.17 After spraying 100 insects with a new chemical, the number of dead insects were counted after each minute.

AFTER	TOTAL NUMBER DEAD
1 minute	10
2 minute	20
3 minute	45
4 minute	63
5 minute	79
6 minute	86
7 minute	91
8 minute	94
9 minute	96
10 minute	98
15 minute	99
21 minute	100

a. Draw a cumulative frequency diagram showing cumulative number of dead insects after each minute.
b. Determine the number of insects that died during each 1 minute interval and draw a histogram.

2.18 Explain why a cumulative frequency diagram cannot be drawn for Problems 2.1 to 2.5. What type of data must you have in order to provide a cumulative frequency diagram?

2.19 Refer to Problem 2.7. Determine the cumulative percents of employees that have worked each number of overtime hours and draw a cumulative frequency diagram (in percents).

2.20 Refer to Problem 2.8.

a. Calculate the cumulative frequencies and draw a cumulative frequency diagram
b. What percent of the time should I expect at least two absences?
c. What is the chance of 20% or more of the class being absent?
d. Discuss the following statements:

"Since on no days were there five absences, the chance of five absences must be 0."

"Since on no days were there seven or more absences, I would never expect to have seven students absent."

2.21 Draw a cumulative frequency diagram for Problem 2.15. Below what value are 40% of the breads? What is the 60th percentile? What is the chance of a bread weighing more than 20 ounces? Provide a scale and a separate table listing every 5th percentile.

2.22 Refer to the data in Problem 2.16. Calculate the relative cumulative frequencies for suture strength. Ninety percent of the sutures are at least how strong?

2.23 Percentiles refer to the X-value for which a particular percent or hundredth of the population lie below. Two additional subdivisions are also frequently used: quartiles and deciles. As the words imply, quartiles (symbolized by Q) are quarters (thus splitting the population into 4 equal parts) and deciles (symbolized by D) are tenths (splitting the population into 10 parts). Subscripts designate the particular quartile or decile (with the first the smallest X-value).

a. What percentile would Q_1 be?
b. What percentile would the third quartile be and what would its subscript be?
c. What percentile is D_3?
d. What percentile is D_6?
e. What would the upper decile be?
f. How can you express the middle 50% of the population in terms of quartiles?

2.24 Refer to Example 1.1.

a. What value of number of tickets is the bottom or lowest quartile?
b. What value of number of tickets is the upper decile?
c. What is the value of D_3?
d. Twelve falls in what decile?
e. Twelve falls in what quartile?

SECTION 2.9

2.25 A collection of different types of tests are to be given to different students. What would you expect the distribution of grades to look like in each of the following cases? Why?

a. An addition test is given to high school freshmen.
b. A college physics test is given to high school freshmen.
c. An addition test is given to two groups of students—50 second graders and 50 high school freshmen.
d. A college physics test is given to 50 second graders and 50 high school freshmen.

2.26 How would you describe the distribution of Problem 2.7

CHAPTER THREE
MATHEMATICAL DESCRIPTIONS

I have already mentioned several times that we will employ mathematics to describe data. Since measurements and data are generally expressed in the form of numbers, the kinds of words we will find most useful to assist us in discussing this data or to aid us in condensing the data should, naturally, also be in the form of numbers. The importance of being able to describe the data and thereby find a useful way to relate to the data should be obvious. As to the necessity of condensing the data, I merely refer you back to Table 1.1 (Example 1.1) and ask you to "tell me about it!" We will now explore some of the kinds of mathematical descriptions.

3.1 GENERAL CHARACTERISTICS

In the last chapter you drew a type of picture. It told you that the data were somehow distributed over several different values, with some being used more frequently than others. If you now look carefully at the distribution (and this will be true of any distribution) you should observe that the pattern has at least *two* very important characteristics. They are:

1. A *center* or central value.
2. A *spread* around that center.

It will be necessary to find mathematical values to describe these two characteristics for virtually all of the sets of data you encounter. Of course, these aren't the only characteristics, but they are the most important and about the most useful. Some other types of descriptive characteristics (from a mathematical approach) will be mentioned, but we will generally avoid them in our work.

We are now about to engage in some mathematics. There will be some basic algebra used to tell you about what is going on and what has to be done. This simplifies the writing. The actual manipulation of the data is arithmetic (mostly addition, multiplication, and squaring). The problems are designed so that if you must work by hand without the aid of *any* computational equipment, there are enough problems available so you can learn and apply what I am talking about without exhausting yourself doing arithmetic. I will also give you some hints and shortcuts to use, since everyone who uses a computer at some time is faced with a broken one (or the batteries of the calculator go dead), and you must resort to old-fashioned techniques. For individuals who have access to calculators or computers I have provided some additional and more realistic sets of data for you to work with.

3.2 CENTRAL VALUES OR AVERAGES

You have already encountered at some point in time the term "average." You probably have computed a baseball "average." You certainly have read about "average" salaries or "average" costs of living. Well, "average" and central values are one and the same. The difficulty is that you have been deceived with the word "average." There is not just one exclusive quantity that can serve as *the* measure of the center of a set of numbers. There are several kinds of averages and they may have different values *for the same set of data.*

We will work with four kinds of averages: *midrange, mode, median,* and *mean.* Each plays its role, although one, the mean, will be used more than any of the others. Some will be easier to calculate than others, several may have the same value for some data. Sometimes one (the mode) may not even exist for some data. I will introduce them in the above order, which is somewhat in order of increasing difficulty.

Midrange

The midrange is perhaps the simplest to calculate. It is halfway between the two extreme values—it is the *arithmetic average* of these two values.

$$\text{Midrange} = \frac{\text{Smallest} + \text{Largest}}{2} \tag{3.1}$$

EXAMPLE 1.1 (Cont.)
We found that the maximum number of tickets issued was 19, therefore the largest *value* (not the largest *frequency*) was 19. The smallest value was the issuance of 1 ticket. Thus

$$\text{Midrange} = \frac{1 + 19}{2} = \frac{20}{2} = 10$$

EXAMPLE 1.2 (Cont.)
The smallest value is the shortest time it takes to get to school or 13.6 minutes. The largest value was 16.7 minutes. Thus

$$\text{Midrange} = \frac{13.6 + 16.7}{2} = \frac{30.3}{2} = 15.15$$

Mode

The mode is often the next easiest to find. It is the X-value that appears most frequently. The main difficulty is that there may not be just one single mode, there may be several—as we have already seen in the last chapter. I have already introduced you to the idea that the value associated with the peak of a distribution or histogram is called a mode (recall Figure 2.18). We have also explored the situation where two modes may exist—a bimodal distribution. When discussing data, it is often not the actual mode that is important but, instead, the number of modes that are present.

EXAMPLE 1.1 (Cont.)
Nine occurs 17 times. No other value occurs as frequently as this and therefore 9 is the mode.

EXAMPLE 1.2 (Cont.)
Here there is no single mode, since 15.1, 16.4, and 16.5 all occur with the same frequency—2. This, as I have said before, may not be unusual. We merely state that there is no single mode.* The condensed histogram of Figure 2.11 does not have a single mode of 16.5.

Median The median is a very simple value to calculate *if* the values have been placed in order of largest to smallest (or smallest to largest) of *if* a histogram has been constructed (a tally sheet will serve equally well). The median corresponds to the middle of the values, *when the numbers have been placed in order*. The way we obtain *the* median depends on whether we have an even or an odd set of numbers. In general, we will try to split the set of values up into two groups, and the median lies midway between the two groups.

If we have an *even* set of values (n is even) when we divide them into two groups, we obtain two middle values. The median then is the arithmetic average of these two middle values. Thus we have half the values greater than (or equal to) the median, and the other half less than (or equal to) the median.

If there is an *odd* set of numbers you cannot split them up evenly. The number left over, which should be in the middle, is the median. For example, given an *odd amount of numbers*, say the five numbers 1, 3, 9, 2, and 6, one of them is the median. Arranging them in order of increasing value, we would have 1, 2, 3, 6, 9. Then, two of the numbers are less than 3 and two are greater than 3, so 3 is the median. The median is *not* the number 9. It should be obvious that if the numbers have just been randomly written down, they could be rewritten again and the middle one in that random set could be any value. But there can *only be one median* to any set of numbers. The way to assure that there is one value is always to place the numbers in numerical order.

If there are 11 numbers, the median is the 6th one when the numbers are placed in order. So you would have 5 numbers, then the median, then 5 more numbers. A quick formula for figuring out which is the median is simply

$$\text{Median} = [(n + 1)/2]\text{th value}$$

* With small sets of data, say less than 40 values, the mode is often not very useful for any purpose.

where n is the number of numbers. (Throughout this text and virtually everywhere data is used, n stands for the *number* of values or the size of the sample.)

So for 11 numbers, the median is the

$$\frac{11 + 1}{2} = \frac{12}{2} = 6\text{th value}$$

For *even* sets of numbers, the median does not have to be one of the values. Here you can split the numbers into two completely separate groups. Given the six numbers 1, 3, 8, 5, 1, 7, again you *first* must put them in numerical order—or 1, 1, 3, 5, 7, 8. Split them into two groups: 1, 1, 3 and 5, 7, 8. The median is halfway between the two middle numbers—or $(3 + 5)/2 = 4$. If the two middle numbers are the same, then obviously the median is that common value. [Note also that we are looking for the value between the third and fourth values, the 3.5th value. Thus, $(n + 1)/2 = (6 + 1)/2 = 7/2 = 3.5$ holds.]

EXAMPLE 1.1 (Cont.)

Since there are 112 numbers, they would be split into two groups of 56 values each. The median would then fall between the 56th and the 57th number. The easiest way to find these numbers is to refer to the tally sheet, frequency table, or even the histogram (Figure 2.4 or 2.5). Start on one end and add up the frequencies until you reach the 56th observation. Record that value and then record the next one (the 57th). If they are identical, that value is the median. If they are different, the median is halfway between them. Here, both the 10s (in fact, the 46th to the 61st are all 10s), so the median is 10.

EXAMPLE 1.2 (Cont.)

We have 20 numbers. The median would be between the 10th and the 11th ones. Looking back at the histogram of Figure 2.9 and counting until we reach the 10th and 11th numbers, I obtain: 10th = 15.6, and 11th = 15.7. Therefore the median is 15.65.

Mean We now come to what is generally regarded as the most important and most commonly used measure of central value. You have probably encountered it by the name of arithmetic average. This is the one you generally think of when you hear the word "average." To compute it, you merely add up all the values and divide by n (the number of values). To write this instruction down mathematically, I will symbolize the mean by $\bar{X}$ (read this as "X bar"). If we call each of the individual numbers Xs, we must be able to distinguish them apart. We do this by calling them X_1, X_2, X_3, and so on. You read X_1 as "X sub 1" (the 1 is a subscript) and you consider this as the first one of the set of numbers. X_2 is the second, X_3 is the third, and X_n is the nth number (whatever the n might be). These are merely substitutes for the actual numbers you have in a set of data. We may also use a little i or a little j as a subscript to stand for any one of the individuals.

So

$$\text{Mean} = \bar{X} = \frac{X_1 + X_2 + X_3 + X_4 + \cdots + X_n}{n} \tag{3.2}$$

The $+ \cdots +$ is used to indicate that we may be adding many more than just the five X-values. If we were adding up 112 values (as we will do in Example 1.1), n would be 112 and I would have to write down 112 different Xs. But you should get the meaning of what you are to do from just the few terms written down.

We will use the operation of adding up a set of numbers quite frequently. Again, for the sake of making the writing easier and simplifying the equations we will encounter, mathematicians use a symbol $\sum$ (a Greek capital sigma) to stand for "Take the sum of a certain set of numbers."

If we let X_i stand for those X's in Equation 3.2, we can rewrite the sum (the numerator) as

$$\sum_{i=1}^{n} X_i = X_1 + X_2 + X_3 + X_4 + \cdots + X_n \tag{3.3}$$

The left part is interpreted as follows: the $\sum$ says you will add up a group of numbers; the types of numbers are those next to the $\sum$, in this case the Xs; the $i = 1$ means the first X you will start with is X_1; the n on top of the $\sum$ says you will keep on adding Xs until you reach the one called X_n. So if n is 112, you will start with X_1 and add up all the Xs up to and including X_n or X_{112}. We can now write the mean as simply

$$\text{Mean} = \bar{X} = \frac{\sum_{i=1}^{n} X_i}{n} \tag{3.4}$$

If you need more work with the subscripts and the summation ($\sum$) notation refer to Appendix B—now. (The next few examples may help, but the work in the appendix may help much more).

EXAMPLE 1.1 (Cont.)

Refer to Table 1.1 for the set of data. n is 112. If we start from the upper left corner and go to the right, X_1 is 14, X_2 is 17, X_3 is 18, ..., X_{15} is 11, X_{16} is 3, ..., X_{110} is 9, X_{111} is 4, X_{112} is 7. What are X_{21}, X_{35}, X_{100}?*

So

$$\begin{aligned}\sum_{i=1}^{112} X_i &= X_1 + X_2 + X_3 + \cdots + X_{110} + X_{111} + X_{112} \\ &= 14 + 17 + 18 + \cdots + 9 + 4 + 7 \\ &= 1132\end{aligned}$$

* *Answer.* $X_{21} = 11$, $X_{35} = 19$, $X_{100} = 10$.

and

$$\text{Mean} = \bar{X} = \frac{\sum_{i=1}^{n} X_i}{n} = \frac{14 + 17 + 18 + \cdots + 9 + 4 + 7}{112} = \frac{1132}{112} = 10.11$$

Often you will find the $\sum$ written without the $i = 1$ below it or the n above it. I will also leave off these symbols. You should realize that if you are adding up a set of numbers it must begin with one value and end with another value. The only times I will include the extra instructions is when they are necessary to avoid confusion (it is possible that you could be starting with $i = 5$ and end with $n = 12$). My analogy would be the instruction "start the car." If I assume it has an automatic transmission, most cars will start with the shift level in neutral or in park, and I do not have to include in my statement where the shift lever must be placed. But my car *might* be an exception and I would then be forced to say what the particular exception is; for example, put the lever in park, pull the lever toward you until you hear a click, then turn on the ignition.

EXAMPLE 1.2 (Cont.)
Here $n = 20$. Refer to Table 1.2 for the individual values. Then

$$\text{Mean} = \bar{X} = \frac{\sum X_i}{n} = \frac{16.5 + 16.5 + \cdots + 14.5 + 15.5}{20} = \frac{310.0}{20} = 15.5$$

Means for grouped data

The means we have calculated for Examples 1.1 and 1.2 were found directly from the original data. Quite often, however, the data have been put into categories, as in a table, tally sheet, or histogram. In these cases we will work with the categories and the frequencies associated with those categories. We spent a considerable amount of time discussing categories in the last chapter. The important idea now is that all the individuals in a particular category are treated *as though they all have the same value*—the X-value of the center of the category or interval to which they belong. Usually a mean calculated from categories will be different than one calculated from the raw data. However, the difference is generally not important.

In Table 3.1 I have listed the individual X-values and their frequencies for each of the X-values of Example 1.1. Each of the X-values represents a separate category, and there are 21 of them. If we take the sum of the frequencies by adding up column 2, we obtain the quantity 112. This is also the total number of observations (the n). Thus, if I use the symbol j to stand for *category* and I have c categories, I can write this as

$$\text{Sum of the frequencies} = \sum_{j=1}^{c} f_j = n$$

or simply

$$\text{Sum of the frequencies} = \sum f = n$$

TABLE 3.1 Calculation of the Mean for the Number of Tickets Issued (Example 1.1)

COLUMN	1	2	3
QUANTITY	X-VALUE	f FREQUENCY	$f \cdot X$ FREQUENCY TIMES X-VALUE
	20	0	0
	19	1	19
	18	1	18
	17	2	34
	16	0	0
	15	5	75
	14	7	98
	13	8	104
	12	13	156
	11	14	154
Data	10	16	160
	9	17	153
	8	10	80
	7	4	28
	6	4	24
	5	1	5
	4	2	8
	3	3	9
	2	3	6
	1	1	1
	0	0	0
Sum		112	1132

$$n = \text{Sum of frequencies} = 112$$

$$\text{Mean} = \bar{X} = \frac{\text{Sum of (frequency times } X)}{\text{Sum of frequency}} = \frac{\sum (f \cdot X)}{\sum f}$$

$$= \frac{\text{Sum of column 3}}{\text{Sum of column 2}} = \frac{1132}{112} = 10.11$$

Before, we added up a quantity every time that it appeared. Since 9 appeared 17 times, we made 17 separate additions of 9 in our summation. But I can also multiply them together to obtain the same 153 (17 times 9 is the same as $9 + 9 + 9 + \cdots + 9 + 9$), and I will have eliminated a large number of additions. Thus, I will multiply each X-value by the number of times it appears (its frequency) and I will add up all of these products. This sum will be the same as the sum of all of the individual values. We write this as

Sum of all of (frequency of a category) times (X-value of the category)

$$= \sum_{j=1}^{c} (f_j \cdot X_j)$$

(We are adding up the *products*, and that is why I am using the parentheses around the $f \cdot X$.) Then, finally,

$$\text{Mean} = \frac{\sum_{j=1}^{c} (f_j \cdot X_j)}{\sum_{j=1}^{c} f_j} = \frac{\sum (f_j \cdot X_j)}{n} \tag{3.5}$$

EXAMPLE 1.1 (Cont.)

Refer to Table 3.1. Here column 3 contains the products of $f \cdot X$. Then

$$\text{Mean} = \frac{(0 \cdot 0) + (1 \cdot 1) + (3 \cdot 2) + (3 \cdot 3) + (2 \cdot 4) + \cdots + (1 \cdot 19) + (0 \cdot 20)}{0 + 1 + 3 + 3 + 2 + 1 + \cdots + 1 + 0}$$

$$= \frac{0 + 1 + 6 + 9 + 8 + \cdots + 19 + 0}{0 + 1 + 3 + 3 + 2 + 1 + \cdots + 1 + 0} = \frac{1132}{112} = 10.11$$

Observe that there were 21 categories (0 to 20), although only 18 had frequencies of 1 or more. This points out that the total number of categories is somewhat arbitrary and really doesn't matter as long as all of the values are counted.

When data have been grouped into wider intervals, we can still follow the same procedure.

In column 1 of Table 3.2 I have listed a group of intervals (which we previously called Set 2.1—see page 31). In column 2 I give the center of the intervals that we will use as the X-values for the various intervals or categories.

TABLE 3.2 Computing the Mean Number of Tickets Using Grouped Data and Given Intervals (Example 1.1)

COLUMN	1	2	3	4
QUANTITY	INTERVAL	CENTER OR *X*-VALUE	FREQUENCY	FREQUENCY TIMES *X*-VALUE
Data	18.5–21.5	20	1	20
	15.5–18.5	17	3	51
	12.5–15.5	14	20	280
	9.5–12.5	11	43	473
	6.5–9.5	8	31	248
	3.5–6.5	5	7	35
	0.5–3.5	2	7	14
Sum			112	1121

$$\text{Mean} = \bar{X} = \frac{\sum (\text{frequency} \cdot X)}{\sum (\text{frequency})} = \frac{\text{Sum of column 4}}{\text{Sum of column 3}} = \frac{1121}{112} = 10.01$$

Column 3 contains the frequencies for each interval, and column 4 has the products of the frequencies times the center or X-values. Then

$$\text{Mean} = \overline{X} = \frac{\sum(\text{frequencies} \cdot X\text{-values})}{\sum(\text{frequencies})}$$

$$= \frac{(7 \cdot 2) + (7 \cdot 5) + \cdots + (3 \cdot 17) + (1 \cdot 20)}{7 + 7 + \cdots + 3 + 1}.$$

$$= \frac{14 + 35 + \cdots + 51 + 20}{112} = \frac{1121}{112} = 10.01$$

As we should have expected, this mean is slightly different than the original 10.11 that we obtained by using all of the numbers. However, the calculations here were simpler, and you should realize that there was only a 0.10 difference between the two means, or an error of 1%. Most of the time a small error of this type is not objectionable, since the data itself are not as precise as the mean we are able to calculate.*

3.3 WHICH AVERAGE TO CHOOSE

I have introduced you to four different numbers that can be used to state what the "average" or typical value for a set of numbers could be. We have seen that they are not necessarily the same. Table 3.3 summarizes the

TABLE 3.3 Summary of Averages for Two Examples

EXAMPLE	1.1	1.2
Midrange	10	15.2
Mode	9	No single mode[a]
Median	10	15.65
Mean	10.11	15.5

[a] Note that with so few values it is especially difficult to identify *a* mode even if we would expect *only one* to be present.

averages that we have computed for two examples. But how do we decide which *one* to use if only one is to be quoted? This is a very important question and is one that constantly pops up. Let me begin to answer this by first exploring the question of how to interpret the different "averages."

You have already seen that the mode can be pictured as the value associated with the *peak* of a distribution. The median was shown to be a point that splits the distribution into two *equal areas*. Finally, you might

* Here we have individual values accurate to at most two digits. The mean we calculated has four digits, two of which must be regarded a little bit skeptically. Generally, we calculate the mean to *one* more significant place than the data is recorded. However, I have gone an extra place mainly to illustrate how good the approximations are.

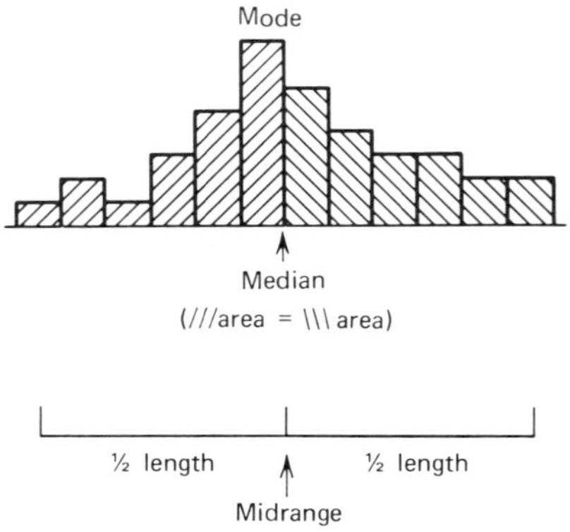

FIGURE 3.1 Geometrical interpretation of the median, midrange, and mode.

observe that the midrange cuts the *length* of the distribution into two equal parts (see Figure 3.1). What about the mean?

Let us look at the histogram of a distribution. In the last chapter I asked you to think of the areas as weights, with the greater heights having greater weight and the lower heights being less heavy. Carrying this analogy one step further, consider the X-axis as a support for these weights. Then the mean is the one value that can act as a balancing point for the entire distribution. So if we place the distribution on a seesaw (Figure 3.2), the spot that allows it to be in a perfect balance, what the physicists call the fulcrum, is also called the *mean*. What significance does this have with respect to our way of interpreting or thinking about the mean? Think about the seesaw. A single point two units away from the fulcrum can counter the effect of two points (or weights) that are only half as far away from the fulcrum (or one unit away). Thus a single point that is some distance from the center of a distribution can significantly alter the value of the *mean* by offsetting the effect of *several* points that are close to the center (see Figure 3.2). Of course, the mean is not the only average strongly influenced by a single very large or very small value—the midrange depends entirely on the extreme values. However, because the midrange is so seriously affected by an individual wild value, it is not regarded as a very important measure of average value. This is particularly true with large sets of data. The median is not affected by wild or peculiar individuals

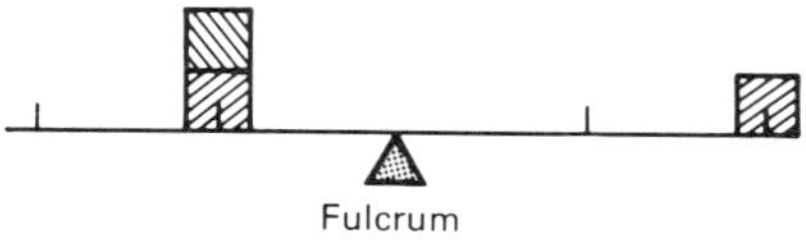

FIGURE 3.2. A seesaw: The fulcrum is the balancing point for a distribution called the mean. Two weights one unit from the fulcrum are balanced by *only* one weight placed two units from the fulcrum on the other side.

or even small sets of wild individuals. For this reason it is usually a good indicator of the typical size of the values.

Statistical theory tells us that for the important application called estimation, the best estimator (the value that best describes the population center or the place from which all these values come from) for *many* important applications is the *mean.* But it *is not always* right and it *isn't always* the closest to the true center. It often does not even act as the best description of the center. But it is the only one that we have discussed that uses all of the numbers, and generally is the one most often quoted and used in statistical calculations.

The most common problem with averages is that encountered with skewed distributions, and skewed distributions, unfortunately, are far from uncommon. If the distribution is symmetrical and has only one mode, all four averages are identical (Figure 3.3*a*). However, when the distribution is skewed, the four averages fall into a pattern usually like that in

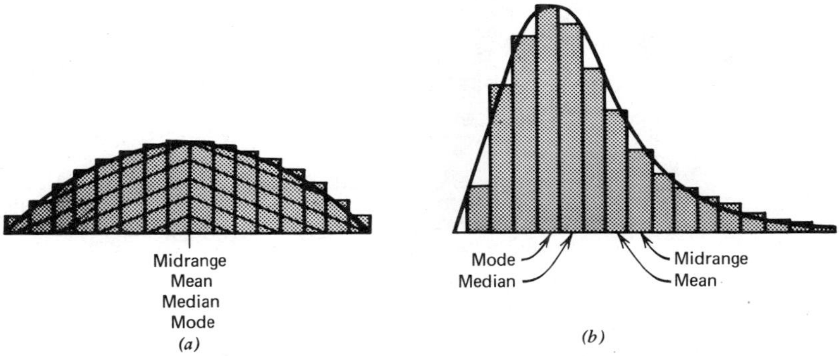

FIGURE 3.3 Relationship of the various "averages": (*a*) For a symmetrical distribution, and (*b*) for a skewed distribution.

Figure 3.3*b*, having an order going from *mode to median* to *mean* to *midrange.* As a result, the word "average" alone can provide misleading information. A good illustration would be the following example.

EXAMPLE 3.1

A company that employs 10 people reported the average income as $19,100. Is this a good figure? The actual incomes were also listed (Table 3.4 and Figure 3.4).

The average that was reported was the mean. What are the other averages?

Median = 7150—quite a bit lower, but do you agree that it is a much more typical value, and presents a much better single description of the collection?

Mode = $6900.

Midrange = $65,600 = this is really high. Is this a "good" value?

TABLE 3.4 Salaries of personnel in dollars

6200	7200
6600	7300
6900	7800
6900	10,000
7100	125,000

The main question always is whether one particular average is better than another as a *description of the data*. Do *not* fall into the easy trap of saying that you like the mean or the midrange "*because they are bigger* than the median." The criterion for saying that you like a particular average in any case should be "because it tells it like it *is*," not because it looks better.

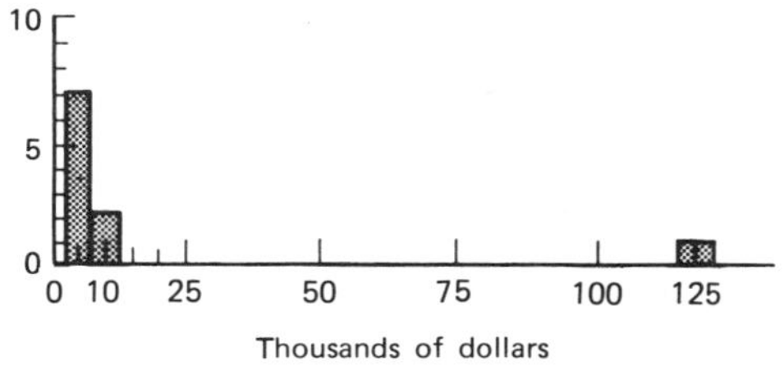

FIGURE 3.4 Grouped histogram for the distribution of salaries of Table 3.4

3.4 SPREAD

The second word that is extremely valuable for describing a set of numbers is the measure of *spread*, or dispersion. Since a distribution has been defined as a collection of numbers that are not identical, we need some way of saying just how different they are. I will present three different measures of spread: *range*, *variance*, and *standard deviation*. A fourth, called the *mean deviation*, will be introduced in the problems, but it is not nearly as important as these three.

Range

This is about the simplest of all descriptive statistics to calculate. It is the total spread of the numbers and is just the difference between the two extreme values.

$$\text{Range} = \text{Largest value} - \text{Smallest value} \tag{3.6}$$

EXAMPLE 1.1 (Cont.)

$$\text{Range} = 19 - 1 = 18$$

EXAMPLE 1.2 (Cont.)

$$\text{Range} = 16.7 - 13.6 = 3.1$$

The range will be particularly useful when working with very small sets of numbers. When there are only four or five values, the range is a very good description of spread. But observe that it is extremely susceptible to individual wild values. Thus, since large sets of numbers tend to have *some* extreme individuals, we frequently ignore the range when we deal with samples greater than about 20 to 25. In addition, the range disregards the middle of the distribution, and thereby treats a distribution shaped like a *U* (with very few values in the middle) no differently than one that is highly concentrated at the center.

3.5 VARIANCE AND STANDARD DEVIATION

The most common and generally most useful measures of the variability or spread of a set of numbers are the *variance* and the *standard deviation*. We will use them somewhat interchangeably during our discussion because the *variance is just the square of the standard deviation*. If one is calculated, I will assume you will be able to easily convert to the other.

$$\text{Variance} = (\text{Standard deviation})^2$$

or

$$\text{Standard deviation} = \sqrt{\text{Variance}} \tag{3.7}$$

We will always calculate the variance. Then, to find the standard deviation, we will take the square root of the variance. A table of squares and square roots is provided in Appendix Table A.8.

The definition of variance that we will use will be:

DEFINITION *The variance is an average of the squares of the differences from the mean.* Mathematically it is

$$\text{Variance} = s^2 = \frac{\sum_{i=1}^{n} (X_i - \overline{X})^2}{n - 1} \tag{3.8}$$

You will see both s^2 and σ^2 (read as sigma squared) used as symbols for the variance at different times. You will also see the two quantities $n - 1$ and n in the denominator. The differences have to do with the notions of sample and population. If we are dealing with a sample, which is what most data is, then we will use the s and the $n - 1$. The few times we are dealing with a complete universe or a theoretical population, we will use σ and N. We will also use μ, the Greek letter mu, to stand for the *population* average or mean. When n is very large (say 100 or more), there is essentially no difference between the two variance values.

So the variance may be written in two ways:

Sample

$$s^2 = \frac{\sum (X - \overline{X})^2}{n - 1}$$

Population

$$\sigma^2 = \frac{\sum (X - \mu)^2}{N}$$

(We will use N for population size and n for sample size.) In this text we will almost always be working with the sample variance or s^2.*

The mathematical definition, Equation 3.8, is not usually the best one to follow when we must actually calculate a variance. There are several other arrangements of the formula that can be used that are often easier (although not necessarily easy). I will work now with two formulas, which follow. *If you are not grouping the numbers* (even to the extent of using a tally sheet), then just *ignore the f_i-values*, since they will all be 1's anyhow.

$$s^2 = \frac{\sum f_i(X_i - \bar{X})^2}{n - 1} \tag{3.9}$$

Equation 3.9 is the same as the definition, and generally it is very tedious to use. Another approach is:

$$s^2 = \frac{\sum f_i(X_i^2) - (\sum f_i X_i)^2/n}{n - 1} \tag{3.10}$$

Equation 3.10 is fine for a calculator. However, the numbers often become very large.

If *the numbers are not organized* this is what you would generally use.

Now let us go through a very simple illustration to see what all those different summations are and what kind of calculations are involved.

EXAMPLE 3.2

I have observed the number of students awake and alert in class at 8:00 A.M. on six consecutive Monday mornings as follows.

6, 9, 8, 8, 12, 9

In categories they would be

CATEGORY i	VALUE X	FREQUENCY f
1	6	1
2	8	2
3	9	2
4	12	1

For example, $X_1 = 6$, $X_4 = 12$, $f_2 = 2$, and $f_4 = 1$.

The mean would be

$$\bar{X} = \frac{\sum_{i=1}^{4} f_i \cdot X_i}{\sum f_i} = \frac{(1 \cdot 6) + (2 \cdot 8) + (2 \cdot 9) + (1 \cdot 12)}{1 + 2 + 2 + 1} = \frac{52}{6} = 8.67$$

and $n = \sum f_i = 6$.

* One of the main concerns here with differences in population and sample variances is also to alert you to why you will find different ways of calculating variances discussed in different books. There is no similar problem with means and medians, since there is usually no difference between the formulas used to calculate population average and the sample averages (at least we will not deal with such cases in this text).

Using Equation 3.9, the variance would be calculated as

$$s^2 = \frac{\sum_{i=1}^{4} f_i(X_i - \bar{X})^2}{n-1}$$

$$= \frac{f_1 \cdot (X_1 - \bar{X})^2 + f_2 \cdot (X_2 - \bar{X})^2 + f_3 \cdot (X_3 - \bar{X})^2 + f_4 \cdot (X_4 - \bar{X})^2}{n-1}$$

$$= \frac{1 \cdot (6 - 8.67)^2 + 2 \cdot (8 - 8.67)^2 + 2 \cdot (9 - 8.67)^2 + 1 \cdot (12 - 8.67)^2}{6-1}$$

$$= \frac{1(-2.67)^2 + 2(-0.67)^2 + 2(0.33)^2 + 1(3.33)^2}{5}$$

$$= \frac{1(7.129) + 2(0.449) + 2(0.109) + 1(11.089)}{5}$$

$$= \frac{19.33}{5} = 3.87$$

The standard deviation is then $s = \sqrt{3.87} = 1.97$. Observe that each term that is added together contains both the product of a difference that has already been squared and a frequency.

With Equation 3.10 we have two separate summations to perform. The first is

$$\sum f_i \cdot X_i^2 = f_1 \cdot X_1^2 + f_2 \cdot X_2^2 + f_3 \cdot X_3^2 + f_4 \cdot X_4^2$$
$$= 1 \cdot (6)^2 + 2 \cdot (8)^2 + 2 \cdot (9)^2 + 1 \cdot (12)^2$$
$$= (1 \cdot 36) + (2 \cdot 64) + (2 \cdot 81) + (1 \cdot 144)$$
$$= 470$$

The second is

$$\sum f_i X_i = f_1 X_1 + f_2 X_2 + f_3 X_3 + f_4 X_4$$
$$= 1 \cdot 6 + 2 \cdot 8 + 2 \cdot 9 + 1 \cdot 12 = 52$$

Then observe that

$$\left(\sum f_i X_i\right)^2 = (52)^2$$

This is because the squaring is done *after* the summation—the exponent is outside the parentheses. The previous summation had the square around the X itself, so that *each term had a square.*

Thus

$$s^2 = \frac{\sum f_i X_i^2 - (\sum f_i X_i)^2/n}{n-1} = \frac{470 - 52^2/6}{6-1} = \frac{470 - 450.67}{5}$$

$$= \frac{19.33}{5} = 3.87$$

So several different formulas will all yield the same result, even though the calculations may be slightly different. I should point out two additional useful things of which to be aware.

1. If the numbers were *not grouped* but just listed, Equation 3.10 would have been easiest to work with (with a few additional calculations). Since all the f_i's are 1's,

$$s^2 = \frac{\sum X_i^2 - (\sum X_i)^2/n}{n-1}$$

$$= \frac{6^2 + 9^2 + 8^2 + 8^2 + 12^2 + 9^2 - (6 + 9 + 8 + 8 + 12 + 9)^2/6}{6-1}$$

$$= \frac{36 + 81 + 64 + 64 + 144 + 81 - (52)^2/6}{5}$$

$$= \frac{470 - 450.67}{5} = 3.87$$

2. You will sometimes see Equation 3.10 rewritten as

$$s^2 = \frac{\sum f_i \cdot X_i^2 - n(\bar{X})^2}{n-1} \tag{3.11}$$

The only difference is in the right-hand term. If you happen to already know the mean this may be easier to use than Equation 3.10.

For actually carrying out the calculation of a variance you should put the data in a table form. The following formats are particularly suitable for the formulas that I have used. In addition, you should be able to use these same tables to compute the mean.

By Equation 3.9.

COLUMN	1	2	3	4	5	6
QUANTITY	X	f	$f \cdot X$	$(X - \bar{X})$	$(X - \bar{X})^2$	$f \cdot (X - \bar{X})^2$
	...	...	...	...	...	...
Individual values	...	...	...	...	...	...
	...	...	...	...	...	...
		Sum	Sum			Sum

$$\text{Mean} = \frac{\sum f \cdot X}{\sum f} = \frac{\text{Sum of column 3}}{\text{Sum of column 2}}, \qquad n = \sum f = \text{sum of column 2}$$

$$\text{Variance} = \frac{\sum f \cdot (X - \bar{X})^2}{n-1} = \frac{\text{Sum of column 6}}{n-1}$$

Where the word *Sum* is stated this means add up the column.
Note that here you must compute the mean *before* you can fill in columns 4, 5, and 6.

By Equation 3.10.

COLUMN	1	2	3	4[a]	5
QUANTITY	X	f	$f \cdot X$	X^2	$f \cdot X^2$
Individual values	...	...	...	...	...
	...	...	...	...	...
		Sum	Sum		Sum

[a] This column may be omitted. I find it easier to square each X-value and then multiply by the f-value. After a while I have memorized most of the smaller squares and find it simpler to then multiply by the frequency rather than taking a peculiar $f \cdot X$-value and multiplying it again by X. It is entirely a personal preference and you can eliminate the X square columns and still reach the last column, which is the most important consideration.

then

$$\text{Mean} = \frac{\text{Sum of column 3}}{\text{Sum of column 2}}$$

$$\text{Variance} = \frac{\sum f \cdot X^2 - (\sum f \cdot X)^2/n}{n - 1}$$

$$= \frac{\text{Sum of column 5} - (\text{Sum of column 3})^2/n}{n - 1}$$

EXAMPLE 1.1 (Cont.)

I have calculated and presented in Table 3.5 the variance using the methods of Equations 3.9 and 3.10. The values and frequencies are in part A of the table. The calculations are in parts B and C. Both methods give answers with differences of less than one in a thousand. Considering the precision of the data itself, these differences are rather minute. You should now see the disadvantages of the first two techniques. The first requires you to square and multiply awfully messy numbers. Squaring a number like 7.11 and then multiplying it even by a 3 is not my idea of spending an enjoyable afternoon. The second method requires you to square or multiply some very large numbers. Again, 156 times 13 is not the nicest kind of calculation to do over and over. In addition, you often must carry out the calculation to six or seven places in order to obtain an accurate answer.*

* There is also a technique called *coding* that can help relieve some of the difficulties inherent in these two methods. Basically, this involves several simple adjustments to the data that make the numbers easier to handle. The most useful facts are:

1. If you subtract a single constant from each and every individual of a set of numbers, then the same constant has been subtracted from the mean. Thus, you can subtract a number from all the values, calculate the mean in the normal way, add the number back, and you have the actual mean.
2. Adding or subtracting a constant from every value in a set of values does *not* change the variance or standard deviation of the set of values. So if you subtract something from each value to make the numbers smaller this will *not* affect the final value of standard deviation.

TABLE 3.5 Calculation of Variance of the Number of Tickets Issued (Example 1.1); Two Different Methods Are Used

	A. DATA		B. BY THE DEFINITION				C. FOR A CALCULATOR		
COLUMN	1	2	3	4	5	6	7	8	9
QUANTITY	X	f	$f \cdot X$	$X - \overline{X}$	$(X - \overline{X})^2$	$f \cdot (X - \overline{X})^2$	$f \cdot X$	X^2	$f \cdot X^2$
	20	0	0	9.89	97.81	0	0	400	0
	19	1	19	8.89	79.03	79.03	19	361	361
	18	1	18	7.89	62.25	62.25	18	324	324
	17	2	34	6.89	47.47	94.94	34	289	578
	16	0	0	5.89	34.69	0	0	256	0
	15	5	75	4.89	23.91	119.55	75	225	1125
	14	7	98	3.89	15.13	105.91	98	196	1372
	13	8	104	2.89	8.35	66.80	104	169	1352
	12	13	156	1.89	3.57	46.41	156	144	1872
	11	14	154	0.89	0.79	11.06	154	121	1694
	10	16	160	−0.11	0.01	0.16	160	100	1600
Data	9	17	153	−1.11	1.23	20.91	153	81	1377
	8	10	80	−2.11	4.45	44.50	80	64	640
	7	4	28	−3.11	9.67	38.68	28	49	196
	6	4	24	−4.11	16.89	67.56	24	36	144
	5	1	5	−5.11	26.11	26.11	5	25	25
	4	2	8	−6.11	37.33	74.66	8	16	32
	3	3	6	−7.11	50.55	151.65	6	9	27
	2	3	6	−8.11	65.77	197.31	6	4	12
	1	1	1	−9.11	82.99	82.99	1	1	1
	0	0	0	−10.11	102.21	0	0	0	0
Sum		112	1132			1290.53	1132		12,732

$$\sum f = n = 112$$

$$\overline{X} = \frac{1132}{112} = 10.11$$

Equation 3.9

$$s^2 = \frac{\sum f \cdot (X - \overline{X})^2}{n - 1} = \frac{\text{Sum of column 6}}{n - 1} = \frac{1290.53}{111} = 11.626$$

Equation 3.10

$$s^2 = \frac{\sum f \cdot X^2 - (\sum f \cdot X)^2/n}{n - 1} = \frac{\text{Sum of 9} - (\text{Sum of 7})^2/n}{n - 1} = \frac{12{,}732 - (1132)^2/112}{111} = \frac{12{,}732 - 1{,}281{,}424/112}{111} = \frac{12{,}732 - 11{,}441.28}{111} = \frac{1290.72}{111} = 11.628$$

Standard deviation

$$s = \sqrt{11.626} = 3.41$$

$$s = \sqrt{11.628} = 3.41$$

3.6 INTERPRETATION OF STANDARD DEVIATION AND VARIANCE*

Now that you have learned how to go through the drudgery of calculating a variance and standard deviation, you probably are saying, "So what?" Well, initially we set out to find some kind of numerical or mathematical way of expressing how much a set of numbers was spread out. We have done just that. Your problem now is to find some way of figuring out if your number is large or small. I recognize that you have no built-in mechanism that tells you how to make sense out of the numerical size of a variance or a standard deviation. I will now try to give you a way of relating the standard deviation to a distribution. This discussion will deal principally with the standard deviation, since it is virtually impossible to give a simple illustration that will provide a meaningful idea of how the variance is related to the distribution or the data.

Interpretation of standard deviation

Over the years you have used many different units to measure different things. You have paid for a hamburger with so many units of cents (perhaps 95); you have filled the gas tank with units of gallons (maybe 12.6); your height has been measured in units of inches (say 69). If you took a number of measurements of one of them, you would end up with one of our distributions and you could compute a mean and a standard deviation. Remember now that the mean and the standard deviation would be in the *same* units as the measurements were. So the mean cost of 10 different hamburgers might be 100 *cents* and the standard deviation 4.3 *cents*. Similarly, you might have filled the gas tank five times with an average (mean) of 10.8 *gallons* and with a standard deviation of 3.5 *gallons*. The height of one person measured on 12 days might have a mean of 69 *inches* and a standard deviation of 0.29 *inches*.

From now on we will freely convert from the units of the distribution (such as the cents, gallons, and inches) to units of standard deviations. We will say that two numbers in a particular distribution are a certain number of standard deviations apart. If we calculate the difference between the two numbers we then merely divide this difference by the standard deviation of the distribution to convert the units. In general,

$$\text{Number of standard deviations} = \frac{X_2 - X_1}{s} = k \qquad (3.12)$$

where X_2 and X_1 are the two different values, s is the standard deviation of the distribution, and k is the number of standard deviations. We can also express the difference in terms of s, or as

$$X_2 - X_1 = ks \qquad (3.13)$$

So we have two ways of talking about differences—the second being in terms of standard deviations.

Often you are pretty certain that you are talking about a population or you are told that the standard deviation you have is the population

* This section may be deferred until Chapter Five and can be read along with Section 5.3.

standard deviation. In these cases the "number of standard deviations" is expressed as $k\sigma$ (or k sigma). So you frequently will hear differences stated as "so many sigma." These merely are the number of standard deviations the two numbers are apart. To convert back to the actual units, you merely multiply the quoted number (k) by the σ of the population (or s of the sample). I will generally talk about sigma and not s and we will use Equation 3.12 or 3.13 with σ in place of s.

Now lets look at a few examples (each refers to data just mentioned).

EXAMPLE 3.3

Two hamburgers that cost 98 cents and 104 cents have a cost difference of 6 cents. Since $\sigma = 4.3$ cents, they are also

$$\frac{X_2 - X_1}{\sigma} = \frac{104 - 98}{4.3} = \frac{6}{4.3} = 1.4$$

standard deviations apart. Therefore, they are 1.4σ (1.4 sigma) apart.

EXAMPLE 3.4

The two times I filled the car this week I got 15.5 gallons and 9.5 gallons. This difference was 6 gallons, or since $\sigma = 3.5$, it is also

$$\frac{15.5 - 9.5}{3.5} = \frac{6}{3.5} = 1.72$$

standard deviations apart. Therefore they are 1.72σ (or 1.72 sigma) apart.

EXAMPLE 3.5

Twice I was measured and the readings were 69 and 69.5 inches. The difference is 0.5 inches or, since $\sigma = 0.29$, it is also

$$\frac{69.5 - 69}{0.29} = \frac{0.5}{0.29} = 1.72$$

standard deviations apart or 1.72σ (1.72 sigma) apart.

I would like to point out two important observations: (1) Examples 3.3 and 3.4 both had differences of the same actual magnitude, 6 units but, because the standard deviations were different, the numbers of sigma were also different. Thus, with different distributions, the same amount of spread may not be the same in terms of standard deviations; (2) Examples 3.4 and 3.5 both have differences that are of different magnitudes (6 and 0.5, respectively) but the differences are both 1.72σ. Thus, in terms of sigma units, *they are the same* and we will find that there is more similarity between these two values than the previous pair we looked at. We will see that expressing certain differences in terms of units of sigma will permit you to make comparisons between different distributions, assist in making a sensible judgment on how the distribution is spread out, and help you determine what would be an excessive or an unreasonable individual value.

For the most part we will talk only about one type of difference—those where one of the X-values is the *mean*. For the previous examples, if we wanted to convert our first measurements into sigma units from the mean, it would have been done as follows.

EXAMPLE 3.3 (Cont.)
If

$$\bar{X} = 100 \qquad X_1 = 95 \qquad \sigma = 4.3$$

then the number X_1 has a difference from the mean that can be expressed as

$$\frac{X_1 - \bar{X}}{\sigma} = \text{number of standard deviations}$$

$$= \frac{95 - 100}{4.3} = \frac{-5}{4.3} = -1.16\sigma$$

(The minus sign says that the X-value is less than the mean, and 95 *is* less than 100).

EXAMPLE 3.4 (Cont.)
If

$$\bar{X} = 10.8 \qquad X_1 = 12.6 \qquad \sigma = 3.5$$

then

$$\text{Number of standard deviations} = \frac{X_1 - \bar{X}}{\sigma} = \frac{12.6 - 10.8}{3.5} = \frac{1.8}{3.5} = 0.51$$

Thus, 12.6 is just about 1/2 a standard deviation greater than the mean.

EXAMPLE 3.5 (Cont.)
If

$$\bar{X} = 69 \qquad X_1 = 69 \qquad \sigma = 0.29$$

then

$$\frac{X_1 - \bar{X}}{\sigma} = \frac{69 - 69}{0.29} = 0$$

The value we got was the mean itself, so it is 0σ from the mean.

Okay, you can express values in terms of sigma or in terms of "so many sigma units from the mean." Now I offer what is often called the *empirical rule*.

EMPIRICAL RULE
If the *distribution* closely resembles *a bell-shaped or mound-shaped distribution** *then approximately*

68% of the individuals are between minus 1σ and plus 1σ from the mean;
95.5% of the individuals are between minus 2σ and plus 2σ from the mean;
99.73% of the individuals are between minus 3σ and plus 3σ from the mean.

Let me use Example 1.1 to explain what I have just stated.

EXAMPLE 1.1 (Cont.)
Recall that the mean was 10.11 and the standard deviation was 3.41. Then

$$\begin{aligned}
\text{Mean plus } 1\sigma &= 10.11 + 1(3.41) = 13.52\\
\text{Mean plus } 2\sigma &= 10.11 + 2(3.41) = 16.93\\
\text{Mean plus } 3\sigma &= 10.11 + 3(3.41) = 20.44\\
\text{Mean minus } 1\sigma &= 10.11 - 1(3.41) = 6.70\\
\text{Mean minus } 2\sigma &= 10.11 - 2(3.41) = 3.29\\
\text{Mean minus } 3\sigma &= 10.11 - 3(3.41) = -0.12
\end{aligned}$$

In Figure 3.5 I have drawn the histogram and I have labeled these points as $+1\sigma$, $+2\sigma$, $+3\sigma$, -1σ, -2σ, and -3σ, respectively. These then can be used interchangeably with the actual values of 13.52, 16.93, and so forth.

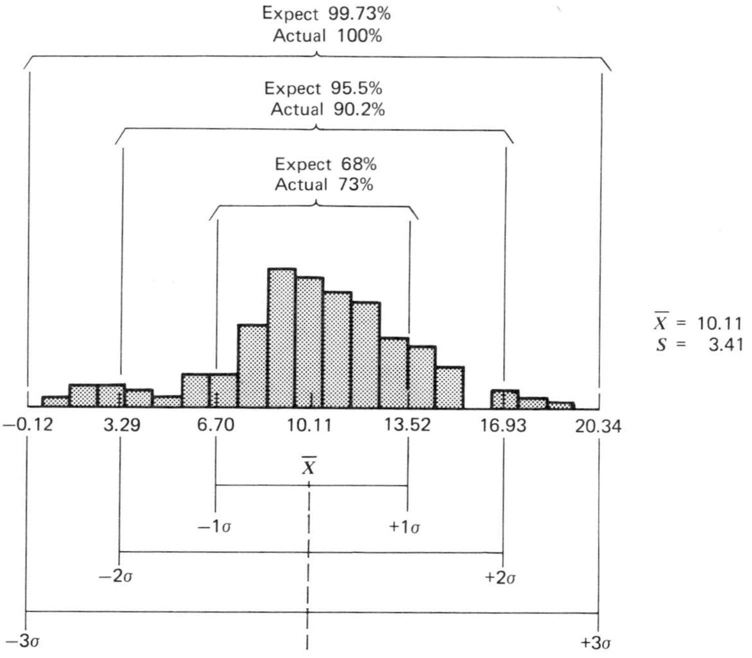

FIGURE 3.5 The plus and minus 1, 2, and 3 sigma limits are shown for Example 1.1.

* Later, in Chapter Five, we will describe the distribution as a normal distribution.

The space between -1σ and $+1\sigma$ is the distance between 6.70 and 13.52 and it includes 82 individual values. Then

$$\frac{82}{112} \times 100 = 73.2\%$$

of the distribution lies between 6.70 and 13.52 or from -1σ to $+1\sigma$. This compares well with the empirical approximation of 68%.

Between -2σ and $+2\sigma$ (from 3.29 to 16.93) are 101 values, or

$$\frac{101}{112} \times 100 = 90.2\%$$

are between $\pm 2\sigma$.* We expect approximately 95.5% to be between $\pm 2\sigma$.

Between $\pm 3\sigma$ (from -0.12 to 20.34) are all 112 values, so

$$\frac{112}{112} \times 100 = 100\%$$

are included. We expect approximately 99.73% of the values here.

As you can see we are quite close in our approximation—the most we are off is about 5%, which is quite acceptable. Here has been the most valuable use of the standard deviation. *With it we can essentially reconstruct the distribution.* Knowing that the mean is 10.11 and the standard deviation is 3.41, I could draw a distribution that practically perfectly fits the distribution we actually had. I can also tell you that rarely should the number of tickets have fallen below 4 or above 16, just about never should the number be less than 0 (which would be absurd anyhow) or more than 20, and the bulk of the time, about two thirds of the time, there would be between 7 and 13 tickets. And I would be pretty close to what actually happened—without all the numbers or the monstrous tables. Hence, we have accomplished a monumental task; we have essentially reduced the entire distribution to *two* numbers—the mean and the standard deviation.

Interpretation of variance

You might now be asking, "Okay, I understand standard deviation, but what's this variance business?" I offer three reasons for studying the variance.

1. We calculate it because we must calculate it before we obtain the standard deviation.
2. There will be several statistical tests that will use it directly. (I won't bother you with them yet.)
3. There is a reason I will try to give you a little insight into now (which you may encounter in more advanced work); in many cases *variances can be added, but standard deviations cannot.*†

* I will now use $\pm$ to mean plus or minus. It will mean that you should consider *both* the plus 2σ and the minus 2σ value. It can be used with any quantity, so that ± 5.2 means $+5.2$ and -5.2.

† The main requirement is that the two values are independent of each other.

To see how this last concept is useful, consider measuring a room with a 12-inch ruler. We may know from past experience that the standard deviation of measurements with this ruler is 0.1 inches (therefore, most—about 95%—of the readings are within ± 0.2 inches). The variance is $(0.1)^2 = 0.01$. A 12-foot room would require 12 independent measurements, each with variance 0.01. If we add these variances we obtain a total variance for the overall room measurement as $12 \times 0.01 = 0.12$. The standard deviation for the sum is therefore $\sqrt{0.12} = 0.34$. If we had added the standard deviation we would have obtained a value of 1.2 inches for the overall variation, or almost four times as large a value as with the addition of variances. There are many practical problems that can actually use this principle.

REVIEW Central value or Average • Midrange • Mode • Median
Mean • Summation Notation • Spread • Range
Standard Deviation = $\sqrt{\text{Variance}}$ • s versus σ • Sigmas • Empirical Rule

$$\bar{X} = \frac{\sum fX}{\sum f} = \frac{\sum fX}{n}$$

$$\text{Variance} = s^2 = \frac{\sum f(X - \bar{X})^2}{n - 1} = \frac{\sum fX^2 - (\sum fX)^2/n}{n - 1}$$

$$k = \frac{X_2 - X_1}{s}$$

Table for Calculations of Mean and Standard Deviation

	X	f	fX	fX^2
Data	...	...	...	...
	...	...	...	...
Sum		x	x	x

PROJECTS

For the projects in Chapter One compute the mean, median, midrange, and the mode, range, and standard deviation of the data. Do you have any additional observations or comments to make?

PROBLEMS

SECTION 3.2

3.1 For the following sets of numbers, calculate the median, midrange, mode (if any), and mean with and without using frequencies or grouping.

a. 5, 4, 2, 9, 13.
b. 2, 3, 5, 8, 8, 12.
c. 1, 0, 9, 27, 3, 8
d. 0, 1, 1, 4, 3, 5, 4, 6.
e. −4, 6, −2, 8, 3, 2, 1, 7, 19, 2, 3, 4, 3.
f. 1, 1, 1, 1, 2, 1, 1, 1, 1, 2.

3.2 If Charlie Brown had an earned run average of 80 runs per game, how many total runs were scored against him in 20 games?

3.3 The number of large ants killed per day in a rural house was

6, 3, 2, 0, 1, 0, 3, 9, 2, 4, 1

Calculate the mean, median, mode, and midrange.

3.4 After an insecticide was sprayed around the foundation and inside the house, the number of ants killed per day was

3, 4, 2, 0, 1, 0, 0, 1, 0, 0, 0, 0, 0, 0, 0, 1, 0, 0, 0, 0, 0, 0

Compare the average number of ants with the situation observed in Problem 3.3.

3.5 The sales in dozens of fresh doughnuts per day during 3 weeks have been

15, 20, 11, 10, 16, 18, 14, 14, 22, 14, 12, 17, 16, 18, 15, 9, 13, 14, 17, 15, 13

Calculate the mean and median sales.

3.6 Twelve people attended "a weight control group" and agreed to eat the identical sets of food. After 4 weeks they reported the following weight loss in pounds (a negative value means that they increased weight)

6, 12, −3, 5, 8, 7, 4, 1, 1, 3, 7, 9

Compute the mean, median, mode, and midrange weight loss for this group of people. How do the averages compare to each other?

3.7 The price of one style of new homes in a development sold during the first 3 months of its opening were

49,990; 48,850; 47,900; 50,280; 48,300; 50,220; 49,340; 50,790; 48,500; 49,200; 50,000; 50,400; 48,650; 50,270

Find the mean, median, mode, and midrange prices.

3.8 The number of deaths in accidents on a major highway each year for the last 23 years were

47, 89, 67, 71, 33, 45, 83, 67, 70, 77, 80, 65, 93, 84, 81, 106, 94, 76, 83, 92, 116, 84, 79

Calculate the mean and median number of accidents.

3.9 The price of 8-foot-long 2 × 4s (boards used in building houses and other items) were priced at 25 different lumber yards and retail stores. They were (in dollars):

1.16	1.29	1.22	1.29	1.15
1.27	1.14	1.37	1.35	1.49
1.45	1.38	1.31	1.40	1.24
0.98	1.42	1.25	1.25	1.33
1.32	1.62	0.88	0.99	1.39

Calculate the average (mean) price.

3.10 Refer to the survey in Example 1.3 (page 9). What are the following averages: Number of hours worked? Rating of cafeteria? Rating for legalizing marijuana?

3.11 Refer to the data in Problem 2.3. To calculate the different "average" number of meals that the people interviewed have eaten, what must be done to the data that were presented? Calculate the four "averages." Which average is easiest to calculate?

3.12 Repeat Problem 3.11 using the numbers obtained in Problem 2.4.

3.13 Refer to Problem 2.5. Can you calculate any type of average? Comment.

3.14 For the data in Problem 2.6, what are the "average" numbers of customers? Which is easiest to calculate? Which seems most realistic?

3.15 Refer to Problem 2.7. What are the various average number of overtime hours worked? Which value seems to be the most appropriate measure of the center of the group of values?

3.16 Refer to Problem 2.8. What are the four "average" numbers of absences? Why are there such differences?

3.17 On a tree farm 2000 trees were planted. After 5 years, 250 were randomly picked and the diameters were measured. The results were grouped as follows.

DIAMETER, INCHES	NUMBER OF TREES
1.75–2.24	7
2.25–2.74	13
2.75–3.24	38
3.25–3.74	86
3.75–4.24	64
4.25–4.74	25
4.75–5.24	14
5.25–5.75	3

Calculate the mean and median diameter of these trees.

3.18 The number of days of hospitalization for 150 patients who underwent a particular operation were as follows.

NUMBER OF DAYS	NUMBER OF PATIENTS
6	0
7	2
8	4
9	13
10	64
11	38
12	13
13	7
14	5
15	3
16	0
17	1

Calculate the mean and median number of days of hospitalization for these patients.

3.19 Group the set of sale prices for the various homes listed in Problem 3.7. What intervals should you use? Calculate the "averages" as before. Are there any differences between the averages obtained from the grouped values and the ones calculated in Problem 3.7?

3.20 Group the number of deaths in traffic accidents listed in Problem 3.8. Calculate the "averages" and compare to the previous values.

3.21 Refer to the prices of 2 × 4s in Problem 3.9. Group the prices and recalculate the mean price.

3.22 Refer to the breaking strength of sutures in Problem 2.16. Calculate the mean breaking strength.

3.23 In Problem 2.15 we examined the weights of bread loaves. Calculate the mean weight (using the intervals you determined in Problem 2.15). You might also calculate the mean without grouping and compare the results and the amount of effort.

*3.24 Refer to Problem 2.17 and calculate the average number of minutes that the insects survived.

3.25 a. Using Table 3.2 find the median. How close is this to the actual median?

*b. When we encounter grouped data, there is one modification to the method of calculating the median that often gives us a value closer to the actual median than we have just used. This method involves determining just how far across the category interval our median should be. In this problem the 46th value is associated with 9.5 and the 98th value with 12.5. We want to know what would go with the 56th value. It would be found by substituting into the formula

$$\text{Median} = X_L + \frac{d}{f_m} L \qquad (3.14)$$

where X_L is the lower value for the group containing the median (the 9.5 here)
L is the width of the interval (3.0 in this case)
f_m is the frequency associated with the group that the median belongs to (43)

d is the difference between $(n + 1)/2$ (or $113/2 = 56.5$) and the cumulative frequency of the next lower category (the last frequency of the previous category, or 45 for this example). Thus, $d = (n + 1)/2 -$ (cumulative frequency of previous category) $= 56.5 - 45 = 11.5$.

Now recalculate the median for Example 1.1 from Table 3.2. How does it compare to the value obtained in part a? How does it compare to the actual value? Discuss just what is happening and why we should often get a better value.

*3.26 In Problem 3.18 the median number of days of hospitalization was calculated using grouped data. Recalculate the median using the technique described in Problem 3.25.

*3.27 Determine the median weight of loaves of bread listed in Problem 2.15 using the method of Problem 2.35. Compare this value to the mean weight calculated in Problem 3.23.

*3.28 Determine the median strength of sutures listed in Problem 2.16. Use the method of Problem 3.25. Compare the median to the mean value computed in Problem 3.22.

*3.29 In Problem 3.17 the median diameter of trees was calculated using groups. Recalculate the median using the technique described in Problem 3.25.

3.30 Quite often we must calculate an average of several sets of values when they carry different weights or degrees of importance. If you want to average together the means of several different-sized samples, you would not regard them as equally important. For example, 20 flashlight batteries had an average life of 150 hours. Another sample of 15 batteries had an average life of 200 hours and a third sample of 5 batteries had an average life of 250 hours. What is the average life of these batteries? We must calculate the average $\bar{X}$ as

$$\bar{X} = \frac{\sum wX}{\sum w} \tag{3.15}$$

where w represents the weight (these are essentially frequencies). Here the weights are 20, 15, and 5, respectively, and the X are 150, 200, and 250. Thus

$$\bar{X} = \frac{20 \cdot 150 + 15 \cdot 200 + 5 \cdot 250}{20 + 15 + 5} = \frac{7250}{40} = 181.2 \text{ hours}$$

If an additional sample of 20 batteries averaged 210 hours, what would be the new overall average life of the batteries?

3.31 A company employs machinists earning \$6 per hour, assemblers earning \$3 per hour, and packers at \$3.50 per hour. If there are 25 machinists, 50 assemblers, and 25 packers, what is the average earning of workers in this company?

3.32 A teacher considers four tests equally important and considers the final exam equivalent to three tests. A student obtains the following grades on the tests: 61, 73, 67, 71, and a 76 on the final exam. What average will the instructor calculate as his average?

SECTION 3.3

3.33 Comment on this statement: "Something is wrong with the way they are teaching our children. Half of our children scored below the National Average! Let's get better teachers."

3.34 The following donations (in dollars) were made to a local charity.

4, 10, 6, 2, 1, 3, 1, 150, 3, 5, 6, 2, 8, 2, 12

Calculate the various averages. Which is the most representative or the "best" average? Explain.

SECTION 3.5 For Problems 3.35 to 3.44 calculate the range, variance, and standard deviation. Use whichever formula is most appropriate for the particular calculation.

3.35 The various sets of numbers of Problem 3.1.
3.36 The number of minutes to school in Example 1.2 (data on page 8).
3.37 The number of ants killed per day in Problem 3.3.
3.38 The number of ants killed after treatment, listed in Problem 3.4.
3.39 Sales of doughnuts given in Problem 3.5.
3.40 Weight losses listed in Problem 3.6.
3.41 Prices of homes in Problem 3.7.
3.42 Deaths listed in Problem 3.8.
3.43 Price of boards in Problem 3.9.
3.44 Days of hospitalization in Problem 3.18.
3.45 Diameters of trees in Problem 3.17.

For Problems 3.46 to 3.51 calculate the range, variance, and standard deviation. The data will be found in previous problems in Chapter Two and the means were calculated in problems in Chapter Three.

3.46 Number of customers in Problem 2.6. Mean in Problem 3.14.
3.47 Number of overtime hours in Problem 2.7. Mean in Problem 3.15.
3.48 Number of absences in Problem 2.8. Mean in Problem 3.16.
3.49 Weights of bread in Problem 2.15. Mean in Problem 3.23.
3.50 Breaking strength of sutures in Problem 2.16. Mean in Problem 3.22.

3.51 In Problem 3.19 the sale prices of homes listed in Problem 3.7 were grouped. Recalculate the standard deviation using the groups and compare the results to the value in Problem 3.41.

3.52 In Problem 3.20 the number of traffic deaths listed in Problem 3.8 were grouped. Recalculate the standard deviation using groups and compare the value to that obtained in Problem 3.42.

3.53 In Problem 3.21 the prices of 2 × 4s listed in Problem 3.9 were grouped. Recalculate the standard deviation using the groups and compare the result to the value obtained in Problem 3.43.

3.54 Refer to the survey data in Example 1.3 and the averages calculated in Problem 3.10. Calculate the standard deviations.

3.55 Another measure of variability that uses all of the data is called the *average deviation*. It is found by computing the mean of the data ($\bar{X}$). You then find the absolute value of the difference between each quantity and the $\bar{X}$ (the absolute value always being a positive value). If you compute the mean of these absolute differences you now have the average or mean deviation. If you refer to the data on page 62, the mean was 8.67. The deviations or differences would be: $6 - 8.67, 9 - 8.67, 8 - 8.67, 8 - 8.67, 12 - 8.67$, and $9 - 8.67$. Therefore they are -2.67, 0.33, -0.67, -0.67, 3.33, and 0.33. The absolute values of these values are 2.67, 0.33, 0.67, 0.67, 3.33, and 0.33. The mean of these last quantities is $8.00/6 = 1.33$. Thus the average deviation is 1.33. The formula for the average deviation is

$$\text{Average deviation} = \text{Mean deviation} = \frac{\sum |X - \bar{X}|}{n} \tag{3.16}$$

Now do the same for Problem 3.1.

3.56 Find the average deviation for the data in Example 1.2. How does this compare to the standard deviation? Should it? Comment on the relative difficulty in computation. (See Problem 3.36).

SECTION 3.6

3.57 The weights of a group of students have a mean of 147 and a variance of 25. How many standard deviations from the mean is 164? 135? 147? Are the weights unusual? What weight is 2 standard deviations from the mean? -3.1 standard deviations? 0.3σ?

3.58 Refer to the data of Example 1.2.

a. What are the 2 and 3σ limits? (See Problem 3.36).
b. Are you willing to accept the necessary assumption about a "bell-shaped" distribution to apply the empirical rule to this data? Explain.
c. Compare the percentages within 1, 2, and 3σ limits to those predicted by the empirical rule. Comment.
d. Would you be surprised if it took you 12 minutes to get to school? How about 17?

3.59 Between what pair of weights do you expect to find about 95% of all the breads listed in Problem 2.15? First look at the histogram to decide if you should apply the empirical rule. (Refer to Problem 3.49 for standard deviation.)

3.60 Calculate the standard deviation of the donations listed in Problem 3.34. The 3σ limits should include about 99.75% of the values. Do they? What change would occur if that \$150 donation were not included? Discuss.

3.61 If the data does not fall into a distribution that resembles a normal or bell-shaped form, the empirical rule is no longer valid. There is a rule called Tchebychev's inequality (also spelled Chebyshev), which states that *no matter what the shape of the distribution at least* $(1 - 1/k^2)\,100\%$ *of the data must fall within* $\pm k$ *standard deviations of the mean.* Thus, within ± 2 standard deviations, or $\pm 2\sigma$, there *must be* at least

$$(1 - 1/2^2)100 = (1 - 1/4)100 = (0.75)(100) = 75\%$$

How do the percents within the 1 and 3σ limits compare to the empirical rule? Can you create a distribution that contains the minimum possible number of points inside 1σ limits?

*3.62 If we have found the range (R) and the standard deviation (σ) of a set of data, an interesting relationship that is sometimes useful to compute is $R/2\sigma$. It is particularly interesting, since it is bounded—it has a largest possible value and a smallest possible value for any given sample size n. Try to compute the maximum and minimum possible $R/2\sigma$ for various values of n. (*Hint*. For minimum $R/2\sigma$ you need to consider n being even and n being odd in separate ways. In all cases try to visualize first just what way the data must be organized to give you various kinds of ratios.)

CHAPTER FOUR PROBABILITY AND DISTRIBUTIONS

So far we have been dealing with descriptive statistics. We now can look at a sample of values from some population and: (1) draw a histogram, which should help us visualize the *distribution* of the values; and (2) calculate the mean and standard deviation of the sample values. But this is usually just the beginning. Generally we are more interested in knowing about *the* population. One important way to choose among the possible populations requires us to look at where the samples *might* have come from. This brings us to the study of probability. We will, however, only examine the aspects of this large field of mathematics that are important in understanding basic statistics.

4.1 PROBABILITY AND STATISTICS

Let us first briefly look at the difference between probability and statistics.

The study of probability assumes that you know the entire universe or population from which you are going to sample. You are then able to determine completely all of the possible samples that can be taken from that population. In addition, you will also be able to state the chances that any particular samples will be chosen.

Statistics assumes that you have only a sample and that you are trying to determine, or at least guess, the parameters or characteristics of the most likely population or source from which the sample was taken.

Figure 4.1 partly illustrates this difference. In Figure 4.1*a* we have a known population. We know all the individuals. We can then determine all of the possible samples that may be taken from that population. In fact, we arrive at a complete distribution of samples. It then becomes possible to determine what probability or chance we might have of obtaining the different samples or values from this population.

In Figure 4.1*b* we have an unknown population. It *may* be the same one as above, but we do not know that. We look at a sample and then try to *infer* what population that sample might have come from, *or* we determine what chance there is that the sample might have come from, for instance, the upper population.

In the next several sections we will examine the concepts of probability. At the end of the chapter we will return to this comparison.

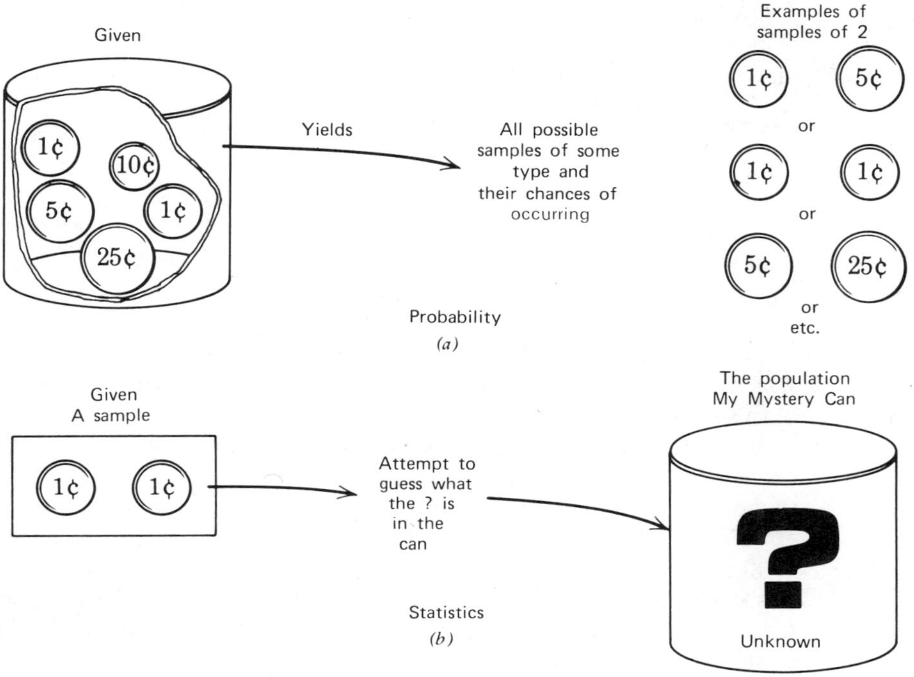

FIGURE 4.1 One distinction between probability and statistics. How the sample relates to the distribution. (*a*) Probability starts with a certain known population and samples are described from that population. You are certain of the population—you know *all* the possible samples. *But* you are uncertain about the *particular* samples. (*b*) Statistics starts with a sample and tries to guess the population. You are certain about the *sample*, *uncertain* about the population.

4.2 POSSIBILITIES AND PROBABILITIES

We have all stated at different times that something has a good probability of happening or a very small probability or low chance of occurring. We have heard statements such as "the chance the Pirates will win the World Series is 60%," or a rain probability index is 50%, or the chance of winning in a particular lottery is only 2%. In each of these cases we have some events that can take place. The Pirates may win or lose or not be in the Series at all. It may rain, be sunny, snow, or do one of several other things. I may win or lose in the lottery with my ticket. Each of these are possible events. When we assign a particular number or mathematical value to our statement of the chance of a particular event taking place, we call the number "the *probability of the event*."

There are several different ways of arriving at this probability value. We will look at three such methods (essentially *guessing*, *estimating* and, for limited situations, *determining mathematically*).

The first is called "subjective" probability; this is really just a statement of someone's *belief*. We may state; "There is a probability of 0.50 or a

50% chance of war in the Middle East in the next 5 years;" the chance of being promoted or getting a certain job is 0.25 or 1 in 4; the probability of me buying a new car is 0.10 or 1 in 10; or the probability of you failing this course is 0.05 or 1 in 20.

Although there are a variety of different ways in which we can interpret these numbers, one such approach is to imply, for example, that in the last case you have a way of sensing that *if you took this course 20 times you might fail it once*. But, hopefully, you will take it only once. The number, therefore, is some kind of a guess on your part of what you think your chance of passing or failing will be. It cannot be proven, since you will not take it 20 times, or even twice. Examine the other *guesses* of probabilities. They also deal with things that either will or will not occur, but for which you will only have one opportunity. We will *not* deal with probabilities of this type.

Another approach involves the use of actual data to estimate what the chances of each occurrence would be. For example, we might observe that out of 10,000 deaths of adults in a particular year, 1800 are from cancer, 5000 are from diseases of the circulatory system (i.e., heart attacks), 1000 are from accidents, and 2200 are from miscellaneous causes. From this kind of information we could state that the proportion of people dying from cancer is 0.18 (1800/10000 = 0.18), the proportion of deaths from circulatory system causes is 0.50 (5000/10000 = 0.50), the proportion from accidents is 0.10, and the proportion from other causes is 0.22. This will then lead us to say that we estimate that the probability of death from cancer is 0.18, from circulatory system diseases 0.50, from accidents 0.10, and other causes 0.22.

The rain or snow probability index so often used by weathermen is another everyday encounter with an estimate based on relative frequencies. The meteorologist studies the particular weather pattern of the past several days. Records of past observations may have shown that in 60 out of the last 200 times when this weather pattern occurred, it was followed by rain. We then arrive at a probability of 60/200 or 0.30 for rain and designate this as a rain probability index of 30% (or 0.30).

The probability of a certain number of telephone calls completed during a particular time, the chances of a particular birth defect in an infant, the probability of success in college, the chance of purchasing a defective light bulb, or the probability of gaining 5 yards or more on an end run in a football game might all be based on relative proportions.

We have now discussed two methods for determining probabilities, and both have involved guessing or *estimating* the relative chances that particular events might occur. The remainder of this section, and most of the chapter, will consider a third method, which is a more mathematically *exact* way of computing probabilities.

Mathematical method

Our first requirement is to be able to prepare a complete listing of *all the possibilities that may occur*. We call these the *outcomes*. Thus, we have the following example.

EXAMPLE 4.1

a. If you are to flip a coin you have two possibilities—heads or tails.
b. A child may be born either a boy or a girl.
c. A biologist might have 5 rats with tumors. If all the rats were operated on and the tumors removed, we might be able to list the weights of all the tumors (e.g., 0.5 gram, 1 gram, 2 grams, 6 grams, and 10 grams).
d. An automobile dealer has 6 cars identical, except that they give various gas mileages (16, 16, 17, 18, 19, and 22 miles per gallon).

After we have listed the possibilities, we want to consider the *chances* that the various possibilities may occur or be picked. The most elementary types of situations are those where all of the individual possibilities or outcomes are *equally likely*. In tossing a balanced coin we would generally agree that the two sides of our coin have an equal chance of appearing every time the coin is flipped. There is essentially the same chance that a given birth will be a boy or a girl. When the biologist picks *a* rat at random, each of the listed weights has the same chance of being observed. If you test drive one of the cars, having randomly selected it, then each of the six possible values of gas mileage has the same chance of occurring (although 2 of those cars give the same value).

Obviously, not all situations have possibilities that are equally likely. For example, if you reexamine the sample data on causes of death, you will see that some are more likely to occur than others. I have a much better chance of dying from a circulatory system disease than from an accident. However, we will consider only those situations where the *individual outcomes* are easily recognized as being equally likely.

Our goal is to determine the *probability* that various outcomes or *groups* of outcomes might occur. Therefore, we must next state or describe what type of occurrence we are interested in. This description is called the *event*. An event may be either an individual outcome (in which case it is called a simple event) or it may consist of any number of outcomes (from none to the entire set of outcomes). Thus we may have reason to describe some of the following events (note that we will often also abbreviate or represent the event by a single letter).

EXAMPLE 4.1 (Cont.)

a. In flipping a coin:
   Event *A* might be "a head."
   Event *B* might be "coin stands on edge."
b. Birth of a child:
   Event *A* will be "a boy."
c. If an animal is operated on:
   Event *A* might be "selecting a rat with a tumor of 2 grams."
   Event *B* might be "selecting a rat with a tumor greater than 3 grams."
d. Cars:
   Event *A* will be "the first car giving 16 miles per gallon."
   Event *B* will be "a car that gives 16 miles per gallon."

EXAMPLE 4.2

If you are one of the 25 students in my class and I randomly select one student each day to solve a problem:

Event A is "you are chosen."
Event B is "picking a student in the front row."

Now we come to the definition of the probability of a particular event.

DEFINITION If all the possible outcomes are considered *equally likely, then the probability of a particular event*, A, is

$$\text{Probability of } A = P(A) = \frac{\text{Number of outcomes included in Event } A}{\text{Total number of possible outcomes}}$$

$$= \frac{n}{N} \tag{4.1}$$

where N = total number of possible outcomes (2 sides, 2 sexes, 5 rats, 6 cars)
and n = the number of outcomes in the particular event.

$P(A)$ is sometimes written as Prob(A) or Pr(A).

Note also that the probability of randomly selecting a single *particular* individual among a collection of equally likely outcomes is $1/N$.

EXAMPLE 4.1 (Cont.)

a. $N = 2$, head or tail. We have one outcome included in Event A ("a head") but *none* included in Event B, since I do not allow the possibility of the coin standing on edge.

$$\text{Probability of a head} = P(A) = \frac{1}{2} = 0.5$$

$$\text{Probability of coin on edge} = P(B) = \frac{0}{2} = 0.0$$

b. $N = 2$ and Event A includes only one outcome. Thus:

$$\text{Probability of a boy} = P(A) = \frac{1}{2} = 0.5$$

c. $N = 5$. Only 1 rat has a 2-gram tumor, but 2 rats are included in the event "greater than 3 grams." Therefore:

$$\text{Probability of rat with 2-gram tumor} = P(A) = \frac{1}{5} = 0.20$$

$$\text{Probability of } P \text{ rat with tumor greater than 3-gram} = P(B) = \frac{2}{5} = 0.40$$

d. $N = 6$. Event A consists of 1 car but Event B contains 2 possible cars.

$$P(\text{first car giving 16 miles per gallon}) = \frac{1}{6} = 0.167$$

$$P(\text{a car with 16 miles per gallon}) = \frac{2}{6} = 0.333$$

EXAMPLE 4.2 (Cont.)
$N = 25$. There are 5 students in the front row.

$$\text{Probability of picking you} = P(A) = \frac{1}{25} = 0.04$$

$$P(\text{picking students in front row}) = P(B) = \frac{5}{25} = 0.20$$

Mathematicians often describe a complete set or collection of all possible quantities of some type as a *space*. Here we are dealing with problems that involve picking samples from some "space" or "collection of possibilities." We therefore call this collection of possibilities a *sample space*. Thus, the sample space for tossing a coin consists of the two possibilities, the sample space of the rats consists of the listing of the weights of the 5 tumors, and the sample space for students in my class is the set of 25 students.

Sometimes we can draw a picture outlining or including the sample space to help us visualize the various possibilities that can occur. One such picture for Example 4.2 is shown in Figure 4.2. The points represent the events that different individual students might be chosen. For convenience, I have drawn in the dots in the way the students sit in class. We can easily see and also provide an outline for the event "a student in the front row is picked."

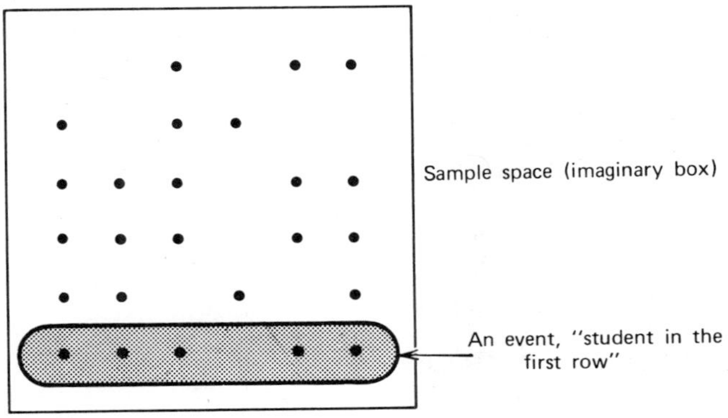

FIGURE 4.2 A sample space for the students in my class. The points have been arranged in accordance with the seating of the students.

Often it is difficult to list all of the possible occurrences. To describe all of the possible bridge hands or even poker hands that could be dealt with a deck of cards is an enormous task. If each student had an equally likely chance of picking any seat to sit in, finding the probability of an empty seat left in the front row would require a listing of *all* the possible ways that seats might be occupied. This, too, is a mammoth undertaking. For those occasions when we need to calculate such probabilities there are certain counting formulas to assist us. These are formulas to determine all possible combinations and all permutations (or arrangements) of various types.

We now have two important and useful ways of defining probability. The first was the empirical or relative frequency method, which used data from a sample. It assumed that we have taken a very large group of trials in arriving at the estimation of the relative frequencies. The last method we discussed was a more exact procedure for arriving at probability values, but it requires certain assumptions and is not always reasonable. Although we will use the two definitions essentially interchangeably, you should realize that the exact method (involving equally likely outcomes), when applicable, provides a stronger foundation for the theoretical concepts of probability (although we will not need to be concerned about such differences).

4.3 SOME RULES OF PROBABILITY

RULE 1 *The probability of an event, $P(A)$, is always some number from 0 to 1 or*

$$0 \leq P(A) \leq 1$$

where $\leq$ means less than or equal to. In addition,

If $P(A) = 0$, *the event is* impossible (certain not to occur)

If $P(A) = 1$, *the event is* certain to occur

To examine these statements, let us return to the definition of probability where $P(A) = n/N$. If the event is impossible it means that there are *no* possibilities (among the overall total number of possibilities) that correspond to the given event. Thus n is 0 and n/N equals 0 or $P(A) = 0$. The probability of my coin standing on its edge is 0, since I have not included this as one of the possibilities that I will accept. The probability that I will choose Albert Einstein to answer a question is also 0, since he is not alive and cannot be in my class. At the other extreme, if the event includes all N individuals, then the probability is $N/N = 1$ and the event also includes the entire sample space.

RULE 2 *The probability that a particular event will not occur is 1 minus the probability that it will occur. This is called the complement of the event and is written as A' (A prime).*

$$P(A') = 1 - P(A) \tag{4.2}$$

The 1 represents the entire group of possibilities. Once we subtract out the group of possibilities associated with our particular event (i.e., heads in flipping our coin, individuals in the first row, or rats with large tumors), we are left with the rest of the group of possibilities. Thus

$$\text{Probability of tail} = 1 - \text{probability (head)} = 1 - \frac{1}{2} = \frac{1}{2}$$

Probability of student *not* in first row = 1 − probability of student *in*

$$\text{front row} = 1 - \frac{5}{25} = 1 - 0.20 = 0.80$$

Probability of not having a large tumor (more than 3 grams)

$$= 1 - P(\text{large tumor}) = 1 - \frac{2}{5} = 1 - 0.4 = 0.6$$

A more complicated event

The areas of statistics that utilize probability often require us to look at somewhat more complex events. For example, samples are not usually single individuals, but the selection of several individuals, often with varying probabilities of selection.

Let us look at another example.

EXAMPLE 4.3

Select 5 coins—2 pennies, 1 nickel, 1 dime, and 1 quarter—and spread them out. We will consider these as our population. Figure 4.3 shows the population of the 5 coins I picked.

Now we will take some samples. One sample could be the 1969 penny and the dime, another could be the nickel and the dime, a third might be the dime and the quarter. A complete list of all of the samples is given in Table 4.1 (this is the sample space for our samples of two). The values of the samples are also listed, so the penny and dime make 11 cents, and so

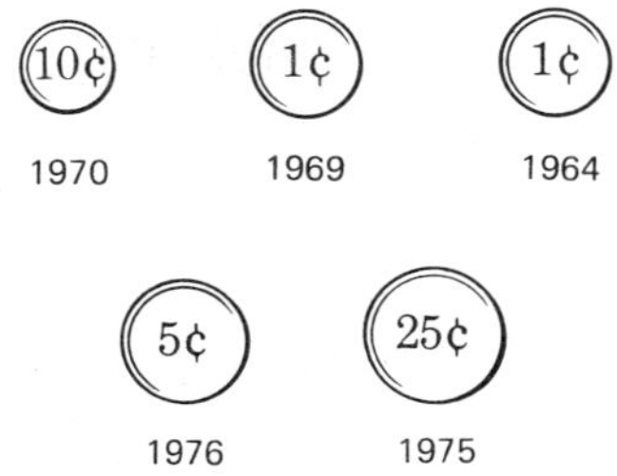

FIGURE 4.3 The population of coins in Example 4.3.

TABLE 4.1 All Possible Samples of Two from the Population of Figure 4.3

FIRST PICK		SECOND PICK	SUM OF THE 2 COINS OR VALUE OF THE SAMPLE (CENTS)
1964 penny	and	1969 penny	2
1964 penny	and	nickel	6
1964 penny	and	dime	11
1964 penny	and	quarter	26
1969 penny	and	1964 penny	2
1969 penny	and	nickel	6
1969 penny	and	dime	11
1969 penny	and	quarter	26
nickel	and	1964 penny	6
nickel	and	1969 penny	6
nickel	and	dime	15
nickel	and	quarter	30
dime	and	1964 penny	11
dime	and	1969 penny	11
dime	and	nickel	15
dime	and	quarter	35
quarter	and	1964 penny	26
quarter	and	1969 penny	26
quarter	and	nickel	30
quarter	and	dime	35

forth, and the possible samples range from 2 cents to 35 cents. Observe that we are picking 2 coins at a time and that we obtain a sum.*

In a similar way we can examine samples of tumors from 2 rats, gas mileage of 2 cars, or we can ask about the probability of 2 heads out of 3 tosses or of 4 boys in a family of 5 children.

In Table 4.1 we have a total of 20 different samples and each has associated with it the value of the "sum of two coins."† We can now draw a histogram or probability distribution for these 20 possible sample values (Figure 4.4), and we will then be better able to visualize the chance of obtaining different sample values. Thus, for example, we find 15 cents is obtained in only two ways and, therefore, $P(15 \text{ cents}) = 2/20 = 0.10$ (or 1 chance in 10). The probability of obtaining 26 cents is 0.20 and the probability of picking 20 cents is 0.00.

* Statisticians describe such values as random variables. In each case, we have some variable (it could be just the values of the individual coins, it could be the sums we have been working with here, or it could involve any other mathematical operation as well). The selection of particular samples is a random operation and thus we obtain *random* values of the *variable* we are measuring. The probability distribution of these samples is really a distribution of random variables.

† This sum is also called a random variable.

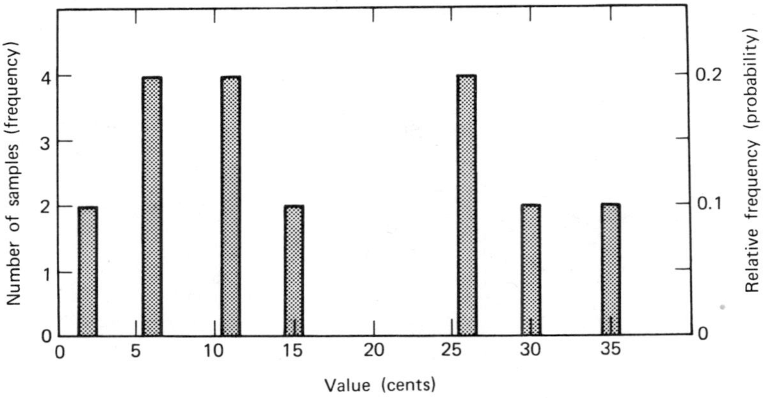

FIGURE 4.4 Histogram for the value of all possible samples of two from the population of Figure 4.2 (samples of Table 4.1).

Next we should look at the way I took the samples from the population, and consider a couple of important concept that I disregarded when I listed those samples. For ease of visualizing we will continue to work with the 5 coins (2 pennies, 1 nickel, 1 dime, and 1 quarter). Each sample could be considered as being taken by first selecting one individual and then, *while keeping that individual out of the population*, picking a second individual. In other words, I did *not replace* the first item I picked. This is a very important distinction to be aware of; if I were to replace the first coin after looking at it, *I would have a different collection* of possible samples. For instance, if the first coin I picked was a quarter and I put it back in, then I *could* pick it again, so one of the samples that would have to be considered now would be 2 *quarters.*

The standard terms used here are:

1. Sampling *without* replacement.
2. Sampling *with* replacement.

Whenever we deal with the first case we *alter our population* as we sample. After we have picked the quarter, the population that remains does not contain the quarter any more (Figure 4.5*a*). So the next pick *cannot* be a quarter. This limits the possible collection of samples of two. How many actually different samples of five could you have picked from this population when you sample without replacement? If you said one, you are correct. After all, there are only five items. After you pick one, you are down to a population of four. If you continue to pick you will finally exhaust the population and so the sample will contain all of the items of the population and the only possible net sample will be 2 pennies, 1 nickel, 1 dime, and 1 quarter, or the 42 cents.

What about sampling with replacement? Figure 4.5*b* shows in effect what occurs. In this case we toss the item back into the population, and

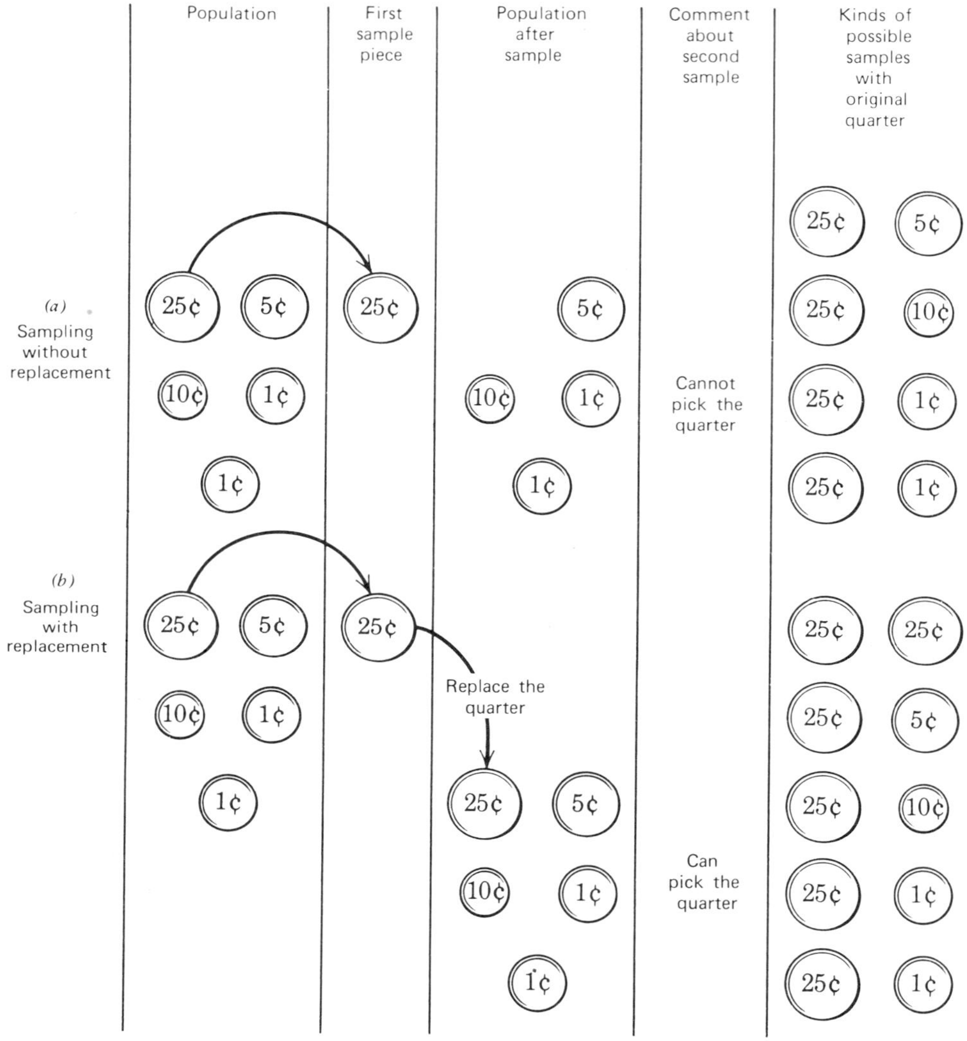

FIGURE 4.5 Sampling with and without replacement. Taking the population of five coins and selecting the quarter on the first pick, we have two possible sets of samples if we (*a*) do not replace the quarter, or (*b*) replace the quarter.

we can pick the same item again. Table 4.2 lists the total collection of possible samples of two from this population when we sample *with* replacement. Observe that all of the previous samples are included *plus* a few additional ones (noted by [a]), namely the five cases of sampling the same item twice. Figure 4.6 shows the histogram of these samples, and you may compare this distribution to the previous one in Figure 4.4 to see the effect of replacing the first item before continuing to sample.

TABLE 4.2 All Possible Samples of Two from the Population of Figure 4.2 (2 pennies, 1 nickel, 1 dime and 1 quarter). This Time Sampling *With* Replacement

FIRST PICK		SECOND PICK	SUM OF THE 2 COINS OR VALUE OF THE SAMPLE (CENTS)
1964 penny[a]	and	1964 penny	2
1964 penny	and	1969 penny	2
1964 penny	and	nickel	6
1964 penny	and	dime	11
1964 penny	and	quarter	26
1969 penny	and	1964 penny	2
1969 penny[a]	and	1969 penny	2
1969 penny	and	nickel	6
1969 penny	and	dime	11
1969 penny	and	quarter	26
nickel	and	1964 penny	6
nickel	and	1969 penny	6
nickel[a]	and	nickel	10
nickel	and	dime	15
nickel	and	quarter	30
dime	and	1964 penny	11
dime	and	1969 penny	11
dime	and	nickel	15
dime[a]	and	dime	20
dime	and	quarter	35
quarter	and	1964 penny	26
quarter	and	1969 penny	26
quarter	and	nickel	30
quarter	and	dime	35
quarter[a]	and	quarter	50

[a] Additional samples because of sampling *with* replacement.

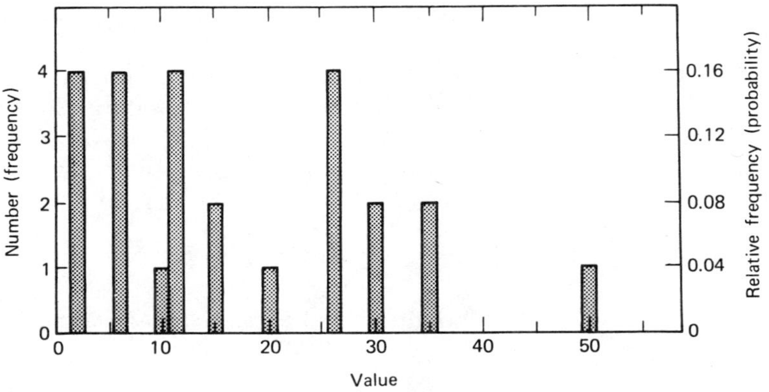

FIGURE 4.6 Samples of two *with* replacement. Compare this histogram to Figure 4.4.

Now consider the population shown in Figure 4.5*a* as a *sample space* for our experiment of picking coins. There have been 5 coins with each equally likely to be selected. When we first picked the quarter, we *reduced our sample space*. We then have a change in the collection of possibilities for any future picks. As a result, the various probabilities are also changed. Thus, if I had originally wanted to consider the event "pick a penny," we would have begun with the probability of a penny as

$$P(\text{penny}) = \frac{2}{5} = 0.4$$

If, however, I revealed that the quarter was already picked, then the probability of selecting the penny is *changed*. We must state the question as: "What is the probability of a penny *given* that we have already picked a quarter?" In symbols we write this as

$$P(\text{penny} \mid \text{already picked a quarter}) = P(A \mid B) = \frac{2}{4} = 0.5$$

where A is the event "pick a penny" and B is the event "pick a quarter." The vertical line is to be read as *given*. The value is 2/4, since there are now 2 pennies *out of a possible 4 coins*.

If we reversed the events we have a different problem. If instead of the quarter I had picked a penny first, my new population would then be reduced to 1 penny, 1 nickel, 1 dime, and 1 quarter. Then the event "pick a quarter, given a penny was already selected" would be

$$P(\text{quarter} \mid \text{already picked a penny}) = P(B \mid A) = \frac{1}{4} = 0.25$$

since we now have 1 quarter out of 4 coins.

Next we must examine the case where the sample *with replacement*—where the *population* is not changed from one sample to the next. Since there is no change in the population, the kinds of samples that may appear are not affected by previous samples and will not affect the samples that may appear later. Thus each individual, and each sample, is independent of every other individual or sample.

In addition, in these cases the given information or given event will *not* change the probabilities of any occurrence. In other words, here we replace the quarter after we observe what it is, and we still have 5 possible coins to choose from. Therefore

$$P(\text{penny} \mid \text{already picked a quarter}) = P(A \mid B) = \frac{2}{5} = 0.4.$$

but this is also $P(\text{penny}) = P(A)$. In general, *if*

$$P(A \mid B) = P(A)$$

that is, *if the knowledge that B has occurred does not change the probability that A will occur, THEN we say the two events A and B are independent.*

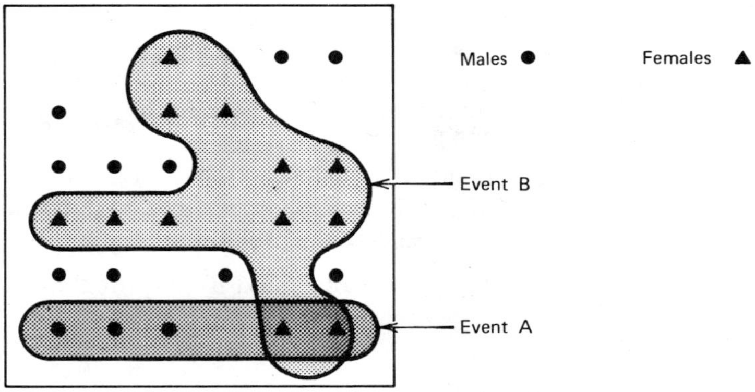

FIGURE 4.7 Sample space for students in my class. I have now indicated which are males and which are females. Two events are illustrated: A is students in the front row, B is a female.

EXAMPLE 4.2 (Cont.)

In Figure 4.7 some more information is now provided, and two events are labeled. I ask "What is the probability of a female being chosen, *given* I will only pick someone in the front row?" We only have 5 students in the first row and 2 are females. Therefore:

$$P(\text{female} \mid \text{first row}) = P(B \mid A) = \frac{2 \text{ females}}{5 \text{ students in front row}} = 0.4$$

Let us turn the problem around again and ask: "If you knew the person chosen would be female (the given part), what is the probability that she is in the first row?" The total N is now reduced to 12 (the 12 females), and you are looking at the area enclosed by Event B. Given that you must remain in this area, there are two females in the front row. Therefore

$$P(\text{first row} \mid \text{female}) = P(A \mid B) = \frac{2 \text{ females in front row}}{12 \text{ female students}} = 0.167$$

4.4 SOME ADDITIONAL PROBABILITY RULES

Multiplication rule and joint probability

There are two other important relationships between events that need to be considered. The first (and more important for our work) involves an *and* statement. The second, which will be discussed in the next section, is the *or* statement.

Whenever we discuss a sample, we are considering the results of several observations. In order to describe the probabilities for samples we must first be able to define the set of probabilities for all of the *single* observations. But the sample is the result of one observation *and* then another *and* perhaps another, and so forth, so that we amass a set of *and* statements.

We are concerned about the probability of my picking you today *and* tomorrow or not today *and* not tomorrow. We ask about the picking of a quarter *and* then a penny. I ask, "What is the probability you will show up for the test *and* receive a 100 on the test?"

These events are called *joint events*. We are not interested in the individual occurrences (the penny or the quarter) but, instead, are concerned about the *pairs of events* that have taken place in a particular way.

In general, the probability of a joint event "A and B" is

Probability of (A and B) = (probability of A)
× (probability of B given that the Event A has occurred)

or

RULE 3
$$P(A \text{ and } B) = P(A) \cdot P(B|A) \tag{4.3}$$

I also can just as easily reverse A and B so that we can have

$$P(A \text{ and } B) = P(B) \cdot P(A|B)$$

Thus, to calculate the probability of picking a quarter *and* then a penny, without replacement, would be solved as follows. Call A the event "pick a quarter" and B the event "pick a penny." Then

$$P(\text{quarter } \textit{and} \text{ penny}) = P(\text{quarter})\, P(\text{penny}|\text{already picked a quarter})$$
$$= \left(\frac{1}{5}\right) \cdot \left(\frac{2}{4}\right) = \frac{2}{20} = 0.10$$

If you return to Table 4.1 and examine all the samples, you will find that out of the 20 samples two (or 10%) include picking a quarter on the first try and a penny on the second try. Note that Figure 4.4 was a histogram of *sums*, not of the various pairs, as listed. We will elaborate on this concept when we introduce the addition rule on page 96.

We can continue this reasoning and see that if we wanted to determine the probability of a quarter followed by a penny followed by a dime, we would calculate $P(\text{quarter and penny}) \times P(\text{dime}|\text{already picked the quarter and penny})$.

The first probability is the value we have just calculated (the 2/20), and the second quantity can be determined by observing that the population now consists of a nickel, a dime, and a penny, or ⑤ ⑩ ①. Thus

$$P(\text{dime}|\text{quarter and penny already picked}) = \frac{1}{3} = 0.333$$

Therefore

$$P(\text{quarter } \textit{and} \text{ penny } \textit{and} \text{ dime})$$
$$= P(\text{quarter and penny}) \cdot P(\text{dime}|\text{quarter and penny})$$
$$= \left(\frac{2}{20}\right) \cdot \left(\frac{1}{3}\right) = \frac{2}{60} = 0.33$$

(Keep in mind that a particular *order* was specified here.)

If I decided to pick one student today and another tomorrow the chance that they will both be female would be

$$P(\text{female today and female tomorrow}) = P(\text{female today}) \cdot P(\text{female tomorrow} \mid \text{picked female today})$$

The probability that I pick a female today is 12/25. That eliminates one female (since I will not pick the same student twice in a row) and one student from which to pick tomorrow. So, the probability that I pick the female tomorrow given I pick someone else today is 11 out of 24 = 11/24. Therefore

$$P(\text{female today and female tomorrow}) = \left(\frac{12}{25}\right) \cdot \left(\frac{11}{24}\right) = \left(\frac{132}{600}\right) = 0.22$$

There is also another way of looking at an *and* statement that does not involve samples but that does involve different events. In Figure 4.7 we examined two events or classifications of students; Event A was a student in the front row and Event B was a female student. Now I ask "What is the probability of the female student in the first row?" We now have the question

$$P(A \text{ and } B) = P(\text{in first row and female}) = P(\text{first row}) \cdot P(\text{female} \mid \text{in first row}) = \left(\frac{5}{25}\right) \cdot \left(\frac{2}{5}\right) = \frac{2}{25}$$

This agrees with the fact that only 2 of the 25 students are actually females in the first row (who can readily be picked out as the individuals located in the sample space drawing where the two events overlap).

Independent cases

So far we have been dealing with dependent situations. That is, the knowledge that one of the events has occurred will affect the chance of the other event occurring. If, instead, the samples are independent, as in the case of sampling with replacement, the probabilities do not change from one pick to the next. Since there is no change in the population, there is no need to work with the "given" portion of the probability statement. In these cases we have $P(A \mid B) = P(A)$ and, therefore,

$$P(A \text{ and } B) = P(A) \cdot P(B) \tag{4.4}$$

Thus, in the case where we replaced the quarter after the first sample, we would now have

$$P(\text{quarter and penny}) = P(\text{quarter}) \cdot P(\text{penny}) = \frac{1}{5} \cdot \frac{2}{5} = \frac{2}{25}$$

This agrees with the listing of samples in Table 4.2 (again noting the order of picking the coins).

Similarly, in picking a quarter, a penny, and a dime *with* replacement, we would have

$$\begin{aligned} P(\text{quarter and penny and dime}) &= P(\text{quarter and penny}) \cdot P(\text{dime}) \\ &= P(\text{quarter}) \cdot P(\text{penny}) \cdot P(\text{dime}) \\ &= \left(\frac{1}{5}\right)\left(\frac{2}{5}\right)\left(\frac{1}{5}\right) = \frac{2}{125} = 0.016 \end{aligned}$$

EXAMPLE 4.1 (Cont.)

a. What is the probability of 2 heads if you flip a coin 2 times (or 2 coins at one time)? Each flip is independent of each previous flip. Thus we have

$$P(\text{head } \textit{and} \text{ head}) = P(\text{head}) \cdot P(\text{head}) = \left(\frac{1}{2}\right) \cdot \left(\frac{1}{2}\right) = \frac{1}{4} = 0.25$$

In a similar way we can extend this to perhaps 4 heads in a row as follows.

$$\begin{aligned} &P(\text{head } \textit{and} \text{ head } \textit{and} \text{ head } \textit{and} \text{ head}) \\ &\quad = P(\text{head}) \cdot P(\text{head}) \cdot P(\text{head}) \cdot P(\text{head}) = \left(\frac{1}{2}\right)\left(\frac{1}{2}\right)\left(\frac{1}{2}\right)\left(\frac{1}{2}\right) = \frac{1}{16} \\ &\quad = 0.0625 \text{ (which is approximately equal to 0.06)} \end{aligned}$$

b. What is the probability of having 3 girls in a family of 3 children? Births are generally considered as independent of each other. Therefore

$$\begin{aligned} P(\text{girl } \textit{and} \text{ girl } \textit{and} \text{ girl}) &= P(\text{girl}) \cdot P(\text{girl}) \cdot P(\text{girl}) \\ &= \left(\frac{1}{2}\right) \cdot \left(\frac{1}{2}\right) \cdot \left(\frac{1}{2}\right) = \frac{1}{8} = 0.125 \end{aligned}$$

Addition rule

The second type of connection between probabilities involves the use of the word *or*. You may have observed while studying the last section that we could not get the value for *P*(quarter and penny) to match the value for *P*(26 cents) in the histogram of Figure 4.4. The probability we obtained for the quarter followed by the penny was 0.1. The probability of 26 cents, in Figure 4.4, is given as 0.2. Why the difference?

The 26 cents can be arrived at either (1) by selecting the quarter *and then* the penny, *or* (2) by selecting the penny *and then* the quarter. So we have two ways of obtaining the 26 cents, and the probability of each method is 0.1. If I add together these two probabilities I would obtain a probability of 0.2, which matches the four cases out of the total of 20 in Table 4.1. In general,

RULE 4 *If two events, A and B, are* mutually exclusive, *then*

$$P(A \text{ or } B) = P(A) + P(B) \tag{4.5}$$

By *mutually exclusive* we mean that no individual outcome belongs to both A and B. Male and female are mutually exclusive, since a person is either male or female but cannot be both. A student may sit in the front row or the second row, but cannot be in both locations at the same time. If you use a sample space to represent the possibilities, such as that in Figure 4.7, then mutually exclusive events are sets of points or areas that do not overlap, with $P(A \text{ and } B) = 0$.

EXAMPLE 4.2 (Cont.)

The probability of picking a student in the first row (Event A) or picking a student in the back row (call this Event B) is

$$P(\text{first row or last row}) = P(\text{first row}) + P(\text{last row}) = \frac{5}{25} + \frac{3}{25} = \frac{8}{25}$$
$$= 0.32$$

EXAMPLE 4.1 (Cont.)

a. What is the probability of 1 head in 2 tosses of a coin? We must be careful here to recognize that the head can come either first *or* second and that the other toss will be a tail. Then

$$P(\text{"head and then a tail" } or \text{ "tail and then head"})$$
$$= P(\text{head and then tail}) + P(\text{tail and then head})$$
$$= \left(\frac{1}{2}\right)\left(\frac{1}{2}\right) + \left(\frac{1}{2}\right)\left(\frac{1}{2}\right) = 0.25 + 0.25 = 0.50$$

Observe that we could also do this by listing out all 4 possibilities for 2 tosses and arrive at the same probability.

b. What is the probability of having 1 girl and 2 boys in a family with 3 children? The girl could be the first, second, or third child. So we have

$$P(\text{girl-boy-boy } or \text{ boy-girl-boy } or \text{ boy-boy-girl})$$
$$= P(\text{girl-boy-boy}) + P(\text{boy-girl-boy}) + P(\text{boy-boy-girl})$$
$$= \left(\frac{1}{2}\cdot\frac{1}{2}\cdot\frac{1}{2}\right) + \left(\frac{1}{2}\cdot\frac{1}{2}\cdot\frac{1}{2}\right) + \left(\frac{1}{2}\cdot\frac{1}{2}\cdot\frac{1}{2}\right)$$
$$= \frac{1}{8} + \frac{1}{8} + \frac{1}{8} = \frac{3}{8} = 0.38$$

Note that each of the events, girl-boy-boy, boy-girl-boy, and boy-boy-girl, are mutually exclusive.

We should not need to delve any deeper into probability for our work here in elementary statistics. Obviously, there is much more that could be discussed.

4.5 PROBABILITY AND STATISTICS AGAIN

We have now spent time looking at some problems of probability. One of the results of our work was to develop a probability distribution. If you look back at the histograms in this chapter you should realize that they do one of the following.

1. Identify all of the possible samples from the original population or distribution.
2. Establish the distribution of samples with the appropriate probabilities for the various samples.

As a consequence, we can say what is likely or unlikely to occur when we take a sample. We cannot tell which sample will be chosen, but we can provide you with a number that will tell you what you probably will or will not select.

Now what about statistics? Here we do not know what the population is, we only have a sample. We put a hand into a container and pick out 2 coins—perhaps 2 pennies (2 cents). Our problem now is to try to guess what the container has as an overall population. This is where the two subjects—probability and statistics—merge. Our usual practice consists of the following steps.

1. We make an assumption about what we *believe* the population looks like. One population we might say we think this sample belongs to could consist of the 2 pennies, 1 nickel, 1 dime, and 1 quarter with which we have been working. Another population we could consider might be one with 24 coins—2 pennies, 10 nickels, 10 dimes, and 2 quarters. There are innumerable other possibilities that could be the source of the sample, but we will only look at these two for the moment.

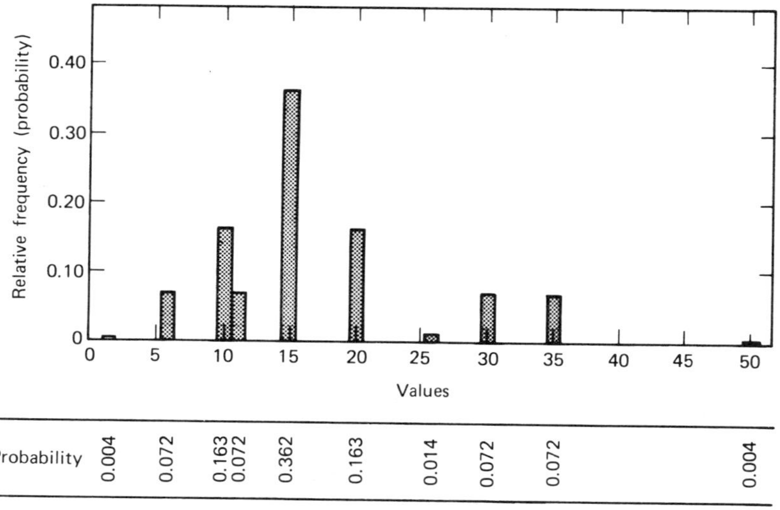

FIGURE 4.8 Distribution of samples of two from a population consisting of 2 pennies, 10 nickels, 10 dimes, and 2 quarters. The exact probabilities are also shown below each possible value.

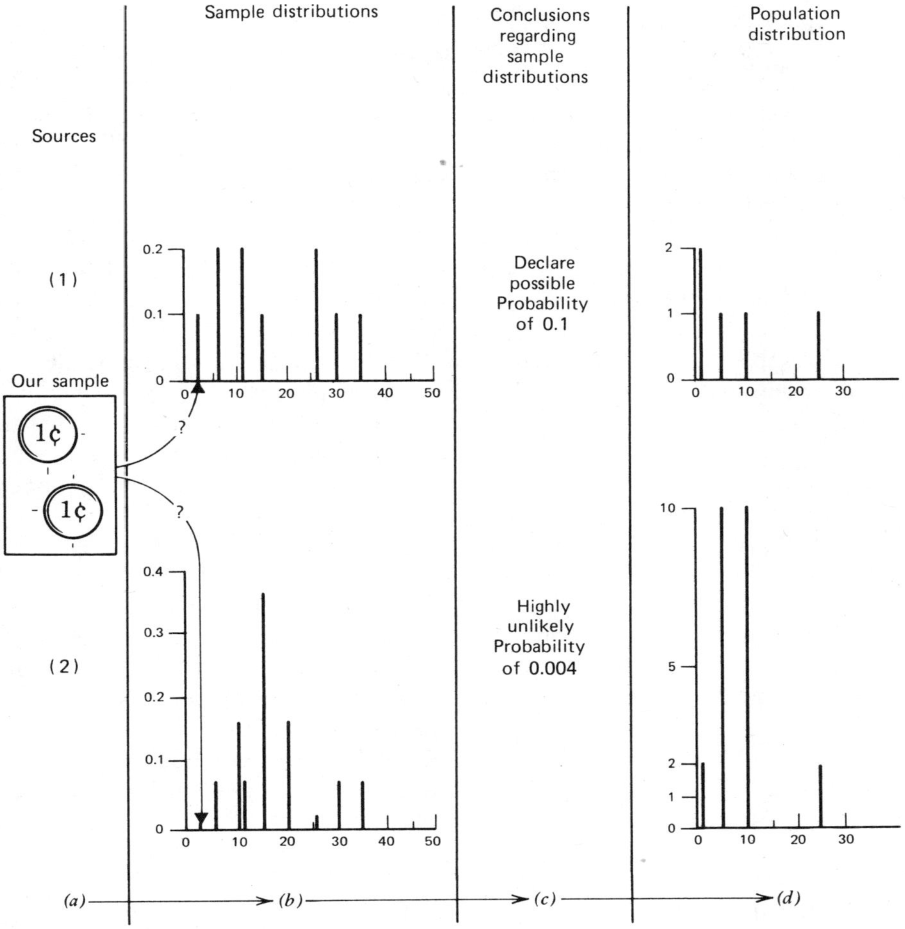

FIGURE 4.9 Comparing a sample to possible sources. (*a*) Look at our sample. (*b*) Look at some possible distributions of samples (1) and (2). (*c*) Compare the sample to the distribution and conclude that with (1) there is a good possibility that the sample came from it and that (2) is a highly unlikely source of the sample. (*d*) Extend our conclusion about the sample and the sample population to the population that generated the sample distribution. Therefore, (1) concludes the population might be the source of the sample and (2) is a highly unlikely source of the sample.

2. We look at the probability distribution of the samples from the theoretical population we believe is the source of our sample.* So we look back at Figure 4.4 (page 89) for the probability distribution of samples from our first possible population. Figure 4.8 shows the distribution of samples from the second population we are also considering.

* We call this theoretical distribution a *probability* distribution because we are graphing probabilities, not data.

3. We compare our *sample* to the probability distribution of the population. We do this by observing what the probability is that we might obtain this particular sample, or a sample this large, or this small, from the distribution. Thus, in Figure 4.4, we would find that the probability of obtaining 2 pennies (2 cents) is only 0.1 or a 1 in 10 chance (actually 2 in 20). This is a moderately small chance, although 3 of the 7 possible samples have this probability associated with them.

The second distribution, Figure 4.8, is even more extreme. Here the probability of picking out 2 pennies is extremely small. The probability of obtaining such a sample is only 0.004, or we would only expect four times out of a *thousand* to have such a sample. Because of the small chance of a sample of this kind coming from the second distribution, we would conclude that *it is highly unlikely that our sample came from a distribution of this kind*. We would extend our statement to say that we believe that the population out of which we picked our coins, the population of coins in our container does *not* consist of 2 pennies, 10 nickels, 10 dimes, and 2 quarters (the population of Figure 4.8).

In other words, we look at the sample and compare it to a theoretical distribution of samples. Depending on how likely it is that our sample may have come from this theoretical distribution of samples, we pass judgment about the population. If we find the sample is not unusual, we would conclude that there is no reason to doubt that the true population is the same as the theoretical population from which the theoretical distribution of samples came. If, however, the sample is an unusual sample, we will say it is highly unlikely that we obtained the sample from the theoretical population. Figure 4.9 summarizes this idea. We will return to this concept several times in later chapters.

REVIEW

Probability—Guess • Estimate • Mathematical
Outcome • Event • Sample Space • Equally Likely

$$P(A) = \frac{n}{N} \quad \bullet \quad 0 \leq P(A) \leq 1 \quad \bullet \quad P(A') = 1 - P(A)$$

Sampling with and without Replacement • Reduced Sample Space $P(A \text{ given } B) = P(A|B)$ • Independence • Mutually Exclusive • And • $P(A \text{ and } B) = P(A)\,P(B|A)$ • Or • $P(A \text{ or } B) = P(A) + P(B)$ Probability Distribution

PROBLEMS

SECTION 4.2

4.1 For the following situation or experiment, what type of probability (subjective, relative frequency, mathematical) can be determined? What value might or should be given?

a. The probability that you will have a broken leg the next time you go skiing.

b. There have been about 20 rotten tomatoes in each bushel of about 250 tomatoes. The probability that a selected tomato is rotten.
c. The probability of a 6 on a single roll of a die.
d. The probability that you will have an accident on the next trip to the store.
e. In dialing a call you have noted that 21 out of the last 85 calls were not completed. The probability of completing a call.

4.2 For the following situations, list all the possibilities. State if they are equally likely. If they are not, state the probability of each possibility or how you might go about determining the probabilities.

a. A die is rolled.
b. A pair of dice are rolled and the sum of the 2 dice are recorded.
c. I attempt to staple together three sheets of paper (consider what happens to the staple)
d. A multiple-choice question is answered with five answers.

4.3 On a true-false test given to a class of 12 students, *describe* all the possibilities of answers to one of the question (number of trues and number of falses).

4.4 On the same test discussed in Problem 4.3, there were nine other questions. Will the probabilities of each possibility for all the different questions be the same? Make up some examples to explain your answer.

4.5 We are to randomly pick one person from a population of which you are a member. What is the probability you will be chosen if the population is (a) 100, (b) 1000, (c) 1 million, (d) 200 million, (e) infinitely large?

SECTION 4.3

4.6 A store has found that sales have been distributed approximately 10% 50 cents; 20% $1; 25% $1.50; 20% $2; 15% $2.50; 10% $3.

a. What is the probability of a sale of $1?
b. What is the probability of a sale of less than $2?
c. What is the probability of a sale of $2 or more?
d. How are parts b and c related?
e. What is the probability of a sale of no more than $1.50?
f. If you knew that a sale was for less than $2, what is the probability it was $1? How and why does this differ from the "probability that the sale was $1?"
g. If we know a sale is at least $2, what is the probability it is $3?

4.7 The area codes for five widely separated states are listed here. For this problem we will only concern ourselves with these 21 area codes.

New York: 212, 315, 516, 518, 716, 607, 914
California: 209, 916, 714, 213, 408, 707, 415, 805

Florida: 305, 904, 813
Kentucky: 502, 606
Colorado: 303

a. Draw a sample space.

For parts b to h consider random selection of an area code.

b. What is the probability I will reach California?
c. What is the probability I will reach Colorado?
d. What is the probability the area code starts with a 2?
e. What is the probability it is odd?
f. What is the probability the second number is 0?
g. What is the probability the second number is not a 0?
h. What is the probability it ends in a 0, 1, 2, or 3?
i. What is the probability it is an odd number given that it is from California?
j. What is the probability it is an odd number given it is from New York?
k. What is the probability it is an odd number given it is from Colorado?

4.8 Six animals are part of an experiment to study heart weight. The actual, but unknown, weights of the animals are as given in the table.

WEIGHT	NUMBER OF ANIMALS	NAMES
4	1	A
5	2	B, C
6	1	D
7	1	E
8	1	F

(I have labeled them for convenience).

Two animals are to be dissected to determine the average weight before the experiment. List all the possible averages and then determine the probabilities of obtaining these average weights.

4.9 Out of 40 people in a room, several are to be questioned about their previous voting. Fifteen had voted Democratic, 14 Republican, and the rest did not vote. What is the probability that an individual selected voted Democratic? Republican? Did not vote?

If I am to pick 3 people, list all the possibilities I can have.

4.10 A pond contains 6 fish. Two weigh 10 ounces, 2 weigh 15 ounces, 1 is 6 ounces, and another is 25 ounces.

a. What is the distribution of all possible samples of total weight for 2 fish?

b. For all samples of 3 fish?
c. If I have to toss back all fish under 12 ounces, what is the distribution of weight for 2 fish?

SECTION 4.4

4.11 Refer to the area codes in Problem 4.7 and determine the following probabilities.

a. The area code starts with a 5 or is from New York.
b. Area code starts with an odd number or is for a state bordering an ocean.
c. Picking a number from New York three times.
d. Picking a number from Colorado twice.
e. Picking a number from New York and with middle number 1.
f. Picking a number that starts with a 7 and is from California.
g. The first and last numbers are the same.

4.12 For the sales distribution in Problem 4.6, determine the following probabilities.

a. What is the probability of 2 sales of \$1 each?
b. How can we have sales that total \$2?
c. What is the probability of 2 sales that total \$2?
d. What is the probability of having 2 sales that total \$2 or less?
e. What is the probability of having 2 sales that total \$3 or more?
f. What is the probability of having 2 sales that total \$3 or less.

4.13 Refer to Problem 4.9. If I pick 3 people, calculate the probability (a) they will all be Democratic, (b) they will all be Republican, or (c) all the other remaining cases.

4.14 Using the *and* and *or* rules, check the probabilities in Problem 4.8.

4.15 A rocket has an electrical unit that must function properly for ignition to take place. The unit is known to work 98% of the time. If a backup unit is available, what is the chance at least one of them will work properly?

4.16 The backup unit in the rocket in Problem 4.15 will not be turned on unless a detector determines that the first unit is not functioning. If this detector fails to find a problem 5% of the time that a failure in the first unit exists, what is the chance that the rocket will work (that at least one of the electrical units will work)?

4.17 A property tax rate increase was proposed and it was found that 60% of the property owners opposed it, while 80% of the non-property owners (renters for example) favored it. If 70% of the voters are property owners, what percent favor the tax increase?

4.18 A shirt is to have 5 buttons sewn on it. If 1% of the buttons are bad and they are automatically sewn on, what is the probability that a shirt will have all good buttons?

4.19 The licensed drivers in a particular state are classified in groups according to age. The proportion of drivers and the accident rate for each group are provided as follows.

GROUP	AGE	PROPORTION OF LICENSES	ACCIDENT RATE
1	17–25	0.155	0.095
2	26–45	0.392	0.041
3	46–65	0.321	0.050
4	Over 65	0.132	0.084

a. What proportion of the licensed drivers were over 65 and had an accident?
b. What proportion of licensed drivers had accidents?
*c. What proportion of drivers who had accidents were under 25 years old? Note here that $P(A \text{ and } B) = P(A|B)P(B) = P(B|A)P(A)$.

4.20 Refer to Problem 4.10. When is this sampling with replacement and when is it sampling without replacement?

4.21 The costs for the textbooks for six of the courses at a college was:

Mathematics	\$12	History	\$10
Chemistry	\$13	German	\$11
English	\$14	Art	\$12

a. A part-time evening student is taking two of these courses. What is the distribution of possible total costs for books?
b. If a student claimed books cost a total of \$24 is this unusual?
c. What do you say if the student claimed books cost \$28?

4.22 If the courses listed in Problem 4.21 were given on the following evenings—mathematics, chemistry, and English on Monday; history, German, and art on Wednesday—how much does this affect the possible distribution of costs for books?

4.23 A book salesperson claims book prices are roughly: \$13 40%; \$12 and \$14 20%; \$11 and \$15 10%. Use this distribution as an alternate population for comparison. Now if a student claims that two books cost \$24, which population do you believe in? How about a cost of \$28?

CHAPTER FIVE NORMAL DISTRIBUTION

In the last chapter I described a difference between the areas of probability and statistics, that is, how they rely on different relationships between samples and populations. Our main purpose in this textbook is to study the topics and concepts of statistics; this implies that we will be concerned mainly with making *inferences or decisions* about the source of the data that we may have gathered or are about to gather.

5.1 THE NEED FOR A THEORETICAL DISTRIBUTION

Recall back to the last chapter how *first* we looked at a theoretical distribution of samples from some populations (in that case, of coins), even *before* we took our sample. We then compared our sample to the collection of *possible* samples that *might* have arisen from that particular population and decided if it was likely or unlikely that our sample might have come from that population. This type of reasoning process is followed throughout the study of statistics. But we need to develop a hypothetical population to compare against; preferably it should be something that can be shown to have some reasonable connection to the source of the actual data.

There are a number of different distributions that can be represented by a mathematical equation. These are called *models*, and we often try to determine if our data came from a population that can be reasonably described by one of these models. We ask if the model and the data are essentially the same, or if a model can accurately duplicate the actual histogram. When this substitution can be made, we can use the model in place of the data, and we often can make many of the useful statements about the data such as probabilities connected to hypotheses and confidence intervals that were mentioned in Chapter One.

One of the most commonly referred to models in statistics is called the *normal* distribution or the *Gaussian* distribution. The name *normal* should not be interpreted as meaning that any other distribution is abnormal. We will be using the term nonnormal, but that should merely convey the fact that the data do not fall on a normal distribution curve. It does not imply that there is anything especially peculiar about the data or the distribution.

The normal distribution is just one of many possible kinds of distributions. It is symmetrical and often described as bell-shaped (see, for example, Figure 5.1). It is particularly important because numerous kinds of data fall into distributions that can essentially be called normal distributions. Many statistical tests *assume* that the source of the data can be considered a normal distribution. Thus, we will continually refer to this kind of distribution.

For those who have covered the empirical rule in Chapter Three, this is the distribution that forms the basis for the rule. We will also find that it will be even more important because the means (averages) of samples will tend to fall into normal distributions.

A second important model is the binomial distribution. We will discuss that model in Chapter Six.

5.2 DISCRETE AND CONTINUOUS DISTRIBUTIONS

One type of classification that distributions can fall into has to do with the kinds of possible values that may be obtained. When we sampled from the 5 coins in my pocket, you only had a small set of definite, distinct, *discrete* possibilities. The number of tickets issued, absences from class, broken eggs in a carton, or deaths from slipping in a bathtub are all illustrations of discrete distributions, because there is a set of exact specific categories into which our individual measurements or samples can fall. The binomial distribution, to be discussed in the next chapter, is such a discrete distribution. Usually, discrete only refers to items that can be counted in whole integer values. However, we can also think of each fractional category as a separate discrete value.

When we discussed the concept of measurement in Chapters One and Two, we observed that most measurements fall into categories only because of the limitation of the instrument we use to make the measurement. We cannot really determine if you had exactly 12 fluid ounces of beer at a temperature of 48 degrees, which you drank in 4 seconds, because in reality there might have been 12.00683142798143 . . . fluid ounces of beer at 47.8314601273 . . . degrees, which you drank in 4.47129814306 . . . seconds. You recorded values that were arrived at by rounding off at some point. The actual measurements or the *possible* values of the distribution may cover an infinity of different values spread out across the number line. We consider this to be in effect a continuous set of possible points, especially if you recognize that the next measurement of fluid ounces perhaps might be 12.00683142798142 . . . or 12.10683142798143 . . . . Then the difference from one true value to another might be only 0.00000000000001 or it could be 0.1. Then we might have over a trillion points to plot if we were to regard *these* values as discrete points; hypothetically we could continue adding more decimal places and thus create the need for more points (every decimal place multiplies the number of points by a factor of 10). When you plot all of these points along the X-axis, you have what is about as close as possible to a continuous line. If at the same time we regard the distribution as a "probability distribution," we can establish or assign a corresponding relative frequency or probability for each of these points and obtain a continuous line for the distribution itself.

Practical cases, however, do not work with the exact points but, instead, with ranges of points and areas of the distribution. In real situations we almost always work with discrete data, although in theory we may be dealing with continuous populations. In this chapter we will be working with a continuous distribution.

5.3 THE NATURE OF THE NORMAL DISTRIBUTION*

This distribution, like all others we deal with, has a mean and a standard deviation. With the normal distribution, if we specify the mean, μ, and nothing else, we are going to be left without a scale.† Figure 5.1 shows a distribution with mean, μ, equal to 50, but neither the variance nor the standard deviation has been specified. This is equivalent to saying that we have a distribution of problems assigned for each week of class, or the lengths of women's skirts, or the ounces lost while on a special diet—all with an average of 50. However, without a variance or standard deviation we *cannot* fill in any other values since we know nothing about the spread

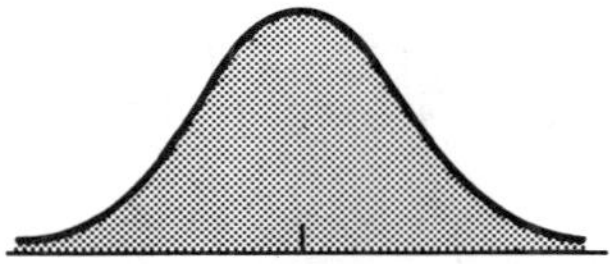

FIGURE 5.1 Normal distribution. Mean $= \mu = 50$. Variance unknown.

of this distribution or the way in which the individuals are dispersed across the scale. Thus we cannot tell if almost all the weeks have 50 problems assigned, if the lengths of skirts are all close to 50, or if the weight loss varied all the way from nothing to 100 ounces. Remember that this distribution is what we based the empirical rule on. At that time we had an estimation of the percent of the data that fell between certain values of sigma. But if we do not know the standard deviation, we are at a loss to know where the limits are to be. In order to work with the complete theoretical distribution it becomes even more vital to know the standard deviation (or variance).

Throughout this chapter I will use a problem involving the distribution of the amount of weight lost by a collection of people using various diets.

As we look at Figure 5.2, the nature of the distribution and its reliance on a value of standard deviation become apparent. Here we can see what is the relative chance that various values have of appearing, since the area between particular sets of values on the scale indicates, as always, the probability of occurrence of those values. Here we see that most people will lose from 45 to 55 ounces, with a few only losing around 40 ounces and some lucky people losing over 60 ounces.

* A review of Section 3.6, would help your work in this section. If you have not covered it you should do so in conjunction with this section.

† Recall that we use μ for population or true theoretical mean. $\bar{X}$ is only used for the mean of a *sample*.

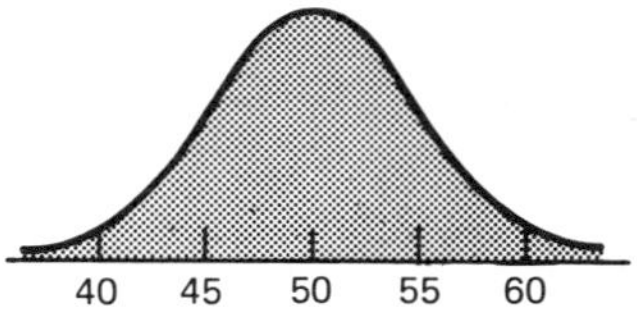

FIGURE 5.2 Normal distribution. Mean = μ = 50. Variance = 25. Standard deviation = 5.

Exactly what portion of the distribution lies between various points on the X-axis and how we will determine these probabilities will be discussed in Section 5.4. There are no simple relationships between the values on the X-scale and the regions of the normal distribution. We will be required to use special tables that will be described shortly.

As we change the variance (and hence the standard deviation), we alter the scale (it may be we changed the diet). Thus, in Figure 5.3, as the variance is increased from 25 to 100, the scale is likewise changed. A point that appears to be approximately the same distance away from the center actually is numerically further away in Figure 5.3. Compare the X-value of 40 in Figure 5.2 with the X-value of 30 in Figure 5.3. They both appear to be

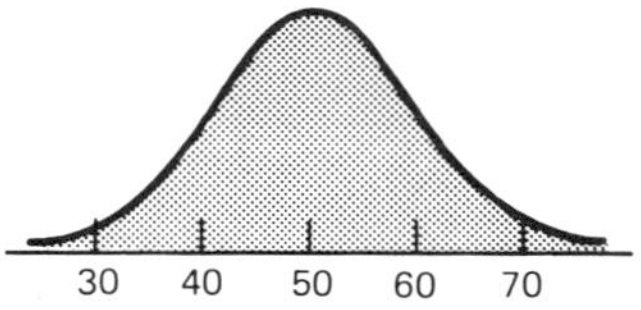

FIGURE 5.3 Normal distribution. Mean = μ = 50. Variance = 100. Standard deviation = 10.

at the same relative locations on the curve. However, in the first case it is only 10 units from the center; in the second case this point is 20 units from the center. When the variance is greater, we find pinpointing the individual loss is more difficult, so that a person with the new diet has the same chance of losing 30 ounces as the old dieter had of losing 40 ounces. The same is true on the high end of the scale.

We can generalize this observation by saying that the greater the variance, the larger the values along the scale. Similarly, if we decrease the variance, we will obtain a smaller scale along the X-axis—the base of the graph. In Figure 5.4 I have drawn a normal distribution that now has a smaller

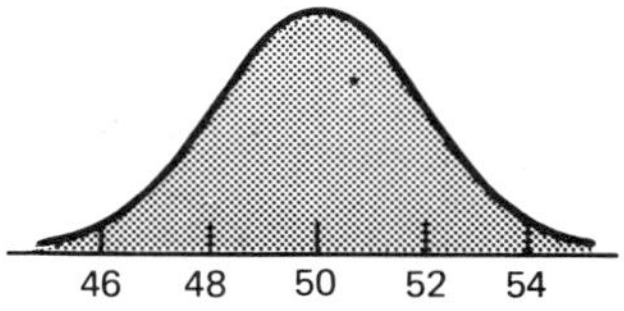

FIGURE 5.4 Normal distribution. Mean = μ = 50. Variance = 4. Standard deviation = 2.

variance (a third diet). Our generalization mentioned above about the variance still holds.

Let us superimpose all three curves on a single set of scales. All of these distributions are probability distributions because they each represent a theoretical distribution. Now we need to reason as follows. The probability of a value falling in a particular region is related to the area of the distribution in that region. The probability that a value falls somewhere in the distribution is 1.00. So the area under an entire normal distribution must be 1.00 square units (area of 1.00). Furthermore, this total area of 1.00 must be the *same for all the distributions.* So the three curves must be drawn in such a way that the areas are *all* the same.

If we use the scale of Figure 5.2 as a standard (the one with a mean of 50 and a standard deviation of 5), when I superimpose the next graph I will have to stretch the curve. This in turn will cause it to shrink in height (see Figure 5.5). The third graph we discussed (the one with the smaller standard deviation—Figure 5.4), because it is on a smaller scale, would be compressed

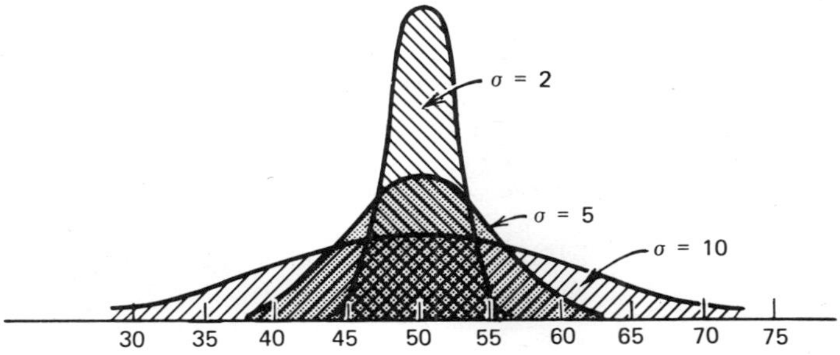

FIGURE 5.5 Normal distributions. Mean = 50. Figures 5.2, 5.3, and 5.4 are superimposed on the same scale. All areas are equal.

inward in order to squeeze it onto this new scale. This would cause the height to increase. Remember, this is because *all three graphs must have the same area.* Let me remind you that these curves are to be interpreted as though they are histograms. The areas between particular sets of values represent relative frequencies with which the particular range of X-values occur. The more compressed curves would have most of their values close to the mean, while the others, which are much more spread out, would have less of a chance of having values that are actually near the mean.

Thus far we have looked at several different normal distributions, but all had the same mean or central value. It is not necessary that the mean be equal to 50. Figure 5.6 shows a normal distribution with mean equal to 100 and variance equal to 25. This does not look much different than Figure 5.2 except for the scale. Figure 5.7 shows a distribution with mean = 0 and variance = 100, and resembles Figure 5.3 except for the scale. Each of these could be a consequence of additional different diets that yield different results. The last diet appears to cause no change, and may just reflect

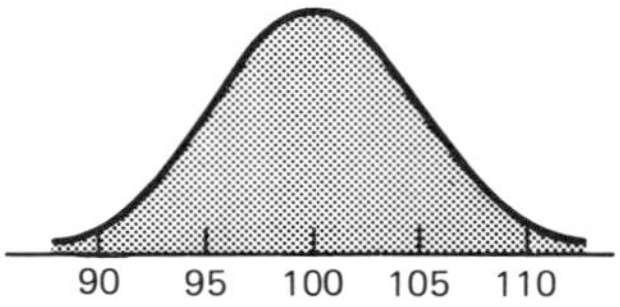

FIGURE 5.6 Normal distribution. Mean = μ = 100. Variance = 25. Standard deviation = 5.

random change in a control or a maintenance diet. (Note that a minus value would indicate that weight was gained, not lost.)

I have taken the pairs of distributions that have identical variances and placed them on the same axis. Figure 5.8 contains both distributions with σ^2 (variance) = 25 (from Figures 5.2 and 5.6). Figure 5.9 contains the two distributions with variance of 100 (from Figures 5.3 and 5.7). I then drew all four of these distributions on a single axis (Figure 5.10). You should now be aware that many different kinds of normal distributions can be drawn—as many different distributions as can arise in real life. They

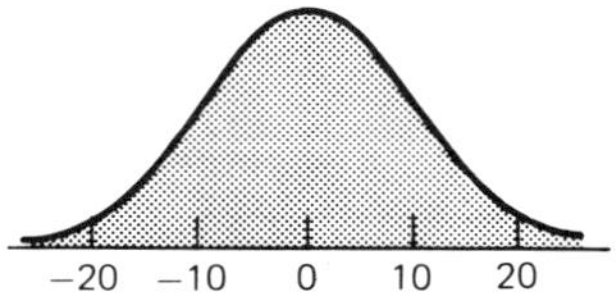

FIGURE 5.7 Normal distribution. Mean = μ = 0. Variance = 100. Standard deviation = 10.

resemble each other in certain ways but, with each new value of μ (mean) or σ^2 (variance), we will effectively be drawing a different picture, or at least using a different scale.

Now if you have been observant, you may have realized that something has been avoided. None of the curves thus far presented have had scales for the Y-axis. There is good reason for this. In order to calculate the heights at each particular X point, we must know what relationship exists between the X and the height. The formula for the normal distribution, however, is extremely complicated, and we will have no real need to determine such values.

Recall now that in a continuous distribution the probability associated with an exact value is zero. However, when we work with practical prob-

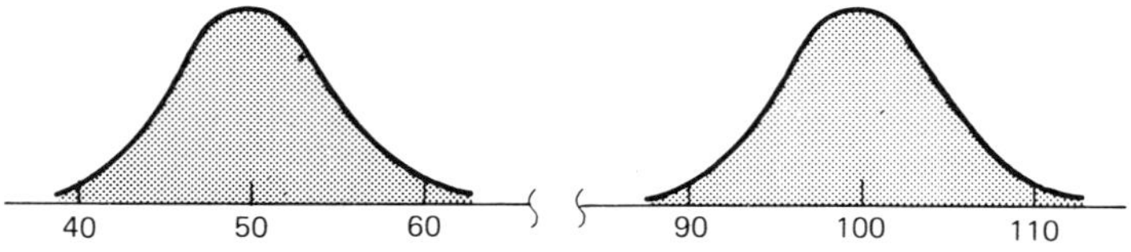

FIGURE 5.8 Two normal distributions having same variance (25) but different means. Figures 5.2 and 5.6 are shown on the same axis.

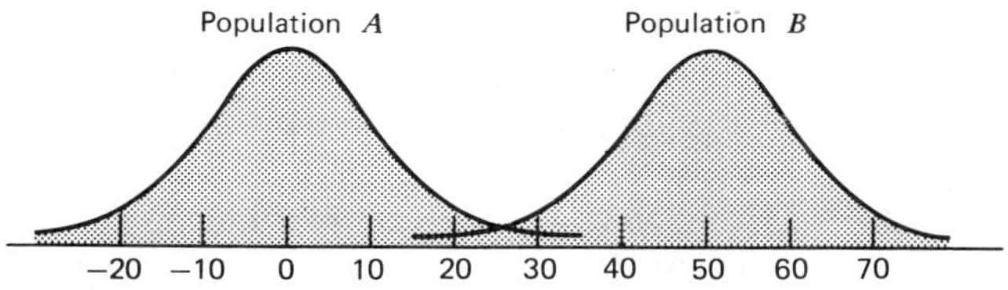

FIGURE 5.9 Two normal distributions with variance of 100 and different means. Figures 5.3 and 5.7 are shown on the same axis.

lems, we work with a scale, and each value has a pair of bounds. These bounds effectively are the class limits we encountered in our work with histograms and measurement data in Chapter Two.

To determine the probability of a particular value or set of values, we are really determining the area under the curve between two limits. This again is because the *area under the curve is equal to the probability* associated with the collection of values between those limits. In addition, we really don't care about determining your weight loss in thousanths of an ounce—the nearest half an ounce is more than satisfactory.

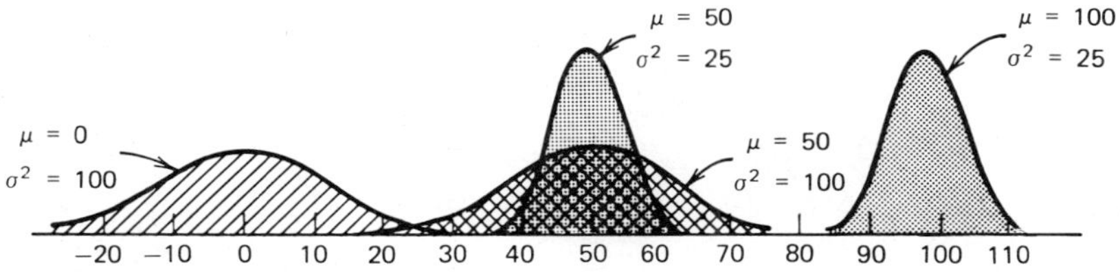

FIGURE 5.10 Distributions of Figures 5.8 and 5.9 on one scale. The scale has been adjusted to allow all four distributions to fit on a single scale. All areas are equal and equal to previous curves.

Whenever we will discuss the probabilities of certain values from normal distributions, we will use the word *between*. In all cases we will be required to specify two points. We will therefore refer to the *probability of a value* (call it X) as that area under the normal curve *between two points*, two points that somehow include the X-value.

In Figure 5.11 we would say that the "probability that the variable X (weight loss) will fall in the *range of possible values* from 0.0 to 0.5 is equal to the area (shaded) under the curve for that range."

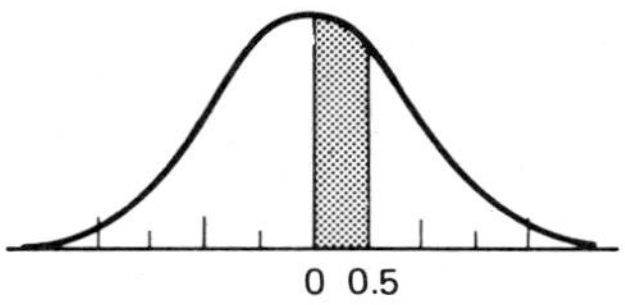

FIGURE 5.11 The probability of a value from 0 to 0.5 is the shaded area.

Recall in Section 3.6 the discussion about interpretation of the standard deviation. At that time we examined the effect of subtracting the mean from an individual value and then dividing by the standard deviation. We found we obtained a new number that could be used to make comparisons of values from different distributions. We will follow the same procedure again, only this time we will give it a name—we will call it *standardizing*. We will now subtract the population mean (μ) and divide by the population standard deviation (σ) to obtain replacements for all the X-values. Thus, we are doing something to each and every individual X-value of the original distribution. We are adjusting them. The effect is:

1. To reduce the distribution center to a *center of* 0.
2. To change the standard deviation to a value of 1 (by dividing by σ).

We have essentially changed the scale, or superimposed another scale on the original one. This new scale reads with a 0 at the mean and has one unit for each standard deviation.

In Figure 5.12 I have taken the curve in Figure 5.4 (mean 50 and variance 25) and superimposed another scale on the same axis. You will soon be

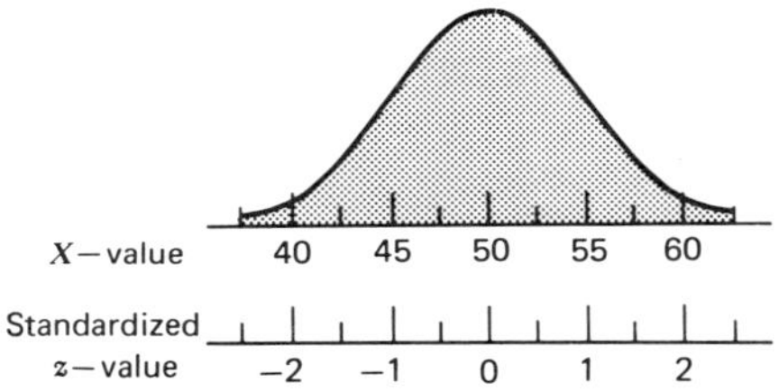

FIGURE 5.12 Relation of X-value to z-value. The difference is merely the scale.

shown how to use a table that will give us the probabilities for the *standardized normal distribution* (the distribution with mean $= 0, \sigma^2 = 1$). Since the area in this standardized normal distribution also corresponds to *our* normal distribution (mean $= 50, \sigma^2 = 25$), we will automatically have probabilities for our distribution.

What I have been saying is that we can adjust the scale on any normal distribution to correspond to the standardized normal distribution. If we then work with the new scale, we can work with a single universal table. This universal scale, as I have said, will possess a central value of 0 and variance or standard deviation of 1. We will label the values on this new scale *z-values*. Thus we will have two scales side by side, and we transfer or change each X-value to a corresponding x-value. The relation between the X-values and the z-values is simply

$$z = \frac{X - \mu}{\sigma} \tag{5.1}$$

where the μ and σ are constants for a particular distribution.

We will make our interpretations using the z-values, and we do this by referring to probabilities associated with the particular z-values. However, since the normal distribution is a continuous distribution, probabilities will always refer to areas. An area must have boundaries. Since we are discussing area under a curve, we need only to specify the end points. One of these must be the z-value. The other end point traditionally is one of two values, either 0 or $-\infty$ ($-\infty$, or negative infinity, means going to the left endlessly). When we use negative infinity, we are considering the case in

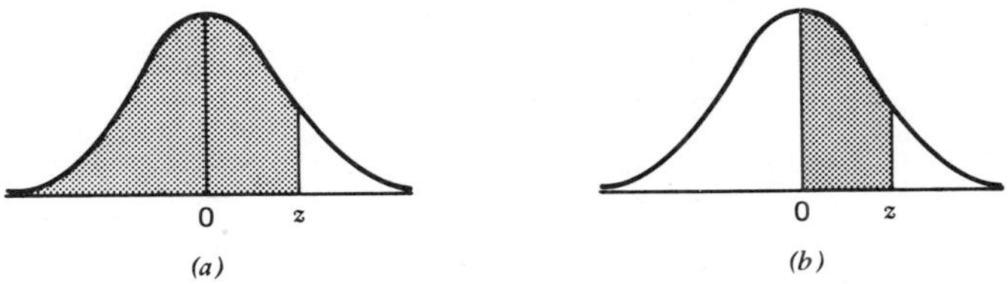

FIGURE 5.13 Cumulative areas for normal distribution. (*a*) True cumulative—lower limit is $-\infty$. (*b*) Area used in many tables—lower limit is 0.

Figure 5.13*a*. This is a true *cumulative probability* value where the value associated with z is the area *up to z*.

More frequently we use tables that provide the area or probability of a value between 0 and z (Figure 5.13*b*). Since the normal distribution is symmetric, we can consider the area from 0 to $+z$ as the same as the area from $-z$ to 0 (see Figure 5.14). Appendix Table A.1a contains a table of values for areas from the mean to z (or from 0 to z). These are listed in column B.

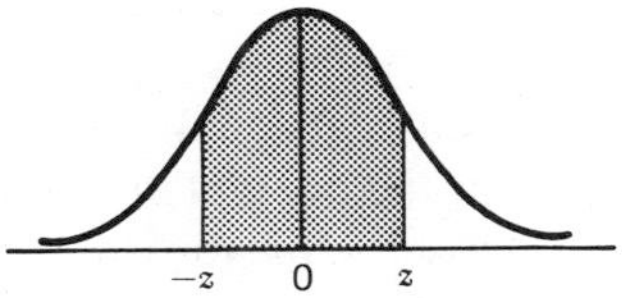

FIGURE 5.14 The area from 0 to $-z$ is the same as from 0 to $+z$.

In column C we also have the areas above $+z$ (the area in the tail of the distribution). Note that this is also the area up to $-z$. (We will gradually develop these concepts, so don't panic.) The table goes up to $z = 2.50$ in intervals of 0.01 and then up to $z = 4.00$ in larger intervals.

5.4 USING THE TABLES

For each of the next nine examples we will explore how you use the tables to find probabilities, or areas, of a normal distribution. We will work these examples for a normal distribution with mean 0 and standard deviation 1, and you can think of the values (the z-values) as weight loss, or just as

z-values. You can think of the examples as discussing chances of having particular weight loss, or you can try to generalize and let the distribution be any type of example you might prefer.

EXAMPLE 5.1

Find the probability of a value being in the area from $z = 0$ to $z = 1.00$.

I have drawn part of the table in Figure 5.15*a*. The left column is for the decimal value of z. In column B we find that the value for the area is 0.3413. This means that 34.13% of the area is in the range from $z = 0$ to $z = 1.00$ and thus there is a 34.13% chance our value is in this area. In Figure 5.15*b* we have the distribution showing the area that we are considering.

A	B
z	Area 0 to z
1.0	0.3413

(*a*) Table

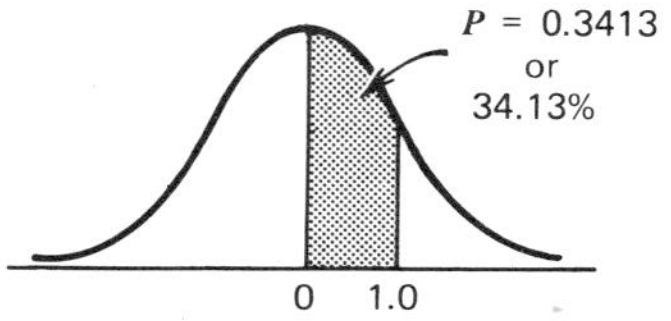

(*b*) Drawing

FIGURE 5.15 Finding the probability (or area) for a z-value of 1.0 (to 0.0).

EXAMPLE 5.2

Find the area between $z = 0$ and $z = 2.34$.

We look down column A until we reach 2.34. Alongside the 2.34 we find in column B that the area between $z = 0$ and $z = 2.34$ is 0.4904 (Figure 5.16*b*). Note that the area between $z = 0$ and $z = -2.34$ would be the same (0.4904) as seen in Figure 5.16*c*.

Now remember that the normal distribution is symmetric and therefore half of the area is above the mean while half is below the mean. Then the area above or below the zero point is 0.5000. We will use this observation

A	B
z	Area 0 to z
2.34	0.4904

(*a*) Table

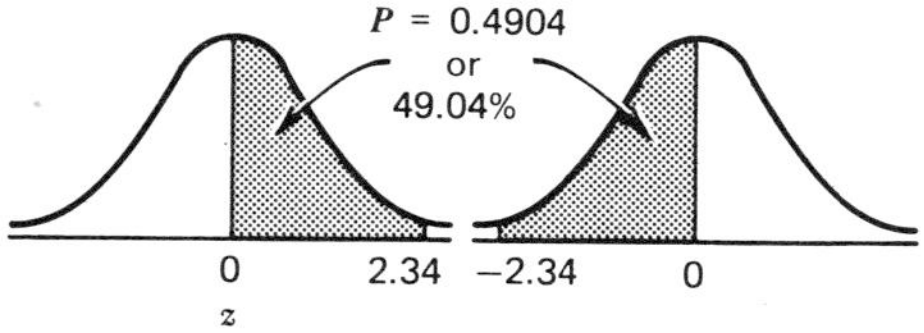

(*b*) z positive (*c*) z negative

FIGURE 5.16 Finding the probability of a value for (1) a two-decimal place z-value, and (2) negative z-value.

to come up with *complete cumulative distributions*. If *z is positive*, say $z = 1.22$, then the total cumulative area is the sum of the amount *below* $z = 0$ and the *amount between* $z = 0$ *and* $z = 1.22$.

EXAMPLE 5.3

Find the area below $z = 1.22$ (the probability of z being less than 1.22).

The area from $z = 0$ to $z = 1.22$ is 0.3888. The total cumulative area $= 0.5000 + 0.3888 = 0.8888$, or we can say 88.88% of the values are below $z = 1.22$ (see Figure 5.17).

A	B
z	Area 0 to z
1.22	0.3888

(*a*)

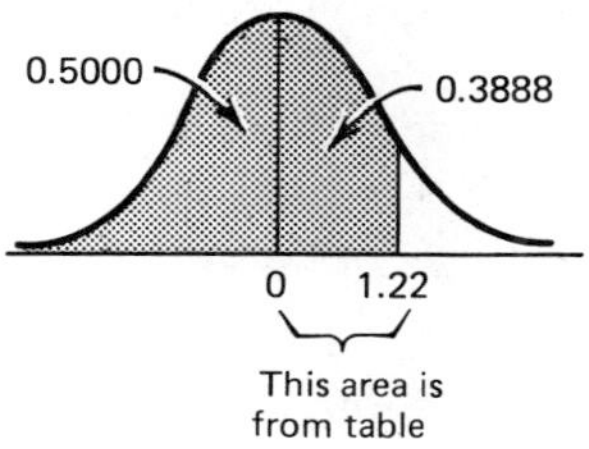

(*b*)

FIGURE 5.17 Finding the cumulative area below a given positive z-value. We must add the two areas.

If we have *z negative* then, for the cumulative area, we really want the area *below* the particular z-value. This time we must realize that we are to *subtract* the area we read in the table from 0.5000. This will leave us the amount in the lower tail and give us the cumulative area we want to know. Our next example will illustrate.

EXAMPLE 5.4

Find the cumulative area for $z = -0.88$ (see Figure 5.18).

A	B	C
z	Area 0 to z	Area beyond z
0.88	0.3106	0.1894

(*a*)

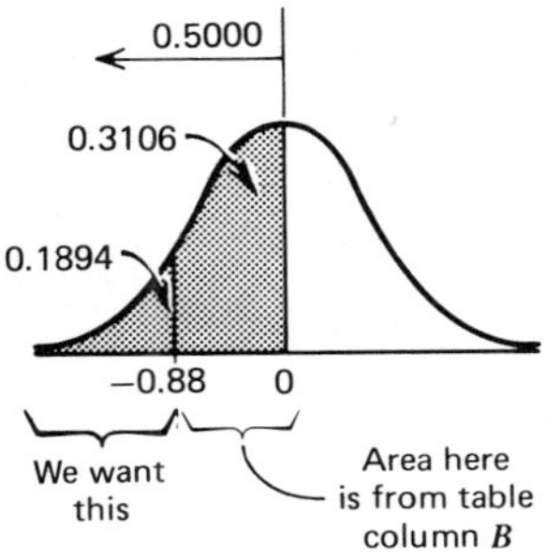

(*b*)

FIGURE 5.18 Finding the cumulative area below a given negative z-value. We must subtract the value in the table from 0.5000, or use column C.

The first way is to look up the area for $z = +0.88$ and obtain in column B a value of 0.3106. If we subtract this from 0.5000 we get 0.1894. This latter value is the cumulative area for $z = -0.88$. Notice that it is the left tail of the distribution that we are actually looking at. But *we can also read this value directly in column C*, since column C is the area *beyond* the z-value.

You should begin to become aware of the usefulness of the little picture. By identifying the approximate locations on the graph, you should be able to determine rapidly the proper procedure needed to calculate the desired area. We will now cover additional major types of problems. They all follow from the previous examples.

EXAMPLE 5.5

Find the area above $z = 1.65$ (see Figure 5.19).

We do one of two things.

1. Determine the area between $z = 0$ and $z = 1.65$; this is 0.4505; if we subtract this from 0.5000 we have our desired value—0.0495.
2. We look in column C to read the value directly.

A	B	C
z	Area 0 to z	Area beyond z
1.65	0.4505	0.0495

(*a*)

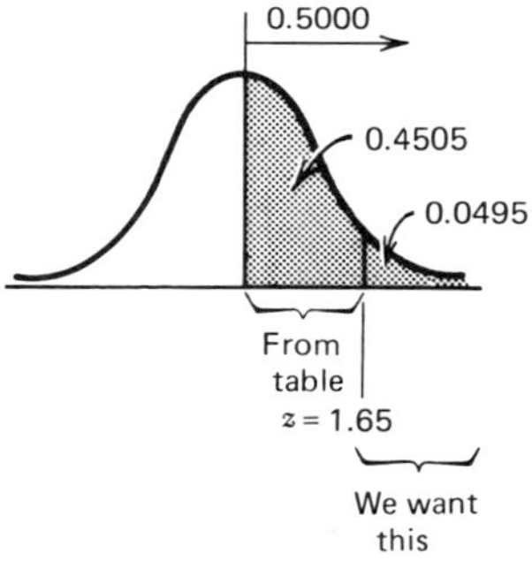

(*b*)

FIGURE 5.19 Finding area of a tail $z = 1.65$.

So just about 5% of the values are above a z of 1.65. This is a very common problem and is referred to as "finding the area in a tail." You should also note that there is no real difference in the technique for finding the area in either of the tails.

I am emphasizing the subtraction method of finding areas in the tails because many tables you may encounter are not as convenient as Table A.1a—they may not have a column C to read the areas in the tail directly.

EXAMPLE 5.6

What is the probability of a value falling between $z = -0.75$ and $z = +1.25$? (See Figure 5.20)

A	B
z	Area 0 to z
0.75	0.2734
1.25	0.3944

(*a*)

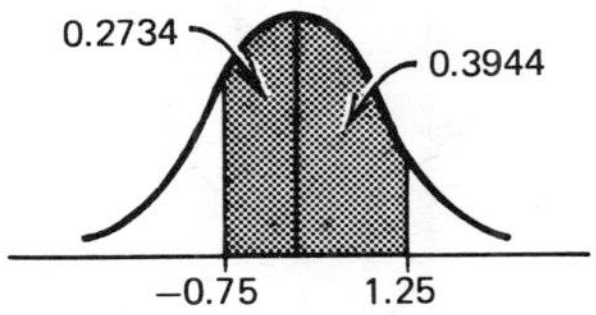

(*b*)

FIGURE 5.20 Finding the area between two z-values. Here we add 0.2734 to 0.3944 to get 0.6678.

This is a problem with *two limits*. We must find the areas for both of these values. First, look at the figure; then decide what we do with the two quantities when we find them. We will add them together to get the total area between the two limits. We obtain $0.2734 + 0.3944 = 0.6678$. So about 2/3 of the values will fall in this range.

EXAMPLE 5.7

Now, what is the probability of a value being between $z = 1.30$ and $z = 1.89$ (see Figure 5.21)?

Look at the figure and observe how the area associated with the larger z-value, the 1.89, extends all the way back to zero *and* also includes *all* of the area up to $z = 1.30$. But we do not care about the area up to $z = 1.30$, so we must subtract this area from the larger one. This is true *whenever the two z-values have the same sign* (both are plus or both are minus z-values). Since for a z-value of 1.89, $p = 0.4706$ and for a z-value of 1.30, $p = 0.4032$, the area between is the difference, or $0.4706 - 0.4032 = 0.0674$.

Let us now reverse the type of problem. We will now try to find z-values between which a given amount of area or percent of individuals lie.

A	B
z	Area 0 to z
1.30	0.4032
1.89	0.4706

(*a*)

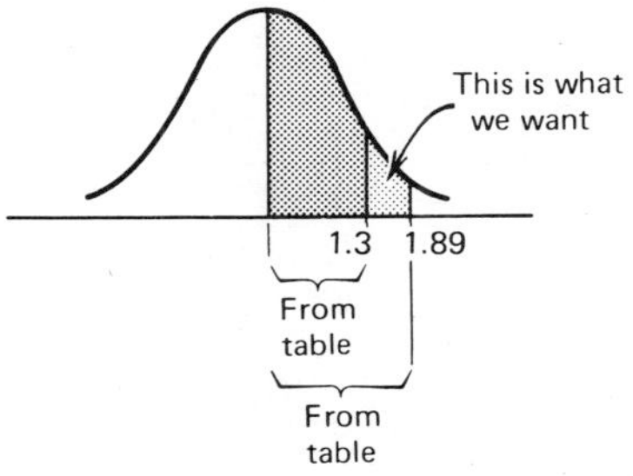

(*b*)

FIGURE 5.21 Finding the area between two z-values. The area is 0.4706 minus 0.4032, or 0.0674.

EXAMPLE 5.8

We are now to answer the question, "17% of the area is enclosed between what value of z and 0?" (Figure 5.22.)

A	B
z	Area 0 to z
0.44	0.1700

(*a*)

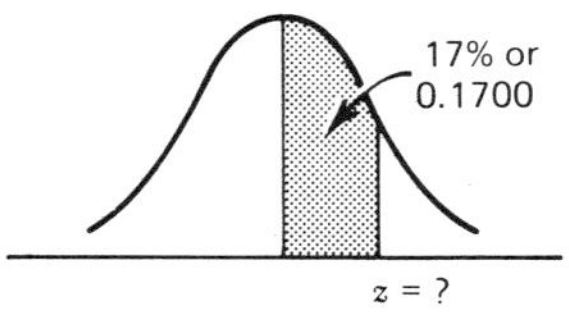

(*b*)

FIGURE 5.22 Given an area of 17% find the corresponding z-value. We work from inside the body of the table to the margins to get a z of 0.44.

Now we work from the *inside* of the table, in column B, to find the probability designated. We then look across to column A for the corresponding z-value. Thus, we look into the table to find 0.1700, since this is equivalent to 17%. In column B we are fortunate to find the exact number 0.1700 (this is not too common an occurrence). In column A we find that it corresponds to a z of 0.44. So 17% of the area lies between $z = 0$ and $z = 0.44$.

EXAMPLE 5.9

What is the z for which 15% of the values are in the upper tail? (Figure 5.23.)

This is a more common problem, involving the determination of the z-value for a percent or probability of values falling in the tail (or tails). Well,

A	B	C
z	Area 0 to z	Area beyond z
1.03	0.3485	0.1515
?	0.3500	0.1500
1.04	0.3508	0.1492

The 1.04 is closer to what we want

Not found in the table

(*a*)

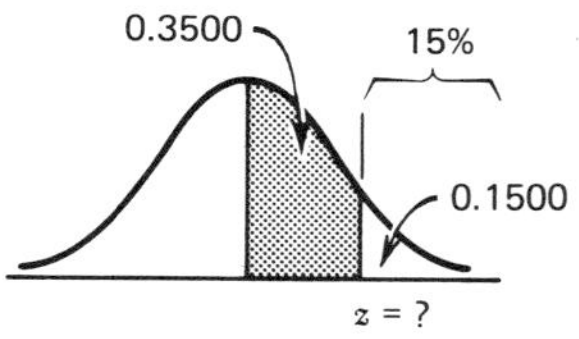

(*b*)

FIGURE 5.23 Given a desired area in the tail, 0.1500, find the z-value. We must find the z-value for 0.5000 minus 0.1500 = 0.3500. We find that 1.04, is closest.

if 15% are to be in the tail, then 50%–15% (0.5000–0.1500) or 35% (0.3500) are included between 0 and z. It is the 0.3500 that we would look up in column B (or the 0.1500 could be looked up in column C). Search as hard as you can, you won't find 0.3500. What do we do?

The simplest approach is to find the two z-values that are nearest to the probability value we have. In this case they are 1.03, which corresponds to 0.3485 (the next lower value), and 1.04, which corresponds to 0.3508 (the next higher value). Since the 0.3508 is closer to the 0.3500 than the 0.3485 is, the z-value that would be closer to the true z-value would be the 1.04.

A more precise method called interpolation can be used, but it is generally unnecessary.

Finally, we need to discuss the cases where we desire to find the two z-values between which a particular percentage of values is included. Generally, we will split the amount evenly between the upper and lower halves. Then, because the split is in equal amounts, the distance from the center to the limits is the same for both directions and, therefore, except for the sign, the z-value will *be the same* in both directions. So we only need to find *one* z-value (in the same manner, as in Example 5.5). In Figure 5.24*a*

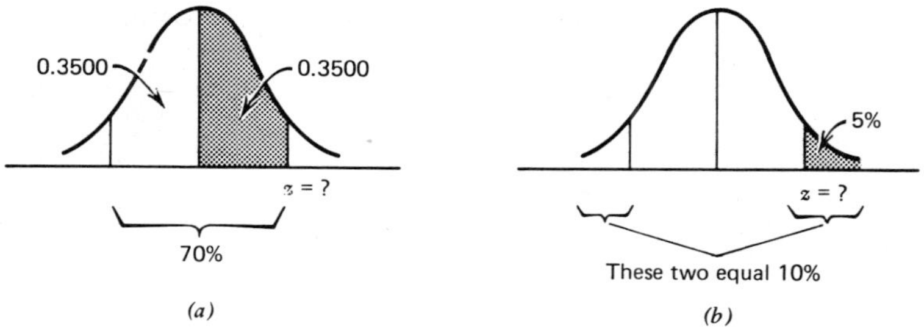

FIGURE 5.24 Finding limits for two-tailed problems. (*a*) Given a percent between the limits, divide the area in two. (*b*) Given a percent in the two tails, divide the area and find the limit for one as before.

you can see that when we ask for the z that encloses 70% of the values, we actually pose the problem of finding the z for which 35% is between 0 and one of the limits. Thus we end up with two limits that are equal in magnitude. Since the z for which 35% of the values lie between it and 0 is 1.04 (from Example 5.9), we would say that the two limits that enclose 70% of the area are $z = +1.04$ and $z = -1.04$.

In a similar manner, to calculate limits for which an equal amount of area is in each tail, say 5% in each tail or a total of 10% outside the limits, we again look at *one* of the limits or one side of the distribution. We would then find the z-value for which 5% were above (or 45% from 0 to z). This limit would be 1.64, and we would say that the two z-values were $z = -1.64$ and $z = +1.64$.*

* If you use Table A.1b, or interpolate, you will find the value to be $z = 1.645$.

Okay, so much for z-values, but what about a typical normal distribution? We already talked about converting individual values from any distribution to z-values or vice-versa, from a z table to another normal distribution. We merely apply the formula (which I repeat here):

$$z = \frac{X - \mu}{\sigma} \tag{5.1}$$

or its counterpart

$$X = z \cdot \sigma + \mu \tag{5.2}$$

to convert in either direction. We can then employ the z table for computing *all* areas or *all* probabilities for *all* normal distributions.

EXAMPLE 5.10

The diet used by the Hillsboro Chapter of the Weightcontrollers Society leads to an average (mean) weight loss of 50 ounces every 2 weeks. The variance among dieters is 100 (then $\sigma = \sqrt{100} = 10$).

What percent of the dieters will lose between 44 and 64 ounces?

First we convert these two limits to z-values using Equation 5.1.

For $X = 44$

$$z = \frac{44 - 50}{10} = \frac{-6}{10} = -0.6$$

and for $X = 64$

$$z = \frac{64 - 50}{10} = \frac{14}{10} = 1.4$$

Next we consult the z table to determine the percent that lies between a z of -0.6 and $+1.4$. We find 0.2257 for the area from 0 to -0.6 and 0.4192

A	B
z	Area 0 to z
0.60	0.2257
1.40	0.4192

(*a*)

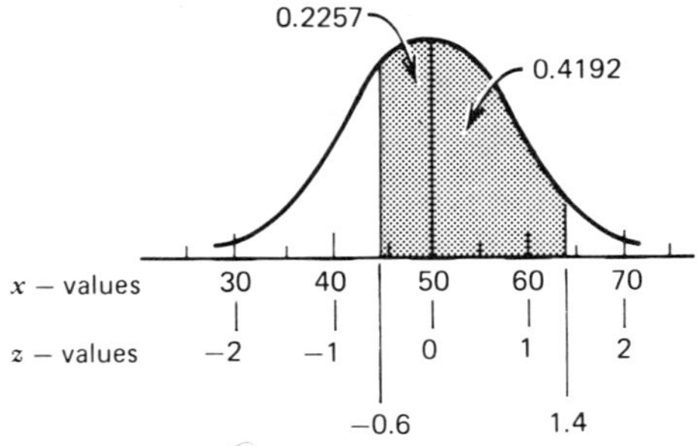

(*b*)

FIGURE 5.25 Given a normal distribution with mean = 50, variance = 100, what is the probability of a value between 44 and 64? (First find the z-values, then consult the table as before to determine the areas.)

from 0 to +1.4. The area between −0.6 and +1.4 is the sum of these two values (see Figure 5.25) and

$$\text{Probability} = 0.2257 + 0.4192 = 0.6449$$

Thus about 64.49% of the dieters should lose between 44 and 64 ounces.

EXAMPLE 5.11

This time let us perform the reverse problem by finding an X-value.

We decide we want to know what minimum weight loss we should guarantee that 90% of the dieters will lose. This question amounts to asking "above what value are 90% of the individuals?" (Note that it also is the same as asking below what value are 10% of the individuals.)

We start by determining the z-value for which 90% of the values are above. This is equivalent to asking for a $-z$ for 0.4000, since we are including the entire upper 50% in our area, along with an amount on the lower end, 40%, which makes up the difference (see Figure 5.26). Consulting Table A.1a, we find that the z for 0.4000 corresponds to an approximate value of 1.28, which is to be −1.28 for our problem (see Figure 5.26 again).

A	B
z	Area 0 to z
1.28	0.3997
1.29	0.4015

0.4000 is closer to this value

(*a*)

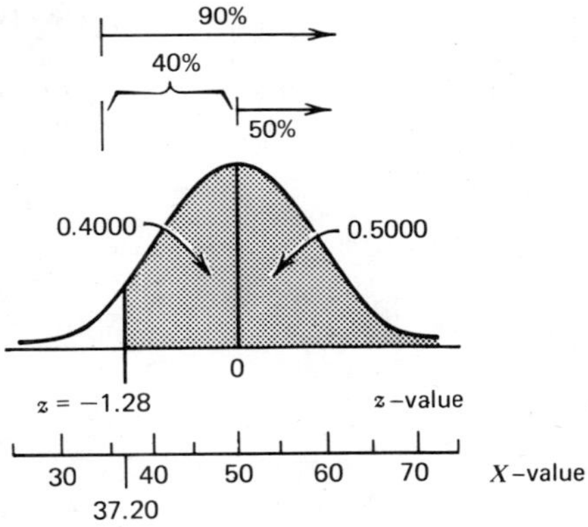

(*b*)

FIGURE 5.26 Above what value is 90% of the distribution? (Where mean = 50, σ = 10.)

Substituting this z-value into Equation 5.2, we will be able to find the X-value, or

$$\begin{aligned} X &= z \cdot \sigma + \mu \\ &= (-1.28)(10) + 50 \\ &= -12.80 + 50 \\ &= 37.20 \end{aligned}$$

Thus 90% of the people should lose more than 37.20 ounces.

The thing to remember is that there are two steps—one is to *find the z for the desired area*; the second is to *find the X-value that corresponds to the z-value* that you obtained.

Two additional observations should be made. Generally we only use a few important z-values. Eventually you may even learn some of them by heart. However, Table A.1b provides a simplified table of *selected* values that are most frequently used. You will probably find this table to be more valuable than the more complete tables (such as Table A.1a).

Finally, I hope you have seen the extreme value of drawing the little pictures of the distribution when working problems. They often assist you in quickly and correctly arriving at answers.

REVIEW Discrete Distribution • Continuous Distribution
Standardized Normal Distribution • $\mu = 0$ • $\sigma = 1$
z-value • Normal Table

$$z = \frac{X - \mu}{\sigma}$$

PROBLEMS

SECTION 5.4

5.1 Find the probability of a value falling in the following portions of a normal distribution.

a. Between $z = 0.20$ and the mean.
b. Between $z = 2.41$ and the mean.
c. Between $z = 1.47$ and the mean.
d. Between $z = -0.42$ and $z = 0.42$
e. Between $z = -0.61$ and $z = 1.35$.
f. Between $z = 0.87$ and $z = 0.65$.
g. Beyond $z = 1.42$.
h. Beyond $z = 0.15$.
i. Smaller than $z = 1.50$.
j. Smaller than $z = -0.50$.
k. Greater than $z = 1.33$.
l. Greater than $z = -0.75$.
m. Not greater than $z = 0.38$
n. Not greater than $z = 1.15$.
o. Less than $z = 1.73$.
p. Less than $z = 0.45$.
q. Less than $z = 1.10$.
r. At least $z = 0.25$.
s. At least $z = -1.11$.
t. No larger than $z = -2.03$.
u. No larger than $z = 0.66$.

v. Equal to $z = 1.0$.
w. Greater than $z = 0.00$.

5.2 Find the z-value or values in which the following areas are contained.

a. The 10% between z and the mean.
b. The 15% beyond z.
c. The middle 25%.
d. The 40% below z.
e. The middle 65%.
f. The 70% below z.
g. The 25% between $-z$ and the mean.
h. The 40% above z.
i. The 80% above z.

5.3 Find the specified proportion of the following normal distributions.

a. $\mu = 25, \sigma = 10$; area between μ and 30.
b. $\mu = 25, \sigma = 10$; area between 20 and 30.
c. $\mu = 25, \sigma = 25$; area between 20 and 30.
d. $\mu = 50, \sigma = 5$; area between 43 and 61.
e. $\mu = 100, \sigma = 15$; area between 98 and 107.
f. $\mu = 2{,}000, \sigma = 150$; area between 2010 and 2090.
g. $\mu = 0.10, \sigma = 0.02$; area between 0.07 and 0.13.
h. $\mu = 35{,}000, \sigma = 200$; area between 35,000 and 36,000.
i. $\mu = 35{,}000, \sigma = 400$; area between 35,500 and 36,000.
j. $\mu = 100{,}000, \sigma = 2{,}500$; area between 95,000 and 99,000.
k. $\mu = 5, \sigma = 20$; area between 0 and 25.

5.4 A test has been shown to result in a distribution of grades with an average (mean) of 65 and a standard deviation of 12. I decide that about 15% of the class should fail on this test. What should be my bottom cutoff for passing? If I decide that only 5% will get As, what is the cutoff for an A?

5.5 My electric bills have averaged $20 a month, with a variance of 9.00 What is the chance that my bill might be more than $24? Ninety-five percent of the time my bill will be at least how much?

5.6 Fire drills have been conducted at a school for several years. The time to evacuate has been approximately normally distributed with a mean of 4 minutes 15 seconds, and a standard deviation of 40 seconds. A phone call is received that a bomb is to go off in 3 1/2 minutes. What is the probability that the building will be fully evacuated? (*Hint*: Convert everything to seconds.) If the caller said 5 minutes, what is the chance of full evacuation?

5.7 The life of a light bulb is normally distributed with an average (mean) life of 100 hours and variance of 36.

a. What percent will last more than 110 hours?
b. What percent will last between 85 and 95 hours?
c. 15% will burn out before what length of time?

5.8 One-pound bars of butter have actually been weighing an average (mean) of 0.98 pound with standard deviation of 0.04 pound. What is the chance of buying a bar of butter that is at least 1 pound?

5.9 A tree I just planted is supposed to grow to 25 feet tall in 5 years (with $\sigma = 3$ feet). What is the chance it will be taller than my house, which is 30 feet high? What is the chance it will not reach the bottom of a window that is 24 1/2 feet high? If the window is 3 feet high, what is the chance that the top of the tree will be in front of the window?

5.10 The family income in Executive Township is $25,000, with a standard deviation of $4000. A sociology researcher randomly picks a person to interview. What is the chance that this person will come from a family with an income of less than $20,000? With an income of over $35,000?

5.11 The time between buses (or subway trains) is 19 minutes, with a standard deviation of 4 minutes. If I just missed the last bus:

a. What percent of the time will I wait between 15 and 20 minutes?
b. Ninety percent of the time I should not expect to wait longer than how many minutes?
c. Less than 5% of the time I will catch a bus as soon as how many minutes?

CHAPTER SIX
BINOMIAL DISTRIBUTION

We will now deal with the second important distribution, one that will also serve as a model for many types of real problems. The normal distribution was a model for continuous or measurement data. We will now turn our attention to one distribution that involves discrete, integer values and limited sets of possible values.

6.1 THE BINOMIAL DISTRIBUTION

In Chapter One we briefly discussed the notion of categorical data. This, in effect, is the simplest form of a measurement scale, since the data are merely classified into one of several categories. Thus, a person may be classed as absent, present, withdrawn, or perhaps excused (the last two really being subcategories of absent). A bill is paid in full, in part, or not at all. A word has 1, 2, 3, 4, . . . letters spelled correctly or incorrectly. And so on. We will look at the simplest case of categorical scales—one with only *two* possible categories.

Consider the problem of checking on a person working at a checkout counter at a supermarket. The price of an item is to be punched into the cash register, and this price may end up being correctly or incorrectly entered. Thus we have two possibilities for a single price. If 25 items are to be purchased, each price may be correct or incorrect. We can check to determine the number of incorrect prices in this order of 25 items (and perhaps we find one incorrect item).

Now let us continue by looking at a series of orders of 25 items each. If we record the number of incorrect prices for each of 10 more orders, perhaps we would have 0, 1, 0, 0, 2, 1, 0, 0, 2, and 3 errors. Each number then serves as a sample. In effect, what I actually am doing is assigning a value to the two conditions; I have assigned a numerical value of 0 for "correct" and 1 for "incorrect," and we have totaled the number of 0s and 1s for all the individual items in a single order. This sum is the numerical value of the sample. As we look at many more sample orders we continue to assume that each individual price may or may not be correct and that each independent entry has the same chance of being correct.

So much for the real data and the real process. Now if the chance of making an error is constant and each entry is independent of each other, a mathematical model exists, a theoretical distribution of sample values that we can derive and later compare with our set of samples. Furthermore, if the theoretical model serves as a satisfactory substitute for the data, we may say that it duplicates the process and we will then be able to determine what the *relative chances* of occurrence are for *all* of the different numbers of incorrect prices possible in an order. Some further examples of observations, measurements, or phenomena that may be classified as having only two possible results follow.

EXAMPLE 6.1

a. A coin may come up *heads* or *tails*.
b. A poll may permit two responses: *in favor of* or *against*.
c. A battery is *defective* (does not work) or *not defective* (works).
d. A traffic light may be *red* or *not red*.
e. A customer may *buy* or *not buy* something.
f. A bill is paid *on time* or *late*.

Now that we have the two possible cases, we must consider what constitutes *a particular sample*. The possible value of the sample will then serve as the X-values of a theoretical distribution, the binomial distribution. It will, therefore, be the variation of the sample values that will most concern us.

The next examples will point out what *can* occur.

EXAMPLE 6.1 (Cont.)

a. If we were to toss a coin 5 times we could get 0, 1, 2, 3, 4, or 5 heads. Then each of these are possible X-values and the list included all possible samples of 5 ($n = 5$). If 4 heads came up, $X = 4$.
b. You poll 50 people; $n = 50$. The possibilities are none in favor, 1 in favor, . . . all the way up to 50 in favor. *Each* case can occur. Some will be highly unlikely, while others will have a good chance of happening. If 15 were in favor, $X = 15$ for the poll.
c. If we try 6 batteries we may find that none work, 1 works, 2 work, 3 work, . . . or all 6 work.
d. You arrive at the traffic light 8 times. It can be red none of the times, 1 time, 2 times, . . . or 8 times.
e. Of 30 customers who enter a store none, 1, 2, 3, . . . on up to 30 may buy something.
f. Of 8 bills sent out 0, 1, 2, 3, 4, 5, 6, 7, or 8 may be paid on time.

So we must realize that a particular sample can have only one value. However, many possible values can occur, although the number of possibilities is limited by the sample size (obviously if you start with 8 bills sent out you don't expect 10 paid.) The recognition of the fact that this variety of values is possible is the beginning of the development of the model called the binomial distribution.

I have said several times that we want to determine a theoretical model that, given some assumptions, might possibly indicate to us the chance of having 23 items correct, or 15 customers buying things. Or we might want to know the entire model: what the entire distribution of heads, people in favor, red lights or defective batteries might likely be.

This leads us to make some assumptions to arrive at the underlying concept that we will use to establish values for the chances or the probabilities. Thus, we must describe the model we will use, remembering that we need probabilities for each X-value in order to establish this hypothetical probability distribution or model.

Our first assumption is that *each of the individuals* in the sample has the *same chance* or the same probability (of falling into one category). I have already said that each and every one of the students is to be considered as having the same chance of being absent. I am assuming each item in the order now has the *same chance* of being correct, each battery has the *same probability* of working, and each time I arrive at the traffic light I have the *same chance* of it being red. *And*, furthermore, this *same* probability is to be extended to *all* the individuals in *all* the possible samples. So when tossing coins the chance of a head occurring will not change during *all* the tosses for the first sample or the second sample or *any* sample. The decision or chance of an incorrect price is *assumed* to be the same for *all* items in *all* different orders.

In addition, each trial, each person polled, and each item looked at must be *independent* of every other trial, person, or item. Thus, when we toss a coin and get a head, we assume that this will not influence our chance of getting a head or a tail on the next trial. When we ask an opinion on a poll, we are assuming that the next person polled has not been influenced by the previous person or will not influence a future individual. As you might observe, this last case requires careful control of the way the data are accumulated, otherwise valid interpretations are virtually impossible to make. (Checking day-to-day temperature could very well not involve independent measurements. I would suspect that the chance of it being hot tomorrow would be greater if it were hot today than if it were cold today.)

Nonetheless, if we use the binomial distribution, we are assuming that independence is present, and not only within the sample, *but* from sample to sample (day to day, and so forth).

Now to identify the probability values. We actually have to assign a value to *both* of the possible occurrences, but knowing one of them we can quickly compute the other:

We often determine the probability either by guessing, by using past data, or by our experience. Or we may use a sample to approximate the value. If we use a sample (including any previous samples) we again say that

$$p = \text{the long-range probability or chance of } \textit{one} \text{ of the occurrences}$$

$$= \frac{\text{Number of times one of the occurrences took place}}{\text{Total number of occurrences}} \tag{6.1}$$

This is an empirical method and we are really *estimating* the true probability of the occurrence, p. Observe that the total number of occurrences need not be just one sample but could include any number of trials or individuals, as we will see in the following examples.

If we have a type of situation where we can provide *exact* probabilities, as we discussed in Chapter 4, then the p that we would use would not be an estimate, and it would be a more correct value than an empirical estimate for describing our probability distribution.

EXAMPLE 6.1 (Cont.)

a. If we have a fair coin, we would say that there is a 50-50 chance of a head appearing. So, in a very large number of tosses, say 10,000, we guess that 5000 should be heads. Therefore we say that $p = 5000/10000 = 0.50$ that a *head* will occur on any given toss. If we make the assumption that there is an equal chance for each occurrence, we can state that this is an *exact* probability value.
b. We may assume for some reason that an individual has a p of 0.60, or a 60% chance of saying "Yes, I am *in favor*." If we have taken a preliminary sample of 250 people where 150 (or 60%) said they were "in favor," then we are estimating p.
c. Out of 1800 batteries tested we found that 720 were defective; then the chance of a *defective* battery would be 0.40. Or we might want to determine what possible situations could occur if we *assumed* that the probability of a bad battery is 0.4.
d. If the light is red a quarter of the time, then the probability of the light being red is 0.25.
e. Perhaps we feel that 20% of the customers purchase something; then $p = 0.20$.
f. From past experience we find that the probability of a bill being paid on time is 0.85.

Recalling Rule 2 of Chapter Four, we can say: *If we know the probability, p, of the occurrence of one of the items, then the probability that the other will happen is 1 minus p*, and is called q. So

$$q = 1 - p \tag{6.2}$$

EXAMPLE 6.1 (Cont.)

a. The probability of a tail is

$$q = 1 - p = 1 - 0.50 = 0.50$$

Observe again that if the two items are equally likely to happen, they have the same probability.
b. The probability of an individual saying "no" is $1 - 0.60 = 0.40$.
c. The chance of a good, nondefective battery is 0.60.

d. If the probability of *red* = 0.25, then the probability of *green* is 0.75 if "green" is interpreted as "not red" and includes the case of an amber light).
e. The probability of *a* person *not* buying is 0.80.
f. The probability of *a* bill *not* being paid on time is 0.15.

Now, we are really not interested in the chance that the price of *apples* is incorrectly entered or *this* battery is bad but, instead, in the chance that 1, *or* 2, *or* . . . , and so forth, items would be incorrect out of the entire order, or that various numbers of batteries are bad. We must now determine what the probabilities are for each possible kind of sample.

The simplest technique for obtaining these probabilities is to consult a table.* Tables of probability values for a number of different combinations of p and n are provided in Appendix A Table A.6. The following examples will illustrate their use.

EXAMPLE 6.1 (Cont.)

a. If you want to know the chance of 3 heads occurring in a set of 5 tosses of a coin (or 3 heads out of 5 tossed coins), look in the table under $n = 5$, $X = 3$, and $p = 0.50$.

n	X	p = 0.50	
5	0		
	1		
	2		
	3	0.3125	← Answer

Thus we would expect that 31.25% of the time we would have 3 heads.

c. What is the chance of having 4, 5, or 6 bad batteries out of a group of 6 batteries? $n = 6$, $p = 0.4$, and we will look up values for $X = 4$, 5, and 6.

n	X	p = 0.40
6		
	4	0.1382
	5	0.0369
	6	0.0041

If I wanted to determine the chance that I would *not* have enough good batteries for my 2-battery flashlight, I would be asking the probability of 5 bad or 6 bad batteries and would have a probability of 0.0369 + 0.0041 = 0.0410, or there is a little over a 4% chance of not having enough good batteries to operate the flashlight.

* The mathematical method is described later in Section 6.2.

d. What is the probability of 0, 1, 2, . . . , 8 red lights? Here I would like the entire distribution, so we will look up all of the possibilities under $n = 8$.

n	X	p 0.25
8	0	0.1001
	1	0.2670
	2	0.3115
	3	0.2076
	4	0.0865
	5	0.0231
	6	0.0038
	7	0.0004
	8	0.0000

You should note that the chance of the light being red 8 times is less than 1 in 10,000. (0.0000 does not mean 0.0000000. . . . It says that it is closer to 0.0000 than to 0.0001.) There is less than a 3% chance of being stopped 5 times and about a 31% chance of being stopped 2 times, and so on.

Now, with these values we can draw a histogram (Figure 6.1). This histogram is our theoretical distribution or model of the number of occurrences when n is 8 and p is 0.25.

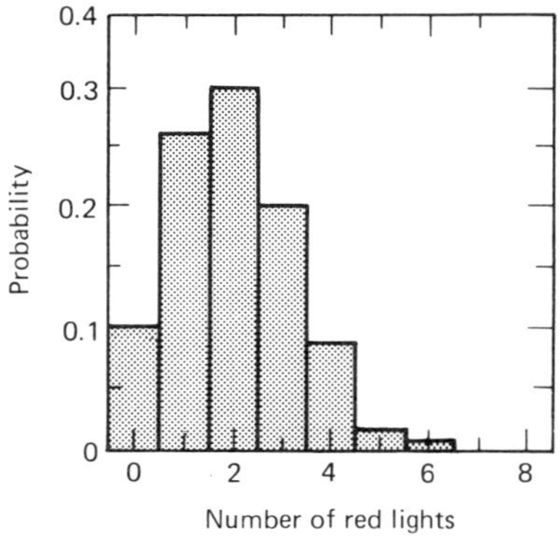

FIGURE 6.1 Distribution of the number of red lights out of a total of eight traffic lights. The probability of a red light is 0.25.

EXAMPLE 6.1 (Cont.)

b. Here we were taking a poll of 50 people and we felt that 60% would be in favor of the issue (or each person had a probability of 0.6 of answering favorably). Now I ask, "What percent of the samples of 50 will show a majority in favor of the issue?"*

* Note that this also begs the question "What proportion of the samples will have a sample p that is greater than 0.50?"

The first step must be to come up with the entire distribution of *number in favor*. But your tables stop at $n = 20$, so they are useless for this problem. I consulted other tables.* In Figure 6.2 I have drawn the distribution. Look at the distribution of *samples*. Some samples will only have 20 or 21 people in favor out of the 50 that are asked. These samples will *not* indicate that the majority is *in favor*. Only those samples with X-values of 26 through 50 will indicate that the majority is *in favor*.

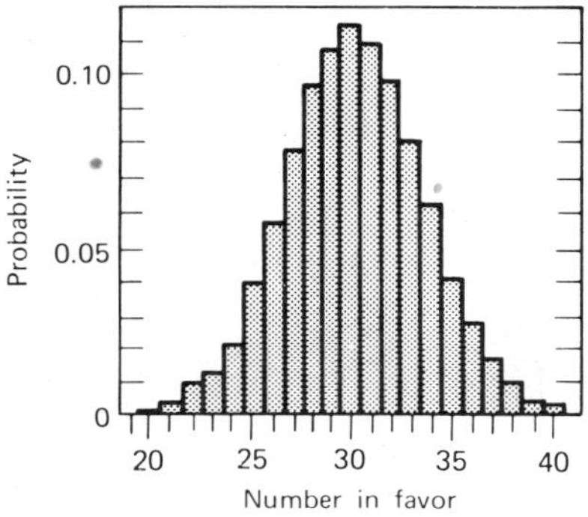

FIGURE 6.2 Distribution of *number of people in favor* of an issue. Fifty people voting, each with a probability of 0.6 being in favor.

To find the probability or percent of times that the sample will show a majority in favor, we merely add up all of the heights above all of the values from 26 to 50, and we find that this is 0.90.† Thus there is a 90% chance that the poll will have the majority in favor of the issue.

If you look further at Figure 6.2 you should see how close the lines on the histogram are. This distribution is beginning to resemble a smooth continuous distribution. With large samples, large n, we will find that this smoothness will become even more pronounced and we will find that we will be able to use some other distributions to approximate the binomial distribution.

Some other observations should be made about the binomial distribution. Figure 6.3 shows a collection of histograms for different values of n and p. Regardless of the n, when $p = 0.5$ the distribution is symmetrical and centered at the middle of the set of possible values. When p is at one of the extremes, either around the 0.1 or the 0.9, the distribution is heavily skewed. However, as the sample size increases, this skewness decreases. When the p is small, the samples consist mostly of small values; when p is large, the samples tend to be mostly large values. These factors are true pretty much regardless of n. As we observed before, as n increases, the

* H. G. Romig, *50–100 Binomial Tables*, New York: John Wiley & Sons, 1963.

† This is most easily found by determining the chance of being defeated. This is found by adding the probability of 19, 20, 21, 22, 23, 24, and 25 being in favor—or $0.000 + 0.002 + 0.003 + 0.005 + 0.010 + 0.015 + 0.025 + 0.040 = 0.100$. Then the probability of a majority is $1.000 - 0.100 = 0.900$.

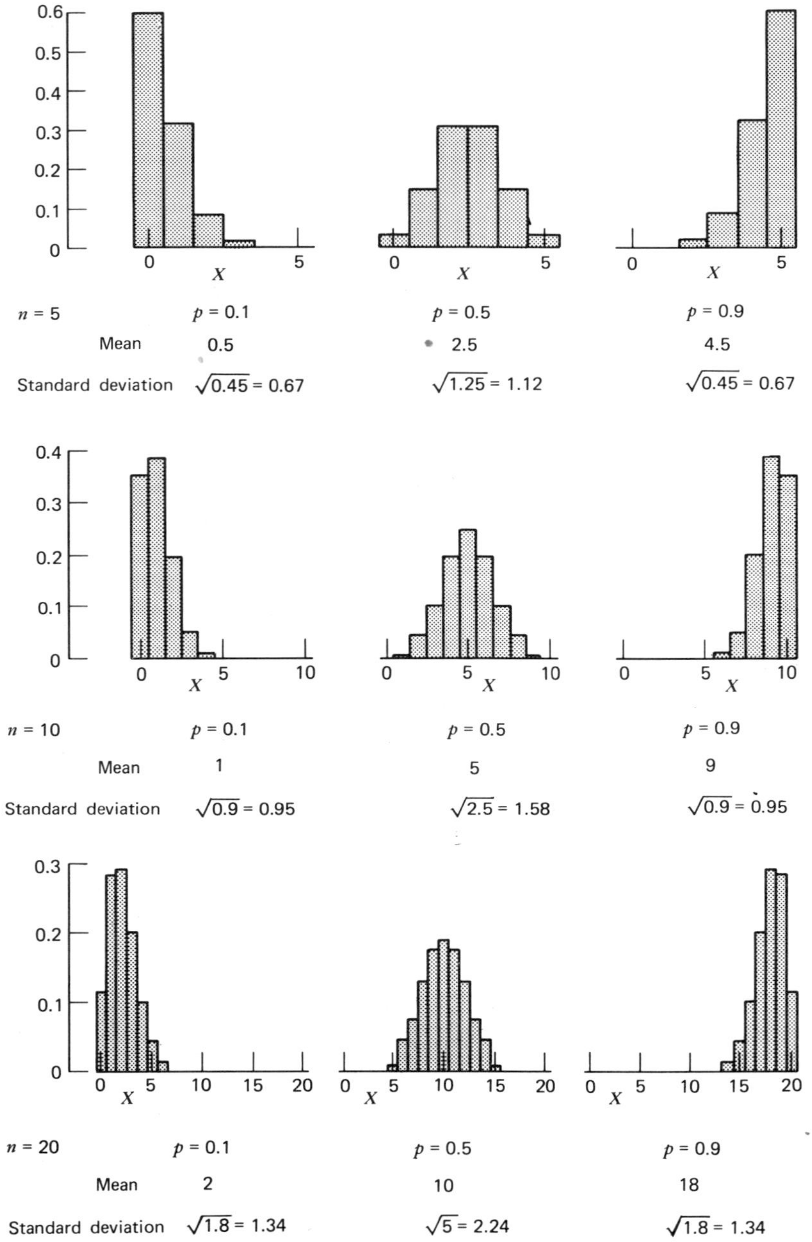

FIGURE 6.3 Binomial distributions for different values of n and p.

distribution begins to smooth out, regardless of p. We just noted that as n increases, the skewness decreases. In addition, the distribution of samples covers much less of the total possible sets of samples. For $p = 0.5$ and $n = 5$ there is at least a 3% chance for the most unlikely possible values to occur (the 0 and the 5), so *all* the values have at least some small chance of being observed (3% chance is small but not impossible). For $n = 20$, however, the span covering those values with at least a 3% chance of being observed goes from 6 to 14, only *half* of the possible values. There is some chance for the other values to appear, but it is much less, and so basically the distribution covers much less of the whole set of *possible* values.

Obviously, the tables do not satisfy all our cases. Of course, you do not have every possible p or every possible n value. Think how large the table would have to be if you wanted every p from 0.01 to 0.99 for just an n of 100. We will have to turn to other means to arrive at these distributions. We will have to calculate them either by hand or with the aid of a calculator or computer, or in some cases by an approximation that will come pretty close to the real distribution.

*6.2 CALCULATING VALUES OF BINOMIAL DISTRIBUTION (OPTIONAL)

You may have noticed that much of what we have done in this chapter is very similar to what was done in Chapter Four. In fact, the actual formula for determining probabilities of various X-values is a simple extension of what was done in that chapter.

If, for example, we want to consider the problem of determining the probability of 4 defective batteries out of the 6 we bought, we may be looking at:

bad $p = 0.4$	bad $p = 0.4$	bad $p = 0.4$	bad $p = 0.4$
	good $p = 0.6$	good $p = 0.6$	

in that specific order. In this case, the probability of the combined event of first bad, second bad, . . . , fifth good, and sixth good is:

$$\begin{aligned} P(\text{bad}) \cdot P(\text{bad}) \cdot P(\text{bad}) \cdot P(\text{bad}) \cdot P(\text{good}) \cdot P(\text{good}) \\ &= (0.4)(0.4)(0.4)(0.4)(0.6)(0.6) \\ &= (0.4)^4(0.6)^2 \\ &= 0.0092 \end{aligned}$$

But this is only one of many possible ways that I could obtain 4 bad and 2 good batteries, and each of those ways would have a probability of

$$(0.4)^4(0.6)^2 = 0.0092.$$

Or, in more general terms, $(p)^X(q)^{n-X}$.

Now how many ways could we have obtained the 4 bad and 2 good batteries? If we consider the package as having a specific order, the first bad battery could be in any of 6 locations. For each of these locations (place number 1, 2, 3, ...) there are 5 locations for the next bad unit. Thus there are a total of $6 \cdot 5 = 30$ possible pairs of locations for the first and second bad batteries. Continuing, we would have 4 locations for the third and 3 locations for the fourth bad battery, or a total of $6 \cdot 5 \cdot 4 \cdot 3 = 360$ possible separately identifiable arrangements of bad batteries. Thus we might have the following arrangement.

1st bad	good	2nd bad	4th bad	good	3rd bad

In general we have $n \cdot (n-1)(n-2) \cdots (n-X+1)$ arrangements.* But now look again at those batteries. Do you care if it is battery number 1, 2, 3, or 4 that ends up in the first location? Of course not. You only care (or for that matter are only able to know) that it is a bad battery. So if we look at the 4 bad batteries we find that we could have placed any of the 4 in the location that we say the first bad battery is. And then any of 3 could be placed in the next location, any of 2 in the third place, and the last one in the last place. We have $4 \cdot 3 \cdot 2 \cdot 1 = 24$ different arrangements of the 4 batteries that cannot be distinguished from each other. This is true for each set of possible arrangements of good and bad items. Thus we obtain

$$\frac{6 \cdot 5 \cdot 4 \cdot 3}{4 \cdot 3 \cdot 2 \cdot 1} = \frac{360}{24} = 15$$

ways if arranging 2 good and 4 bad batteries (when all we are concerned about is good or bad). The overall probability of 4 bad out of 6 items would be this order or any of the other 14 orders, and we would therefore have

$$15(0.0092) = 0.1380$$

which is what we obtained before.

In general we have

$$\text{Prob}(X \text{ occurrences out of } n) = P(X)$$
$$= \frac{n \cdot (n-1) \cdot (n-2) \cdots (n-X+1)}{X \cdot (X-1) \cdot (X-2) \cdots (3) \cdot (2) \cdot (1)} (p)^X (q)^{n-X} \qquad (6.3)\dagger$$

* This term is also called the number of permutations of n things taken X at a time.

† The first part of this expression

$$\frac{n \cdot (n-1) \cdot (n-2) \cdots (n-X+1)}{X \cdot (X-1) \cdot (X-2) \cdots (3) \cdot (2) \cdot (1)}$$

is called the combination of n things taken X at a time and is abbreviated C_X^n or $\binom{n}{X}$. It is also written and calculated as

$$C_X^n = \frac{(n)!}{(X!)(n-X)!}$$

where $n! = n \cdot (n-1) \cdot (n-2) \cdots (2) \cdot (1)$ and is called n-factorial.

This is very tiresome to calculate for even a few values, especially if p is not an easy number to handle or the n is fairly large.

EXAMPLE 6.1 (Cont.)

a. Find the probability of 3 heads out of 5 tosses of a coin.

$$n = 5, X = 3, p = 0.5, q = 0.5, n - X + 1 = 5 - 3 + 1 = 3$$

$$P(\text{3 heads out of 5 tosses}) = \frac{5 \cdot 4 \cdot 3}{3 \cdot 2 \cdot 1}\left(\frac{1}{2}\right)^3\left(\frac{1}{2}\right)^2 = 10\left(\frac{1}{8}\right)\left(\frac{1}{4}\right) = 0.3125$$

d. What is the probability that you will get stopped by 5 traffic lights? n is 8 (since we spoke of driving past 8 lights), X is the 5 we are looking at, p we found to be 0.25, and q is 0.75. Then $n - X + 1 = 8 - 5 + 1 = 4$ and

$$\text{Probability of 5 out of 8} = \frac{8 \cdot 7 \cdot 6 \cdot 5 \cdot 4}{5 \cdot 4 \cdot 3 \cdot 2 \cdot 1}(0.25)^5(0.75)^3$$

$$= \frac{6720}{120}(0.0009766)(0.421875) = 0.0231$$

I think you can appreciate the magnitude of work involved in calculating just one simple probability value. For example to arrive at one of the probability values for Example 6.1b, the poll, would involve an enormous amount of computations.

6.3 MEANS AND STANDARD DEVIATIONS

The next thing we should determine about these distributions are the descriptive statistics—the *mean* and the *standard deviation* (or variance). Since we are talking about a theoretical distribution, we will use *population* symbols. We will call the mean by the letter μ (the Greek letter called mu), and the standard deviation by σ (sigma). Then the variance will be σ^2 (sigma squared).* We will not derive the following formulas for calculating

---

You may have observed that in the previous problem we also had 2 good batteries which might have been considered as 1st good, 2nd good, or 2nd good, 1st good. Thus, there were two arrangements of the good batteries. The last formula we have looked at approaches the problem by considering that there are $n = 6$ batteries which can all be arranged in $6! = 6! = 6 \cdot 5 \cdot 4 \cdot 3 \cdot 2 \cdot 1 = 720$ different ways. But, as we described before, these 720 arrangements can be divided into groups of bad batteries that cannot be distinguished apart. These groups consist of $X! = 4! = 4 \cdot 3 \cdot 2 \cdot 1 = 24$ arrangements. In addition, there are $(n - X)! = (6 - 4)! = 2! = 2 \cdot 1 = 2$ arrangements of the good batteries that also cannot be distinguished. Taking the good and bad batteries together, we are left with a total of $720/(24)(2) = 15$ different arrangements—the same result as before.

* The fundamental formula used to calculate this variance would be

$$\sigma^2 = \frac{\sum (X - \mu)^2}{n}$$

Note that we use μ not $\bar{X}$ as the mean and n not $n - 1$ for the denominator. These are used because we are working with a population, *not* a sample. Refer to page 61 for our original discussion of this idea.

the mean and standard deviation of a binomial distribution, but state them as follows. Mean

$$\mu = n \cdot p \tag{6.4}$$

Variance

$$\sigma^2 = n \cdot p \cdot q = n \cdot p \cdot (1 - p) \tag{6.5}$$

Standard deviation

$$\sigma = \sqrt{n \cdot p \cdot q} = \sqrt{n \cdot p \cdot (1 - p)} \tag{6.6}$$

Using Equations 6.4 to 6.6, we can quickly find the two values that describe the distribution, given only the population, p, and the number of items n. The mean, μ, and the variance, σ^2, are considered fixed constants for the particular distribution and are established once we know n and p. Note that once you find the mean ($n \cdot p$), to find the variance you only have to multiply the mean by $1 - p$ (or by q).

EXAMPLE 6.2 (Cont.)

a. The mean number of heads in five tosses is

$$\mu = n \cdot p = 5 \cdot (0.5) = 2.5$$

The variance is

$$\sigma^2 = n \cdot p \cdot q = 5(0.5)(0.5) = 1.25$$

and the standard deviation is

$$\sigma = \sqrt{1.25} = 1.12$$

b. The mean number of "yes" votes is

$$n \cdot p = (50)(0.60) = 30$$

So on the average 30 people will vote for the issue.
The variance

$$\sigma^2 = n \cdot p \cdot q = (50)(0.60)(0.40) = 12$$

and the standard deviation

$$\sigma = \sqrt{12} = 3.46$$

c. The mean number of defective batteries per package of 6 is

$$n \cdot p = (6)(0.4) = 2.4$$

and

$$\sigma^2 = n \cdot p \cdot q = (6)(0.4)(0.6) = 1.44$$

or

$$\sigma = 1.20$$

6.4 NORMAL APPROXIMATION TO THE BINOMIAL DISTRIBUTION

Now that you have an ability to compute the mean and standard deviation relatively easily, you should be able to reproduce closely the distribution using the empirical rule, or the normal distribution right? Wrong. Only sometimes. Look at Figure 6.3 again. I have provided the means and standard deviations for each of the distributions. Go only one standard deviation to the left of the $n = 5$, $p = 0.1$ distribution, and where are you? You are out of the distribution because it is so heavily skewed.

However, the distributions near the center (those with p around 0.5) or those with large values of n tend to have a form that begins to approach the form of a normal distribution. We now would like to take advantage of this similarity, especially when the probability of occurrence, p, is not an easy number with which to work or is not in a table (such as 0.34), or when n is large.

In general, if you are given the n and p of a binomial distribution, then whenever:

1. The mean values, $n \cdot p$ and $n \cdot (1 - p) = n \cdot q$ are both greater than 10.
2. The variance or $n \cdot p \cdot q = n \cdot p \cdot (1 - p)$ is also greater than 10.

we can replace the binomial by a normal distribution with the *same mean and variance*.* Thus we can refer to the normal distribution table to determine the nature of a binomial distribution.

The first thing you should realize is that the binomial was a discrete distribution, with exact probabilities for each circumstance or sample. This is not the case with the normal distribution. The normal distribution is a continuous distribution, where at best we can provide probabilities associated with intervals. So we have to ask, just what is the purpose of having the distribution in the first place? Usually you will find that we will not be too concerned with knowing what is the chance that a particular value will occur. We will generally be more interested in ranges or sets of values that would include various percentages of the distribution. Thus we will desire to know where are the upper 5%, the lower 10%, or the middle 95% of the individuals. We found these ranges separately for both the binomial and the normal distributions. We will now go on the assumption that both are essentially the same. Look back to Figure 6.2 and you will see

* When the mean is less than 10 we often can use another distribution, called the Poisson, to approximate the binomial. We will not consider the Poisson distribution here. You may also encounter other rules which may be equally acceptable.

that if we were to draw a smooth curve through all of the peaks of that distribution, we would have one that looks pretty much like a normal distribution.

EXAMPLE 6.1b (Cont.)
A poll of 50 persons ($n = 50$) consists of answers *in favor* or *against*. The probability of being in favor is believed to be 0.60. Therefore,

$$n \cdot p = (50)(0.6) = 30$$

This is larger than 10, so the normal distribution should be a good approximation (in addition, the variance $= n \cdot p \cdot (1 - p) = (50)(0.6)(0.4) = 12$, which is also larger than 10).

We will now use a normal distribution with mean, $\mu = 30$ and variance, $\sigma^2 = 12$. The standard deviation will be $\sqrt{12} = 3.46$.

The one question we had previously asked was what is the chance of a majority being in favor of the issue, or what is the probability of a value being *greater than 25*. We can ask the question again, only this time using the probabilities related to a normal distribution, or the area above or beyond the value corresponding to 25. Thus we ask for the area that includes all the values from 26 to 50.

Figure 6.4*a* shows the binomial distribution for $n = 50$ and $p = 0.60$. The shaded area shows the probability that our poll will have 26 or more people stating yes. Now observe carefully the scale on the X-axis and notice that the line separating the two areas does not fall exactly on the 26. The actual separation is halfway between the 25 and 26 or at 25.5.

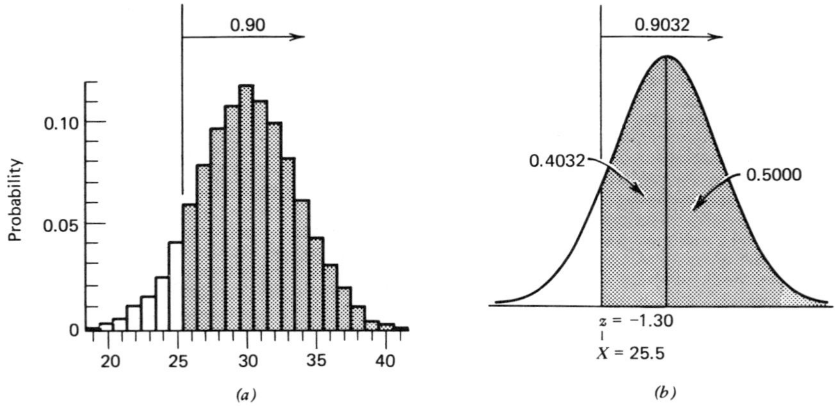

FIGURE 6.4 Comparison of the binomial distribution and the normal approximation. (*a*) The actual binomial distribution of the number in favor (Example 6.1b). The shaded area is the portion of samples that would indicate a majority in favor. (*b*) The normal approximation. Observe how close the proportions (or probabilities) are.

Since we are now going to substitute a normal distribution for this binomial distribution, we need to realize first that the binomial distribution is a discrete distribution for which we looked up areas associated with each discrete X-value. On the other hand, the normal distribution is a continuous distribution and we need to use a pair of X-values located at the ends of an interval to determine a probability value (or area). Therefore, when we approximate the binomial distribution with a normal distribution, we will use the X-value, or 25.5, as the lower end of the area of the normal distribution (which is also equivalent to the lower end of the category 26 on this binomial distribution). If you examine the graph in Figure 6.4*a* you will find that the probability of 26 is approximately the same as the area under the normal distribution in Figure 6.4*b* from 25.5 to 26.5.

The extra 0.5 that we add on to the interval of the normal distribution is called a continuity correction. It should be added on each end of an interval that has an X-value to be used in the calculation of an area of a normal distribution. Thus if we asked the percent between 26 and 30 we would actually want the area of the normal distribution that lies between 25.5 and 30.5, and two extra segments of 0.5 (or one entire integer) would be added to the interval.

Since we need to refer to the normal distribution tables for areas, we will use a z that corresponds to the X of 25.5. Using Equation 5.1, we find that for $X = 25.5$, $\mu = 30$, and $\sigma = 3.46$

$$z = \frac{X - \mu}{\sigma} = \frac{25.5 - 30}{3.46} = \frac{-4.5}{3.46} = -1.30$$

Looking in Table A.1a for the area above $z = -1.30$, which also corresponds to that area below $+1.30$, we find the area between $z = 0$ and $z = -1.30$ is 0.4032, so that the total area above $z = -1.30$ would be $0.4032 + 0.5000 = 0.9032$ (see Figure 6.4).

In Section 6.1, when we worked with the exact binomial distribution, we obtained a probability of 0.90, so we are off by a mere 0.3%, and notice how much easier the calculations are.

So whenever we want to use a normal distribution in place of a binomial, and that will probably be most of the time we run into a binomial, we will work with the standardized normal distribution table by substituting

$$n \cdot p \text{ for } \mu \text{ (the mean)}$$

and

$$\sqrt{n \cdot p \cdot (1 - p)} \qquad \text{for } \sigma \text{ (the standard deviation)}$$

The net effect is to greatly simplify our computational work and also provide a meaningful application of the normal distribution.

REVIEW Independence • p • $q = 1 - p$
$\mu = np$ • $\sigma^2 = n \cdot p \cdot q = n \cdot p \cdot (1 - p)$
Normal Approximation

PROBLEMS

SECTION 6.1

6.1 In what different ways could you classify the following as binomial distributions?

a. Rolling a pair of dice.
b. Making a call with a telephone.
c. Spraying bugs with an insecticide.
d. Starting a car.

6.2 In what way might the assumption that p is the same for all individuals be incorrect for the following examples?

a. Absences in a particular class.
b. Students' score on a test.
c. Matches igniting when struck.
d. Babies crying in the nursery of a hospital.
e. Insect being killed by a bug killer.
f. Being stopped by a red light.

6.3 How does the question of replacement or nonreplacement affect the assumptions of a binomial distribution? (For example, consider polling students for their attitudes on some issue.)

6.4 In flipping a coin it generally is assumed that there is a 50-50 chance of a head. What effect does the possibility of the coin standing on edge or rolling under the table have on this? How can you still describe this as a binomial distribution?

6.5 On a roulette wheel there are 18 reds, 18 blacks, and 2 noncolors (0 and 00). In what way can we still regard this as a binomial distribution?

6.6 Using the tables, determine the following probabilities.

a. For $p = 0.5$ and $n = 6$ what is the probability that $X = 3$?
b. For $p = 0.5$ and $n = 6$ what is the probability that X is less than 2?
c. For $p = 0.4$ and $n = 8$ what is the probability of $X = 6$?
d. For $p = 0.4$ and $n = 8$ what is the probability that X is at least 5?
e. For $p = 0.6$ and $n = 12$ what is the probability that X is between 4 and 8?
f. For $p = 0.6$ and $n = 9$ what is the probability that X is greater than 7?
g. For $p = 0.8$ and $n = 12$ what is the probability that $X = 8$, $X = 2$, X is greater than 10?

h. For $p = 0.5$ and $n = 10$ what is the probability that $X = 0$, X is less than 2, X is 3 or greater?
i. For $p = 0.10$ and $n = 20$ what is the probability that X is 5 or larger?

6.7 A bag of seeds states that 70% of the seeds will germinate.

a. What percentage will not germinate?
b. If I plant 5 seeds, what is the chance that none of the seeds will *not* germinate?
c. If I plant 10 seeds what is the chance that all of the seeds will germinate?
d. For 10 seeds, what is the probability of: all germinating, 9 germinating, 8 germinating, . . . , none germinating?
e. If I need at least 5 plants for an experiment, what is the chance that I will not have 5 plants if I plant 10 seeds?
f. If I need 5 plants and plant 15 seeds, what is my chance now of having enough plants?

6.8 A study at a large university revealed that students in class were really thinking about sex almost 60% of the time. What is the probability that at a given-time more than 15 out of the 20 students in my class are thinking about sex? What is the chance that a majority of the students are not thinking about sex (and I hope thinking about the lecture)?

6.9 On a true-false test, if you just guess the answer, we can say that your chance of getting a correct answer is 0.5.

a. If there are only 4 questions what is the probability of getting all correct by guessing?
b. If there are 20 questions what is the probability of getting all correct by guessing?
c. If just passing on a 20 question test is 15 correct, what is the probability of passing by just guessing?
d. What is the probability of failing the test by just guessing?
e. If an A is 19 or 20 correct, what is the probability of getting an A?

6.10 On the true-false test I assume that a person who understands the material has a 0.8 chance of getting a correct answer.

a. What is the probability of getting all answers correct if there are only 4 questions (and you understand the material)?
b. What is the probability of getting all answers correct on a 20-question test?
c. What is the probability of passing a 20-question test (if 15 is passing)?
d. What is the probability of getting an A (if an A is 19 or 20 correct)?

6.11 It has been assumed at State College that 20% of the students will fail to complete the first year. Seventeen students are admitted from XYZ High School.

a. What is the probability that exactly 3 students will fail to complete the first year?
b. What is the probability that 9 or more students will not complete the first year?
c. What is the probability that exactly 16 students will complete the first year?
d. What is the probability that a majority of students will complete the first year?

SECTION 6.2

In Problems 6.12 to 6.14 use the method of Section 6.2.

6.12 For $n = 3$ and $p = 0.2$ determine the probability of none, 1, 2, or 3 successes. Compare your results to the ones in Table A.6.

6.13 For $n = 5$ and $p = 0.3$ determine the probability of 0, 1, 2, 3, 4, and 5. Compare your results to those in Table A.6.

6.14 If the probability that a spark plug does not operate properly is 0.05, determine the probabilities for no bad plugs, 1, 2, 3, . . . , 6 bad plugs in a 6-cylinder engine (car with 6 spark plugs).

SECTION 6.3

6.15 The probability of a particular seed germinating was 0.7 (Problem 6.7).

a. If I plant 5 seeds what is the average number of seeds I would expect to germinate?
b. If I plant 10 seeds how many should I expect to germinate? What is the standard deviation of the number of seeds germinating?
c. If I were to plant a bag of 10,000 seeds, how many should I expect to germinate?
d. Assume I plant 10,000 seeds. Using the empirical rule, 95% of the time I should have what possible numbers of seeds germinating?

6.16 If 60% of the time students are thinking about sex (see Problem 6.8),

a. What would be the average number of students in my class of 25 that are thinking about sex?
b. What would be the average number thinking about sex in a large lecture hall with 200 students? What are the 2 sigma limits for the number of students thinking about sex (in the lecture hall)?

6.17 What is the average number of correct answers for a student who guesses on a 20-question true-false test (Problem 6.9)? How would this student do on a 100-question test? What is the standard deviation of correct answers on a 100-question test?

6.18 If a person taking a 100-question test understands the material (and therefore the probability of passing is 0.8) how many correct answers should he or she expect to get? What will the standard deviation of the number of correct answers be?

6.19 If 200 students from XYZ High School go to State College, how many would be expected to pass (see Problem 6.11)? Over the course of 100 years the guidance department would expect that the number who would pass would be within what limits for about 99 of those years?

SECTION 6.4

6.20 If a person knows the material on a true-false test, what is the chance he or she will fail if failing is less than 65 correct (see Problem 6.18)?

6.21 If I plant 10,000 seeds (Problem 6.15) what is the probability that more than 7050 will sprout? Is the correction for the discreteness of the distribution really necessary in this problem (would you really care about 1 seed)?

6.22 What is the chance that the majority of the students in a large lecture are thinking about sex (see Problem 6.16)?

6.23 If there are 100 questions on a true-false test, what is the chance that someone who guesses would get at least 65 correct (see Problem 6.17)?

CHAPTER SEVEN SAMPLING AND ESTIMATION

Once again I must make a vital inquiry and ask: "What generally is the most important question about any problem?" It should be: "What is unknown and how can we better know it?" As a way of exploring this peculiar question I offer you my own personal Mystery Can (Figure 7.1) and I pose a problem. Look at that can and tell me what question you should ask about it.

FIGURE 7.1 The Mystery Can—and the big question, "What is inside?"

I hope the question is, "What's in the——can?'

And I answer, "Take a peek, take a pick, or take a guess, but only take *one*."

7.1 WHAT IS SAMPLING?

We have here a fundamental issue that concerns all fields of endeavor and that provides a basic reason for the existence of the subject of statistics. This is because the can is merely a symbol of much of what you do not know. You desire to know fully what is inside that can, but neither I nor the can nor anyone else will reveal fully the contents of that can.

You do not have to limit the concept or idea of this symbol to things that can fit inside it. I prefer to think of the symbol as capable of engulfing whatever population you are interested in. It might contain a million nails, or a million people. It might have all the coins that have ever been produced or that might ever be produced. We might pretend it holds an infinite population of measurements of the length of the Brooklyn Bridge; dishes broken each week in a busy restaurant; number of scratches on a new car; or the weight loss by dieters on a new diet. Or it might only be a collection of pieces of cardboard with numbers on them. Needless to say it contains something, and I will identify that something as our *parent population*. I will claim that it contains the complete population out of which we will draw a sample.

So let's take a sample. Let's also say that the population stands for the number of residents in one-family houses. The population then is the complete distribution of *all* possible numbers of occupants of one-family houses.

I stick my hand into the can and pull out a number—this in turn symbolizes knocking on a door and asking how many people live in that house. You can do it also. "Sampling Pieces," Appendix Table A.9, lists the values and the order of sampling that I am working from. In that table I am starting from the upper right corner and going to the left as I sample individuals. You can make the pieces by following the instructions in Table A.9 and then placing them in a plastic bag or a can. You will use these pieces in future exercises, and right now we will use the values to represent our sampling problem.

So we take the first sample.

It is a 6.

Okay, now what?

Well, you know more than you did before. You know there is a 6 in the population. You know that a house may contain 6 residents. That is useful, valuable information. *But are you willing to guess what the population is* or what the distribution of *all* houses is?

I hope you said *no*. If you didn't think it over.

Now, remember one thing. Whenever you take a single measurement of any type, you have only picked one out of the Mystery Can. You are not ready to do justice in estimating or guessing what is inside that can—not unless you know other information that you have not disclosed. Unfortunately, too many people arrive at conclusions after looking at that first value (and do not have additional relevant knowledge to aid them).

With only the one value as a sample, we can only make a poor guess at where the center of the population lies. We can say absolutely *nothing about its spread* or shape. Without an understanding of the spread or shape of the distribution, we are at a loss to say anything about the kind of number that should appear next.

So we knock at another door. We pick another value from the can. What do we get this time?

8.

Now we know considerably more. We can talk about both a mean *and* a variance. We can say that the mean is

$$\frac{6+8}{2} = 7$$

and the variance is

$$\frac{(6-7)^2 + (8-7)^2}{(2-1)} = \frac{1+1}{1} = 2$$

(The standard deviation is therefore 1.4.) Can we guess now what the population in the can is? We can try, but I wouldn't be too *confident* about my statement.

So we go to a third and a fourth house and see what we get. Here we find 3 and 5 people. Now the mean is

$$\frac{22}{4} = 5.5$$

and the variance is

$$\frac{(0.5)^2 + (2.5)^2 + (-2.5)^2 + (-0.5)^2}{3} = \frac{13.0}{3} = 4.33$$

(The standard deviation is 2.08.) Now will you venture a guess about that can? How confident are you about what is in that can or what the population actually is?

You should be more confident now than after any of the previous observations. The main point, however, is that you still have no way of measuring or estimating what your confidence should be. What should you give in the way of odds that you will get 5 people in the next house or at any other time? What is the mean and the variance of the population? How close to the population values might I reasonably expect this sample mean and variance to be? These are questions that we will now begin to try to answer.

7.2 DISTRIBUTIONS OF SAMPLE VALUES

Whenever we take samples from a distribution, we create another new distribution. In Chapter Four we looked at some probability distributions and distributions of samples from them. The binomial distribution was a similar type of distribution. We will find that some of the new distributions will be extremely important.

In order to explore the concept of sampling more fully we need to start with some kind of a distribution from which we will take samples. Let us consider that our Mystery Can contains a population. We can also think of this population's distribution as a histogram where all the individuals have been stacked at their appropriate X-values. Now we remove a set of individuals. We sample at random and look at 4 values (see Figure 7.2).

6, 8, 3, and 5.

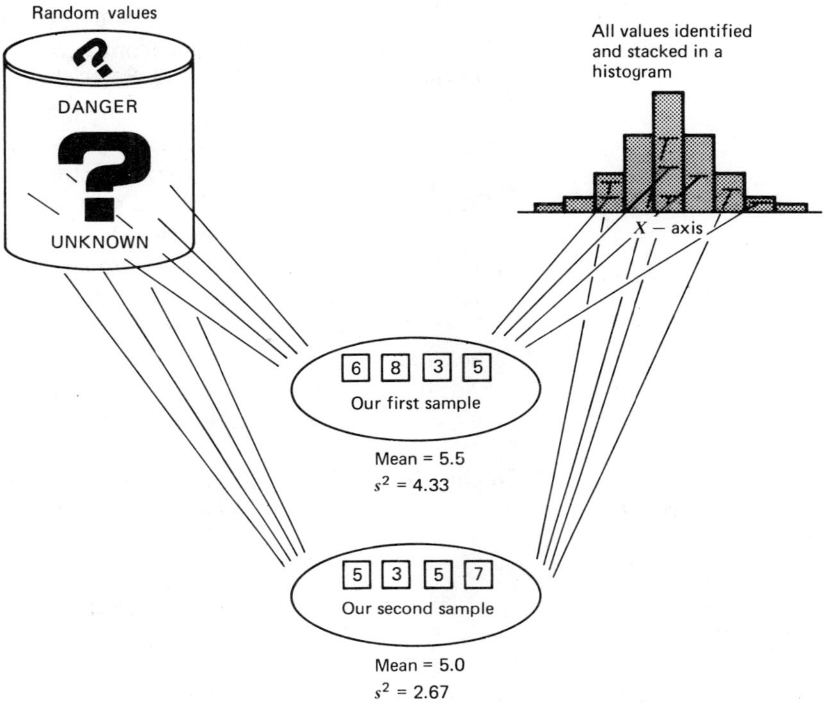

FIGURE 7.2 Taking samples from a distribution. When you observe a random distribution, you can also think of the individuals stacked at their appropriate location on the X-axis.

We can observe the mean, median, variance, and standard deviation of this sample (mean = 5.5, variance = 4.33). (We can divide by n and get one estimate of σ^2 or divide by $n - 1$ and get s^2, and each will provide a different number.) This is just one of the many possible samples we could take. We could take a second sample and get

5, 3, 5, 7

This time we have a mean of 5.0 and a variance (s^2) of 2.67, a median of 5.0, and so forth. Both the mean and the variance are different from the mean and variance of the previous sample We should not be surprised that these descriptive statistics are different, since the samples themselves are different.

If we look at a third sample of 4 what do we obtain?

6, 5, 4, 6

The mean of this third sample is 5.25 and the variance is 0.92.

We are now building up a little set of means and variances, each different. Any of the three is as good a sample from the population as any other. You can take your pick. If we now continue to sample and look at 50 samples of 4 individuals each, we might get the whole collection of different means and variances in Table 7.1. Look them over to see how many are the same. Remember that they all came from a common population, a single source, yet there are so many different kinds of samples that have been generated.

Now that we have sets of numbers we can draw histograms, calculate some more descriptive statistics (means, etc.) and, in general, look at some *new* distributions. The distribution of the means of these samples is drawn in Figure 7.3. For the sake of discussion I have also drawn the distribution of the individuals. For comparison sake I will tell you that the *population mean, variance, and standard deviation* are 5.0, 2.94, and 1.713, respectively, and that the distribution is symmetrical.

Observe that the histogram for the sample means is centered at about the same point or location as the mean of the population. Soon we will find that when we look at sample means (the averages of the samples), in the long run, these sample means will average the same as the population

TABLE 7.1 Samples from an Unknown population. Various Statistics are Computed for each Sample. Two Estimators of Variance and Standard Deviation are Computed and are to be Discussed in the Text.

INDIVIDUAL X-VALUES IN SAMPLES OF FOUR	MEANS $\bar{X}$	VARIANCE		STANDARD DEVIATION		MEDIAN	MIDRANGE
		s^2 (USING $n-1$)	$\hat{\sigma}^2$ (USING n)	s	$\hat{\sigma}$		
6 8 3 5	5.50	4.33	3.25	2.08	1.80	5.5	5.5
5 3 5 7	5.00	2.67	2.00	1.63	1.41	5.0	5.0
6 5 4 6	5.25	0.92	0.69	0.96	0.83	5.5	5.0
6 0 3 3	3.00	6.00	4.50	2.45	2.12	3.0	3.0
4 7 3 8	5.50	5.67	4.25	2.38	2.06	6.0	5.5
3 9 8 5	6.25	7.58	5.69	2.75	2.39	6.5	6.0
6 5 6 5	5.50	0.33	0.25	0.57	0.50	5.5	5.5
9 7 5 4	6.25	4.92	3.69	2.22	1.92	6.0	6.5
1 5 4 5	3.75	3.58	2.69	1.89	1.63	4.5	3.0
5 7 2 6	5.00	4.67	3.50	2.16	1.87	5.5	4.5
5 3 7 3	4.50	3.67	2.75	1.91	1.66	4.0	5.0
7 7 3 6	5.75	3.58	2.69	1.89	1.64	6.5	5.0
5 7 6 4	5.50	1.67	1.25	1.29	1.12	5.5	5.5
6 7 5 7	6.25	0.92	0.69	0.95	0.82	6.5	6.0
4 8 3 4	4.75	4.92	3.69	2.22	1.92	4.0	5.5
2 5 5 4	4.00	2.00	1.50	1.41	1.22	4.5	3.5
5 4 6 6	5.25	0.92	0.69	0.96	0.83	5.5	5.0
7 6 5 1	4.75	6.92	5.19	2.63	2.28	5.5	4.0
3 5 8 8	6.00	6.00	4.50	2.45	2.12	6.5	5.5
7 4 3 4	4.50	3.00	2.25	1.73	1.50	4.0	5.0
6 6 5 4	5.25	0.92	0.69	0.96	0.83	5.5	5.0
7 7 4 5	5.75	2.25	1.69	1.50	1.30	6.0	5.5
6 5 9 6	6.50	3.00	2.25	1.73	1.50	6.0	7.0
6 4 6 5	5.25	0.92	0.69	0.96	0.83	5.5	5.0
2 5 4 6	4.25	2.92	2.19	1.71	1.48	4.5	4.0

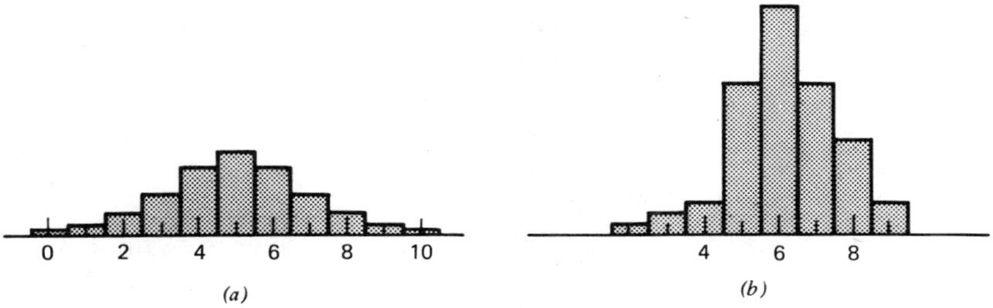

FIGURE 7.3 Distribution of samples from the given population. (*a*) Distribution of parent population. (*b*) Distribution of means of samples of 4 items.

mean. Two ideas are at work here. First, we are discussing the concept of estimation. The sample mean is being used to *estimate* the population mean. For many problems it happens to be the best estimate that we can use for guessing what the true population mean is.

The second observation in this case is that the estimate is *not biased* (is unbiased). When I said that on the average the distribution of sample means has the same center as the true population mean, I meant it was an

TABLE 7.1 (Continued)

4 4 8 6	5.50	3.67	2.75	1.92	1.66	5.0	6.0
7 4 6 6	5.75	1.58	1.19	1.26	1.09	6.0	5.5
5 3 6 6	5.00	2.00	1.50	1.41	1.22	5.5	4.5
4 4 3 7	4.50	3.00	2.25	1.73	1.50	4.0	5.0
3 4 2 5	3.50	1.67	1.25	1.29	1.12	3.5	3.5
5 8 4 4	5.25	3.58	2.69	1.89	1.64	4.5	6.0
5 5 3 6	4.75	1.58	1.19	1.26	1.09	5.0	4.5
4 5 5 4	4.50	0.33	0.25	0.57	0.50	4.5	4.5
6 4 5 6	5.25	0.92	0.69	0.96	0.83	5.5	5.0
4 8 4 5	5.25	3.58	2.69	1.89	1.64	4.5	6.0
8 4 7 3	5.50	5.67	4.25	2.38	2.06	5.5	5.5
4 5 7 7	5.75	2.25	1.69	1.50	1.30	6.0	5.5
3 5 7 2	4.25	4.92	3.69	2.22	1.92	4.0	4.5
6 4 3 2	3.75	2.92	2.19	1.71	1.48	3.5	4.0
6 3 5 3	4.25	2.25	1.69	1.50	1.30	4.0	4.5
5 5 7 5	5.50	1.00	0.75	1.00	0.87	5.0	6.0
5 6 5 5	5.25	0.25	0.19	0.50	0.44	5.0	5.5
4 6 10 6	6.50	6.33	4.75	2.52	2.18	6.0	7.0
2 6 5 6	4.75	3.58	2.69	1.89	1.64	5.5	4.0
3 4 4 6	4.25	1.58	1.19	1.26	1.09	4.0	4.5
2 7 4 4	4.25	4.25	3.19	2.06	1.79	4.0	4.5
5 2 4 7	4.50	4.33	3.25	2.08	1.80	4.5	4.5
4 6 5 6	5.25	0.92	0.69	0.96	0.83	5.5	5.0
4 2 6 5	4.25	2.92	2.19	1.71	1.48	4.5	5.0
1 3 5 5	3.50	3.67	2.75	1.91	1.66	4.0	3.0

unbiased estimate. This is a very good condition; we usually strive to find and use unbiased estimators. In this way, although we do not expect to get an exact estimate on any one individual sample, on the average, in the long run, we will obtain the true population mean. Unfortunately, an individual sample mean is not necessarily even close to the true population mean.*

As with means, we calculate sample variances in order to estimate what the population variance would be. We already calculated values of the variance for our first three samples. They were all different. In Table 7.1 the variance for each of the 50 samples is given, again with many different values. I have provided two different columns for the variance measures, one labeled s^2 (found by dividing by $n - 1$) and the other labeled $\hat{\sigma}^2$ (sigma-hat-squared—it is the estimate of σ^2 found by dividing by n).

Both sets of values are taken from a normal distribution but, if you were to draw a histogram, you would find they are not normal. They are skewed to the right. However, s^2 is unbiased. The average of all the s^2s is the same as the population σ^2 (remember we only have 50 samples and therefore we will usually be slightly off). The average of $\hat{\sigma}^2$ is considerably lower, or biased; therefore, since we generally want to estimate the population σ^2, we use $n - 1$ and calculate s^2.†

7.3 DISTRIBUTIONS OF SAMPLE MEANS

In Figure 7.3 we looked at the distribution of means of samples of 4. We also have the distribution of the individuals, the parent population. As you could readily see, the individuals are spread out much more than the sampled means.

We still have some additional important observations. The first is simply that:

> "When we take samples from a normal distribution, the means of the samples (the sample means) are *also* normally distributed."

* Another important characteristic of estimators is the spread among a collection of these estimators taken from the same population. In Table 7.1 we have 50 samples of 4 items. We calculated the means of these samples and have observed that they are not all the same. The distribution has a spread and we could calculate the standard deviation of these means—arriving at a value of about 0.81. Also listed are the sample medians and sample midranges from these same samples. If you graph the three sets of sample values you will find that the distribution of means has the smallest spread. If you calculate the standard deviation for the medians and midranges you will obtain values of approximately 0.89 for medians and 0.92 for midranges. Thus the mean has the smallest variation amongst these averages.

† You might note that I have said nothing about standard deviation. Oddly enough, even though s^2 is an unbiased estimate of σ^2, s is *not* an unbiased estimate of σ. This arises because by taking squares or square roots you change the shape of the distributions and thus alter the means. For example, given the five numbers 1, 2, 3, 4, and 5 and their squares 1, 4, 9, 16, and 25, note the means of the two groups. The first has a mean of 3 and the second a mean of 11. The question now is whether the square root of the "mean of the squares" equals the mean of the units. But, $\sqrt{11} = 3.32$, not 3. So it isn't true.

There do exist correction factors that you multiply s by to make it unbiased, but the difference is generally so small that we can essentially disregard it.

Thus, the histogram of the sample means in Figure 7.3*a* appears to resemble a bell-shaped normal distribution. This fact will allow us later to be able to describe fully the distribution of sample means. But we are still lacking one important number before we can determine the complete distribution—the variance (or standard deviation). All we can tell from a brief look at the histogram is that the distribution of averages is narrower, or smaller, than the total population.

In Figure 7.4*a* I have redrawn the histogram for the entire population. If we think of the parts of the histogram that make up this distribution as the pieces in our Mystery Can, then we can consider that a random sample is simply a set of points or individuals from this population (see Figure 7.2 also). Do you remember what the distribution (histogram) represents when we use relative frequencies as the vertical axis; or if we consider the areas of the histogram? They stand for *probabilities* or chances that we will obtain the particular values when we pick at random from the distribution. Assume we select 1 chip at random (we pick 1 house). The chance that it could have an extreme value, say a 10, from our population is very, very small. Here it has a probability of 0.005, or 1 chance in 200 of being picked, since there is only one 10 in our distribution of 200 items. Now if by chance we did pick a 10 we would say that it was quite unusual. But let us pick a second time and ask the question, "What is the chance that both cases are 10" or "What is the chance that the average of our selections will be a 10?" If we consider this as sampling with replacement (and we toss the chip back, or we say that the next house we see has the same chance of having 10 people as the one we already looked at), the chance of both coming out as 10 is the product of the probabilities of the individual samples. The probability of both coming out as 10 is *0.000025* or about 1 *chance in 40,000* ($0.005 \cdot 0.005 = 0.000025$). Similarly, the probability of a 9 coming up on a single try is about 0.015 (about 1.5% of the time, since there are only three 9s). Now how about the chance of an *average* of two being 9? It can be shown that this would be about 0.000725.* The chance goes from about 1 in 66 for a single 9 to about 1 in 1380 for an average of 9. So if we look at averages of two selections from a population we should not expect to see as many large values for the averages as we would see for the individuals. If you happen to get one very large value on the first trial you will be likely to get something to balance it on the next trial.

Look at the histogram of samples of 2 from our population of houses (Figure 7.4b). As you can see, no averages of two were up around 8, 9, 10, or as small as 2, 1, or 0.

A similar argument can be applied to larger samples. Thus, for a sample of 4 to have an average as large as 10 would require four 10s, and the prob-

* Here the probability would be equal to

$$
\begin{aligned}
\text{Probability (9 and 9)} + \text{Probability (8 and 10)} + \text{Probability (10 and 8)} \\
&= (0.015)(0.015) + (0.05)(0.005) + (0.005)(0.05) \\
&= 0.000225 + 0.00025 + 0.00025 \\
&= 0.000725
\end{aligned}
$$

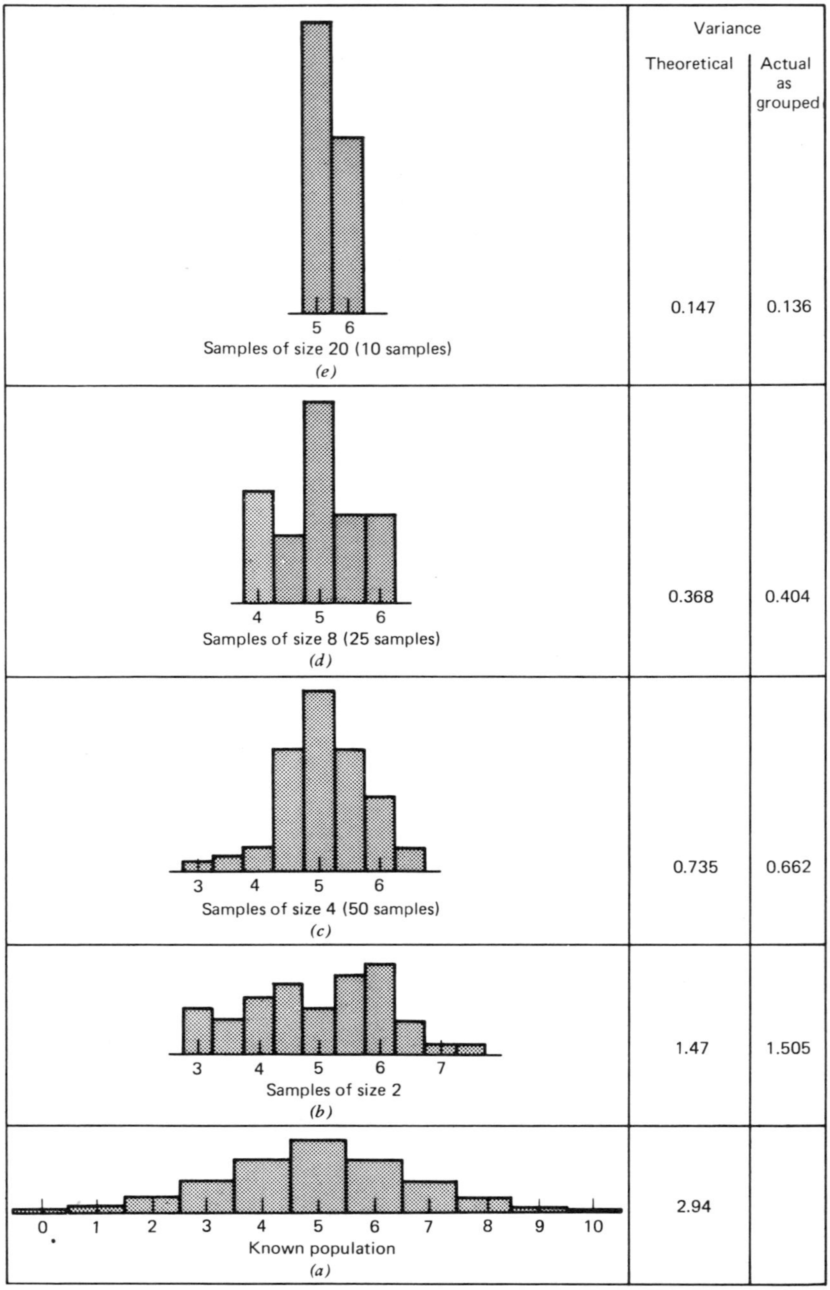

FIGURE 7.4 Distributions of sample means from a known population. Observe the shrinkage of the distributions as sample size increases.

ability of such an occurrence would be in the order of 0.0000000006 or 1 chance in 1,600,000,000. This would be something like the probability of a meteorite striking your roof and putting a hole in it.

Looking at larger and larger samples, the distribution of their means will tend to shrink more and more. We could find many ways to get an average of 7 with say, 8 individuals, but we would find that the probabilities of each way would be so small that the chance of getting an average of 7 would still be very small. Inspect Figure 7.4 and try to see how the distribution gets narrower as the sample size increases. To go to the ultimate extreme, if we were to look at an infinitely large sample (as large as possible), in effect, we would be looking at the entire population, and there would be only *one* average, and obviously no spread.

There is a theoretical relationship between the variance of the population and the variance of the distribution of samples. It is

$$\sigma_{\bar{X}}^2 = \frac{\sigma^2}{n} \qquad \text{or} \qquad \sigma_{\bar{X}} = \frac{\sigma}{\sqrt{n}} \tag{7.1}$$

where $\sigma_{\bar{X}}^2$ is the variance of the sample means—the variance of the new distribution. This is read as "sigma-sub-X-bar square." The square root of this value, $\sigma_{\bar{X}}$, is often referred to as the *standard error of the mean*

σ^2 is the variance of the population itself

n is the sample size

Note that if n is 1, $\sigma_{\bar{X}}^2 = \sigma^2$. So you could say that the population is actually a distribution of samples of size 1.

Since σ^2 equals 2.94, we would expect the variance of the distribution of samples of 2 to be

$$\sigma_{\bar{X}}^2 = \frac{\sigma^2}{n} = \frac{2.94}{2} = 1.47$$

This compares to an s^2 of 1.505 that I calculated for 50 samples of 2 individuals taken from the population. If we had taken another 50 or 100 samples we would have found that the s^2 would begin to be closer to 1.47 and the shape of the curve (histogram) would be closer to a normal distribution.

Continuing in this fashion, alongside the distributions in Figure 7.4 I have shown the theoretical and the actual variances for some sets of random samples from the population. Observe how close most of the actual and theoretical or expected variances are to each other.

The importance of the formula, as well as the whole idea of these distributions of samples, is that the samples are spread out much less than the population itself. And we can state *how much* they are expected to be spread out. This will permit us to put limits around our estimates and allow us to make statements about how confident we are in our estimations (and say how approximate our approximations are).

Now we have the complete set of information necessary to describe a distribution of sample means. We know:

1. The mean of the sample means (the overall mean) is the same as the population mean.
2. The variance of the distribution of means is σ^2/n.
3. If the population is normal, so is the distribution of sample means.

But there is still an even more significant, yet not so obvious fact about the distribution of sample means. In Figure 7.5 we have the histograms of two populations that are quite far from being considered normal distributions. One is rectangular or flat and the other is a highly skewed triangular distribution. A large number of samples of 4 were randomly drawn from these two distributions and the means of each of the samples was calculated.

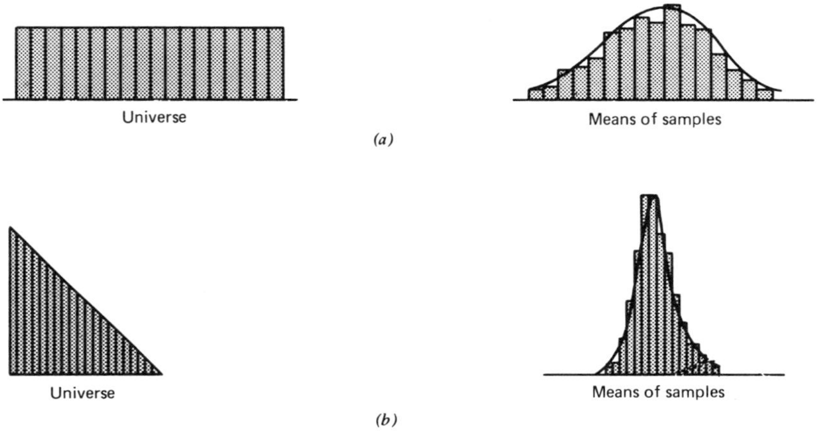

FIGURE 7.5 Illustration of how sample averages of four tend toward normal distributions even when the universe or population is far from normal. (*a*) Rectangular universe. (*b*) Right triangular universe.

Histograms for the means of the samples were then drawn, and I have also superimposed a smooth normal curve on top of the histogram. Observe how extremely close the distribution of sample means approaches a normal distribution. Thus we come to the central limit theorem.

Central Limit Theorem *For large samples from almost any population, the distribution of means can be approximated by a normal distribution (or approaches a normal distribution) whose mean is the same as the population, but whose variance is σ^2/n.*

Although by large samples we generally mean samples with about 30 or more individuals, often we find that samples with only 4 values will be sufficient. As a result, for approximate analysis of most problems, we will find it unnecessary to worry about the actual population shape. However, one criterion is that the population must have a finite variance. This may seem strange, but there are some weird distributions. There is even one called the Cauchy distribution that has practical applications but does not have a mean.

Let us look at an example to see how to use our new information.

EXAMPLE 1.1 (Cont.)

Let's return to our traffic ticket problem. I will now reveal that the individual numbers actually represented tickets issued on 28 different days.

Thus, since we had 112 readings, we now recognize them as 28 days with 4 readings on each day, or 28 samples of size four (see Table 7.2).

TABLE 7.2 Number of Tickets Issued. Samples of Four with the Means and Ranges for each day of February

DAY	1	2	3	4	5	6	7	8	9	10	11	12	13	14
Values	14	17	18	14	12	15	12	10	11	14	11	9	14	13
	11	3	12	8	10	13	11	8	6	9	3	7	12	11
	13	15	11	11	14	13	19	10	11	9	10	12	13	11
	9	10	11	9	9	11	9	6	9	1	2	8	9	2
Mean of four	11.75	11.25	13.00	10.50	11.25	13.00	12.75	8.50	9.25	8.25	6.50	9.00	12.00	9.25

DAY	15	16	17	18	19	20	21	22	23	24	25	26	27	28
Values	15	15	9	10	10	12	14	10	12	10	8	13	14	12
	4	11	5	8	10	13	2	7	9	9	9	12	10	10
	11	8	12	10	9	15	13	9	12	12	10	12	11	17
	8	10	8	10	8	9	6	3	6	7	8	9	4	7
Mean of four	9.50	11.00	8.50	9.50	9.25	12.25	8.75	7.25	9.75	9.50	8.75	11.50	9.75	11.50

Since the population variance is 11.62 (see page 66) and $n = 4$ (the samples have 4 individuals each), we *estimate* the variance of the sample means to be

$$\sigma_{\bar{X}}^2 = \frac{\sigma^2}{n} = \frac{11.62}{4} = 2.90$$

I have drawn the distribution of the $\bar{X}$s in Figure 7.6. Calculating the actual variance of the *sample means*, we obtain

$$\sigma_{\bar{X}}^2 = 2.97$$

This is extremely close to the theoretical value we estimated from the overall σ^2 (the 2.90 above).

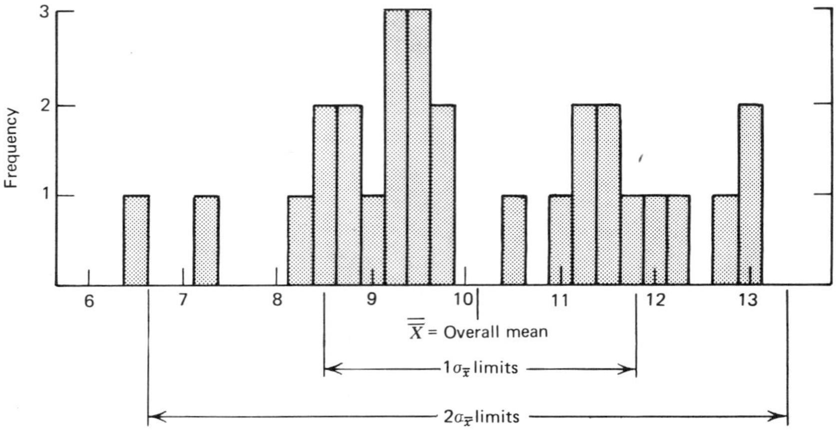

FIGURE 7.6 Distribution of 28 means of samples of four listed in Table 7.2. Assuming a normal distribution, we have one- and two-standard deviation limits, where $\sigma_{\overline{X}} = \sigma/\sqrt{n} = 3.41/2 = 1.70$. Within one $\sigma_{\overline{X}}$ limit we expect 68% of averages—we have 20/28, or 71%. Within two $\sigma_{\overline{X}}$ limits we expect 95% of averages—we have 27/28, or 96%.

Next, we should recall that the distribution of samples is supposed to be approximately a normal distribution. We also know that in a normal distribution essentially 95% of the values fall within the $\pm 2\sigma$ limits. Then, substituting $\sigma_{\overline{X}}$ $(=\sqrt{2.90} = 1.70)$ we should be able to say that we should expect that about 95% of the $\overline{X}$s should be within $\pm 2\sigma_{\overline{X}} = \pm 2(1.70) = \pm 3.40$ of the overall mean. The 1 and 2 sigma limits have been drawn on the distribution in Figure 7.6.

We find that within the $\pm 2\sigma_{\overline{X}}$ limits (or from 6.71 to 13.51) there are all but 1 of the sample means. We would expect 95% of the samples to be within these limits, and we actually have 27/28, or about 96%.

The 1 sigma limits are 8.41 to 11.81, and here we find 20 of the 28 averages, or 71%. We expect about 68%. So in both cases we have reasonably good agreement between the actual and the theoretical numbers of sample means that lie within some of the limits of the theoretical normal distributions.

The most significant part of this discussion, of which you should be aware, is that we are now finding a way of predicting where certain sets of averages should be located. Because certain important distributions shrink as the sample sizes increase, we will find that our ability to make predictions within narrow limits will increase as our sample sizes increase. This will be explored and used many times over in most of the future chapters. In addition, we have now determined what the distribution of sample means *should* look like, so we have another theoretical probability distribution to compare samples of data against.

Since the standard error, $\sigma_{\overline{X}}$, is used quite frequently, a chart has been prepared for you to rapidly find the standard error when you are given n and σ. This chart is in the appendix and is labeled Appendix Table A.5.

7.4 INTERVAL ESTIMATION

One of the major goals of estimation is guessing the parameters of the parent universe or population. Remember that these parameters are fixed constants for the particular population, although the sample statistics we calculate are not constant. Generally we want to choose what we believe to be the best possible guess for the mean, μ, and the variance, σ^2, of the parent distribution. For many situations, especially those dealing with populations believed to be normal, the best estimators to use for these quantities are the sample means and the sample variances. These are point estimators because they only establish a single value or point out of all the possible values that could arise.

Look at the first individual value from our sampling can. It is a 6. At this point it would be our best estimate of the actual population mean. Later we will see how taking a sample will improve on our ability to make estimates. However, we will first concentrate on individuals.

First we should go back and admit that we have not found the true mean with this value. We know that the population has a mean of 5.0 and also has a standard deviation, $\sigma = 1.71$. We can use these values of the parameters of the population to help us draw particular intervals on the population. For example, we know that 95% of values should fall within a range from -1.96σ to $+1.96\sigma$. In this population the interval would be from

$$\mu - 1.96\sigma = 5 - 1.96(1.71) = 5 - 3.4 = 1.6$$

$$\mu + 1.96\sigma = 5 + 1.96\,(1.71) = 5 + 3.4 = 8.4$$

The one value that we picked was a 6 and it falls inside this interval of 1.6 to 8.4 (see Figure 7.7). So if we originally thought that the population mean

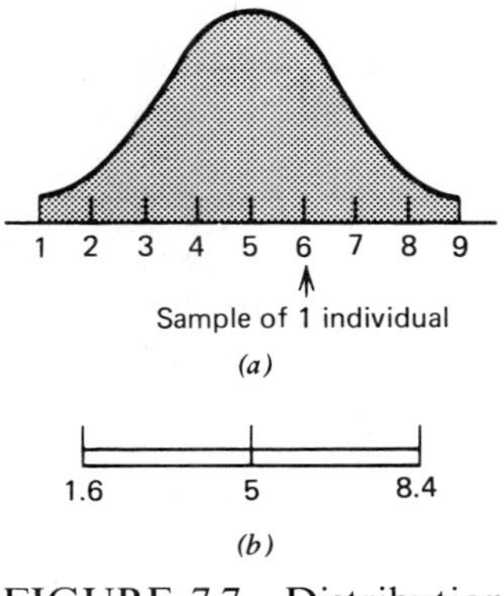

FIGURE 7.7 Distribution of individuals. (*a*) Distribution drawn around *true mean.* This is the theoretical population. (*b*) Interval drawn around true population mean. This interval would include 95% of the individuals that would be randomly picked.

was 5.0, then this sample would be a reasonable one from that population. However, other populations could also have generated this individual. In fact, we could have an interval of $\pm 1.96\sigma = \pm 1.96\,(1.71) = \pm 3.4$ drawn on a population whose mean, μ, was as *low* as $6 - 3.4 = 2.6$. If we took a population with mean of 2.6 our individual value of 6 would have fallen in the 95% interval. Figure 7.8*a* shows this lower population with its 95% interval and the value of 6.

At the other extreme, a population whose mean is as high as $6 + 3.4 = 9.4$ could also have a value of 6 fall within its 95% limits or interval (see Figure 7.8*b*).

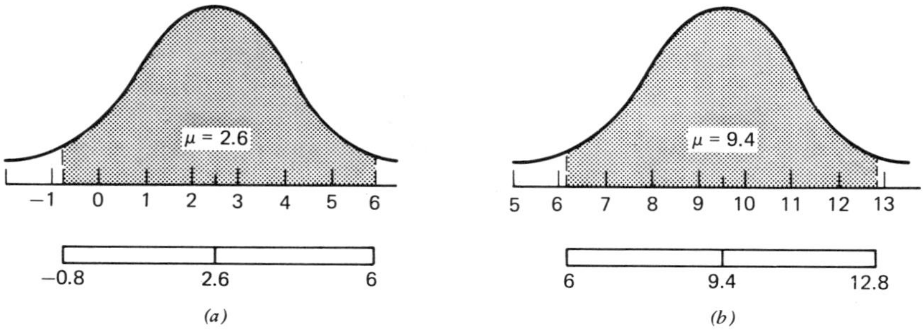

FIGURE 7.8 The two extreme theoretical populations from which an individual value of 6 would fall within the interval that includes 95% of the values. (*a*) The lowest population mean, $\mu = 2.6$. (*b*) The highest population mean, $\mu = 9.4$.

Furthermore, any of the populations where the mean is larger than 2.6 and less than 9.4 could be a likely source of our first individual. All of the $\pm 1.96\sigma$ or 95% intervals for all of these populations include this value of 6. Now if we turn the problem around and look at the sample value itself, we say it most probably comes from *one of the normal distributions* with a mean, μ, between 2.6 and 9.4. We then draw an interval around the individual value. The interval is from 2.6 to 9.4 and we call it a 95% confidence interval.

We call this interval a *confidence* interval because we can express a particular amount, level, or degree of confidence that the interval drawn in this way will actually cross or include the population mean. The 95% figure indicates that *if we were to calculate such confidence intervals for a large number of samples,* 95% *of these intervals would actually include (or cross) the true mean.*

The second individual is an 8 (see Table 7.1), and the 95% confidence interval would go from 4.6 to 11.4. The third is a 3 and the 95% confidence interval is from −0.4 to 6.4. The fourth is a 5 and the 95% confidence interval is from 1.6 to 8.4. All of these intervals cross the true population mean. Figure 7.9 shows the population and these four 95% confidence intervals.

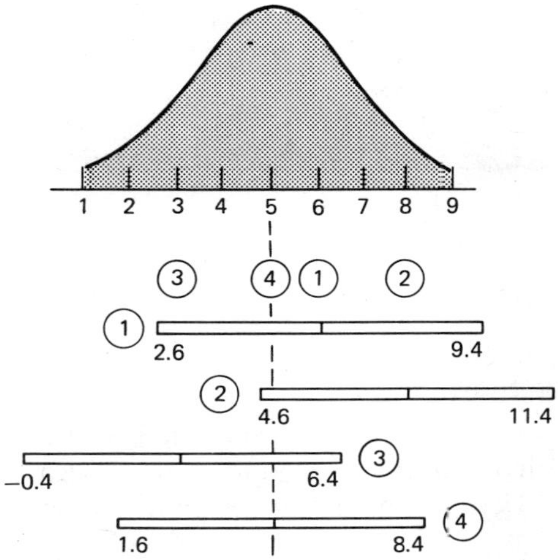

FIGURE 7.9 Ninety-five percent confidence intervals drawn around each of the first four individuals. Observe how all four intervals cross the population mean.

To see further what is occurring, reexamine Table 7.1 and continue looking at the individuals. It isn't until you hit the 17th individual (4th row, 2nd individual) that you obtain a value, the number 0, or an empty house, for which a 95% confidence interval would be from −3.4 to +3.4 and which *doesn't include the true mean.* Of course, if we used only this single individual, we would have made an error in asserting that the interval included the true mean. Unfortunately, we would never really have known that particular confidence interval was incorrect.

Now examine the entire set of individuals in Table 7.1. How many 95% confidence intervals would and would not cross the true mean of 5? You should find 192 that cross this mean and 8 that do not cross it or, 96% of the intervals would include the true population mean.

What if I want to be more sure of myself? What if I want to draw an interval such that I will be 99% sure that it will include the true population mean? Let us again consider individual single values. We must now find the z-values for which 99% of the individuals fall between (instead of the 95% limits, as before). We find that $z = \pm 2.58$ standard deviations, or for this example the limits will be $\pm(2.58)(1.71) = \pm 4.4$. This is a larger interval than before.

Repeating the procedure we just followed for the 95% confidence interval, we would now find that our first individual, the 6, could have come from a population whose mean was as small as $6 - 4.4 = 1.6$ (see Figure 7.10*a*) or as large as $6 + 4.4 = 10.4$ (see Figure 7.10*b*).

In a similar way, we can draw 99% confidence intervals for all of the 200 individuals in Table 7.1. We would then find that *only* 2, or 1%, of

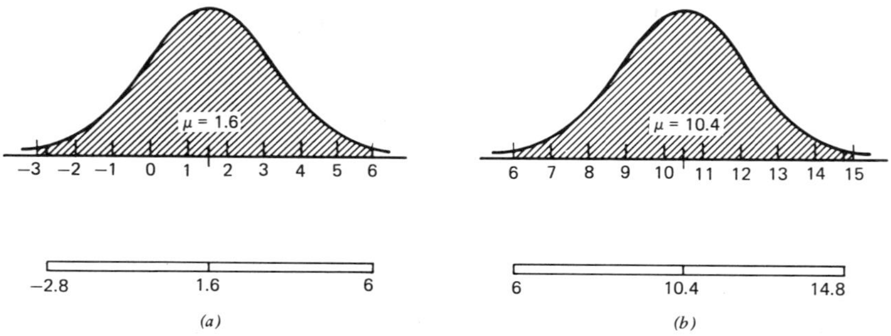

FIGURE 7.10 The two extreme theoretical populations from which an individual value of 6 would fall within the interval, now for 99% of the individuals (99% confidence). (*a*) Lowest population mean, $\mu = 1.6$. (*b*) Highest population mean, $\mu = 10.4$.

the intervals do not cross the population mean. Try to find them. The fact that about 99% of the intervals will cross the mean is why we call this interval a 99% confidence interval.

When we increase from 95% to 99% confidence (we also call this the confidence *level*), we are increasing the size of the interval (see Figure 7.11). Unfortunately, by increasing the interval, we make it harder to be precise or exact about *where* the mean might be. This is the penalty for having a higher level of confidence (say 99% instead of 95%).

Now let us look at samples. Our first sample was a sample of 4. Since we are dealing with a normal distribution, our best estimate of the population mean is the sample mean. This is:

$$\text{Estimate of } \mu = \hat{\mu} \text{ (read this as mu-hat)}$$

$$= \overline{X} = \frac{6 + 8 + 3 + 5}{4} = 5.5$$

We have refined our estimate of the population mean, but this is still just a point estimate.

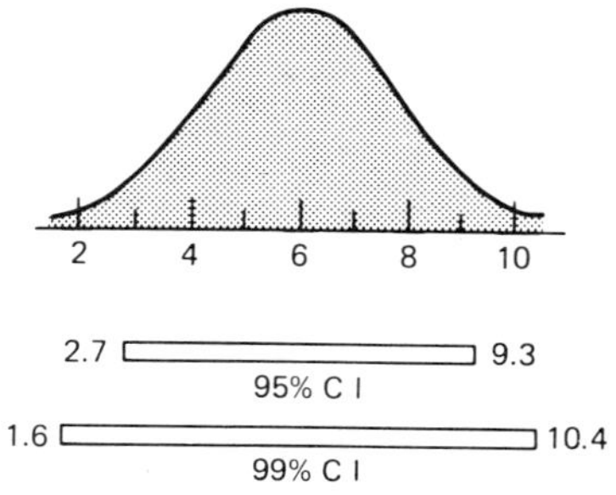

FIGURE 7.11 The effect of increasing "confidence" is to increase the limits.

Again, the most reasonable way of expressing our unsureness is to place an interval around that single point. Thus we come to confidence intervals for sample means.

In the last section we examined the distribution of sample means and found that sample means tend to be closer together than individuals. So we should expect that a confidence interval for samples would be smaller than one for individuals.

We also found that the standard deviation among means, the standard error of the mean, was $\sigma_{\bar{X}} = \sigma/\sqrt{n}$.

Since the means of samples are also normally distributed, 95% of the samples should fall within $\pm 1.96\sigma_{\bar{X}}$ of the population mean.

For our true mean of 5.0 and samples of 4, we would have

$$\sigma_{\bar{X}} = \frac{\sigma}{\sqrt{4}} = \frac{1.71}{2} = 0.855$$

Then 95% of the sample means would fall in a range of $\pm(1.96)(0.855) = \pm 1.7$, or from

$$5 - 1.7 = 3.3 \qquad \text{to} \qquad 5 + 1.7 = 6.7$$

The size of this interval is *half* the size of our previous intervals. The size reduction occurs because we are now using samples of 4 (and as we increase the size of the sample, the standard error is decreased).

The sample value, $\bar{X} = 5.5$, falls within the interval 3.3 to 6.7 and would therefore be a reasonable sample.

However, as with our examination of the *possible* population an individual might have come from, we can look at other possibilities for the sample.

This sample could be the upper limit of a 95% interval from a population with a mean of 3.8. This is found as $\mu = \bar{X} - 1.96\sigma_{\bar{X}}$ or $5.5 - 1.7 = 3.8$ (see Figure 7.12*a*).

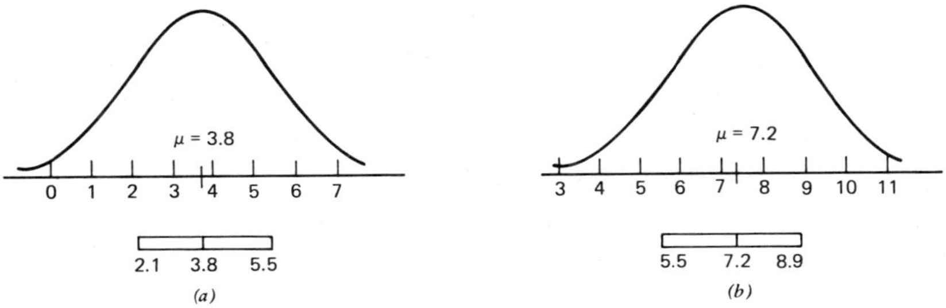

FIGURE 7.12 Two possible extreme populations from which samples of 4 with sample means of 5.5 might have come. Intervals are drawn to include 95% of the possible sample means. (*a*) Lowest population mean, $\mu = 3.8$. (*b*) Highest population mean, $\mu = 7.2$. Observe again that the intervals are half the size of the previous intervals for individuals.

The sample could also have come from a population with a mean, μ, as large as 7.2. Here we calculate $\mu = \overline{X} + 1.96\sigma_{\overline{X}} = 5.5 + 1.7 = 7.2$ (see Figure 7.12*b*). Thus we find two extremes for possible populations from which this *sample* might have originated, and we call this interval the 95% confidence interval for our mean (from 3.8 to 7.2).

We interpret the confidence interval for samples in the same way as we did for individuals. Take repeated samples from the same population and draw 95% confidence intervals for each sample. Then we say that 95% of the confidence intervals for samples will contain the true population center μ. In Figure 7.13 I have drawn the intervals for the first 10 samples of 4. One of these, the 4th, does not reach the true population mean. Bear in mind that if this was the only sample we had taken, these would have been the only limits (the only confidence intervals) that we could have drawn, and we would not have included the true center in our interval. We would have erred. We must expect to make errors sometimes, and our choice of 95% limits forces us to be prepared for about 5 wrong intervals out of every 100 intervals we calculate.

If you were to continue through the entire set of samples of 4 in Table 7.1 you would find that there is only the 1 interval out of all 50 that does not cross the population mean (the 4th one mentioned above).

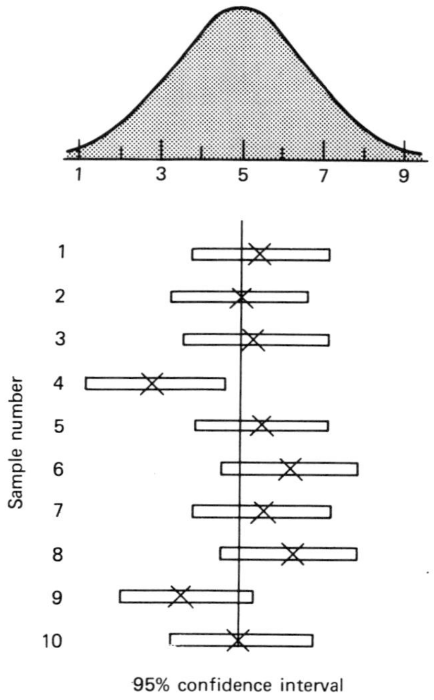

FIGURE 7.13 Ninety-five percent confidence intervals drawn for 10 sample means of 4 from a given population. Observe how one interval, the fourth, does not cross the true mean.

If we were to increase our confidence interval to a 99% level, we would have limits of $\overline{X} - 2.58\sigma_{\overline{X}}$ and $\overline{X} + 2.58\sigma_{\overline{X}}$ or $\overline{X} - 2.2$ and $\overline{X} + 2.2$. Our 99% confidence interval for the first sample of 4 would be from

$$5.5 - 2.2 = 3.3 \quad \text{to} \quad 5.5 + 2.2 = 7.7$$

All of the 99% confidence intervals that might be drawn around the samples of 4 cross the true mean.

We can now see how much taking a sample increases our ability to estimate the location of a population mean. The 99% confidence interval for samples of 4 was even smaller in width than the 95% confidence interval for individuals. And we can still improve on this. By taking larger and larger samples, we can get as small a confidence interval as we might desire. This is illustrated in Figure 7.14, where confidence intervals

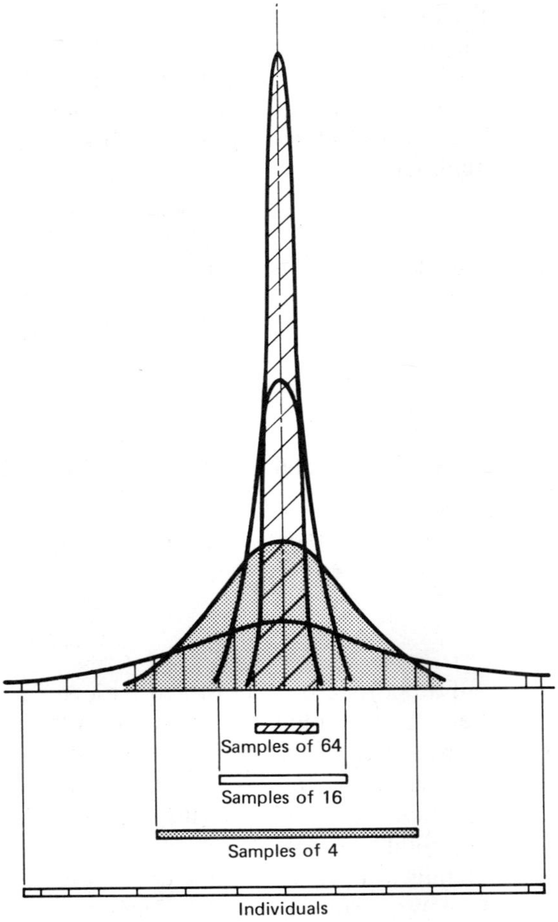

FIGURE 7.14 The effect of sample size on confidence intervals. Distributions of individuals, samples of 4, 16, and 64 are drawn with the respective 95% confidence intervals.

are drawn for individuals and for samples of 4, 16, and 64, all from the same population. In each case the size of the interval is cut in half as the sample is increased by a factor of 4. Remember: $\sigma_{\bar{X}} = \sigma/\sqrt{n}$, so as n gets larger, $\sigma_{\bar{X}}$ must get smaller (and the factor that it decreases by is the square root of n).

In almost any type of measurement where the instrument has been calibrated or tested with a known standard, we may determine a value for the average variation in readings. In other words, we can arrive at a known standard deviation for readings on that instrument. The instrument is then used to obtain a sample of readings for which we calculate a mean. We may then provide a confidence interval around the mean of the measurements obtained.

EXAMPLE 7.1

An instrument for measuring cholesterol level in the blood has been found to have a standard deviation of 10 milligrams per 100 millimeters. Six tests were conducted on one male to estimate the true cholesterol level in his blood. The readings had a mean of 206 milligrams per 100 milliliter. Then

$$\bar{X} = 206, \qquad \sigma = 10$$

The 95% confidence interval for the mean is from

$$\bar{X} - z\frac{\sigma}{\sqrt{n}} \quad \text{to} \quad \bar{X} + z\frac{\sigma}{\sqrt{n}}$$

or

$$206 - 1.96\frac{(10)}{\sqrt{6}} \quad \text{to} \quad 206 + 1.96\frac{(10)}{\sqrt{6}}$$

$$206 - 8 = 198 \quad \text{to} \quad 206 + 8 = 214$$

or

$$198 \text{ to } 214$$

If a cholesterol level of 210 indicates we should go on a special diet, then when we report the cholesterol level not as 206 but in a 95% confidence interval of 198 to 214 we indicate that there is good reason to suggest that the level is high (the interval includes the value of 210).

A different instrument with a standard deviation of only 3 milligrams per 100 milliliters was also used. Here you should note that even a *single* reading will provide a smaller confidence interval. The single reading in the new instrument was 208. Then the 95% confidence interval is

$$\bar{X} - z\frac{\sigma}{\sqrt{1}} = \bar{X} - z\sigma \quad \text{to} \quad \bar{X} + z\frac{\sigma}{\sqrt{1}} = \bar{X} + z\sigma$$

or

$$208 - 1.96(3) = 208 - 6 = 202 \quad \text{to} \quad 208 + 1.96(3) = 208 + 6 = 214$$

This interval also includes the critical value of 210.

REVIEW

Sampling • Population • Estimation • Bias
Distribution of sample means • Central Limit theorem

$\sigma_{\bar{X}} = \dfrac{\sigma}{\sqrt{n}}$ • Standard Error of the Mean

Point Estimation • Interval Estimation • Confidence Interval

PROBLEMS

SECTION 7.2

7.1 Take an additional 5 samples of 4 values from the sampling distribution in the Appendix Table A.9. What are the means and the variances of these samples?

7.2 Table 7.1 contains two sets of estimates of the variance, σ^2. One column is of s^2 and the other $\hat{\sigma}^2$. Calculate the mean of both sets of variance values. The true σ^2 is 2.94. Which average variance is closer to the true variance (is less biased)? Draw a histogram for both sets of estimates of variance.

7.3 Repeat Problem 7.2 for the estimates of the standard deviation. The true value is 1.713.

7.4 Determine the distribution of all possible samples of 2 for the sampling pieces (in Table 7.1). First list all the possible combinations. Knowing that the frequency of each number is as listed below. determine the probability (or relative frequency) of each possible sample. (Consider the samples as *with* replacement.) Calculate the mean, variance, and standard deviation of the distribution of samples. There are a total of 200 pieces. The frequencies of 0, 1, 2, 3, 4, 5, 6, 7, 8, 9, and 10 are 1, 3, 10, 23, 39, 48, 39, 23, 10, 3 and 1, respectively.

7.5 A die has 6 faces with 1 to 6 dots on a face. A fair die will have an equal chance for any given number to occur. Calculate the mean and variance for the distribution of individual values. Draw the histogram for the population of individual values. How would you describe this population?

7.6 If you are to roll 2 dice, list all the possible sums and means of 2 dice. Draw a histogram for the means. Calculate the mean, variance, and standard deviation for these means of two. Calculate the theoretical variance and standard deviation of the means if the central limit theorem holds for the samples (see Problem 7.5 for the population variance).

7.7 For the given population, calculate the theoretical mean, variance, and standard deviation for the distribution of sample means (the standard error of the mean).

a. $\mu = 0$, $\sigma = 10$, sample size 8.
b. $\mu = 10$, $\sigma = 50$, sample size 20.
c. $\mu = 1000$, $\sigma = 25$, sample size 100.
d. $\mu = -50$, $\sigma = 10$, sample size 5.
e. $\mu = 550$, $\sigma = 40$, sample size 16.
f. $\mu = 200$, $\sigma = 30$, sample size 10.
g. $\mu = 0.50$, $\sigma = 0.05$, sample size 3.

SECTION 7.4

7.8 For each of the theoretical distributions of samples in Problem 7.7, calculate the interval within which you expect approximately 95% of the sample $\overline{X}$s to fall (you are to determine two numbers).

7.9 Assume that samples are taken from a normal distribution. For the following populations, determine the standard error of the mean and the percentages of sample means that should fall in the specified areas.

a. $\mu = 0$, $\sigma = 10$, sample size 4, percent between 0 and 2.
b. $\mu = 0$, $\sigma = 10$, sample size 16, percent between 0 and 2.
c. $\mu = 20$, $\sigma = 10$, sample size 25, percent between 15 and 20.
d. $\mu = 100$, $\sigma = 30$, sample size 100, percent between 105 and 110.
e. $\mu = 5$, $\sigma = 0.10$, sample size 8, percent between 4.95 and 5.05
f. $\mu = 20{,}000$, $\sigma = 250$, sample size 10, percent between 19,000 and 20,100,
g. $\mu = 1500$, $\sigma = 500$, sample size 50, percent between 1400 and 1550

7.10 Using the random number table, Appendix Table A.7, calculate the mean of 4 digits. Do this for about 10 sets of 4. Draw a histogram for them. How close to a normal distribution is it? What are the mean and standard deviation of this distribution? What would you expect the mean should be? What would you expect the standard deviation to be? (*Hint*: 10% of the values should be in each category, so if you choose just 10 values the frequency of each would be expected to be 1. Also, this is a theoretical distribution, so use n, not $n - 1$.) What should the standard error be? How close is the standard error of the distribution to the theoretical value?

7.11 If 1-pound bars of butter have been averaging 0.98 pound with $\sigma = 0.04$, what is the chance that an inspector from the Weights and Measures Department would find that the average weight of 8 pounds of butter would be less than 0.96 pound?

7.12 If a barber takes an average of 14 minutes to give a haircut with $\sigma = 3$ minutes:

a. What would he expect the total time for giving 24 haircuts to be?
b. What would the standard error for the average time for giving 24 haircuts be?
c. Ninety-five percent of the time the average time for giving 24 haircuts would fall within what limits?
d. If you multiply the limits for the average by n (or 24) you will obtain limits for the sum (the total time for 24 haircuts). Within what limits would 95% of the total times for 24 haircuts fall? (This can also be developed by finding σ for Σx.)

7.13 Baking potatoes were weighed and found to average 0.50 pound, with standard deviation of 0.08 pound. A banquet of 100 was planned. How many pounds of potatoes should be purchased to provide 95% assurance that there will be enough potatoes, if each person gets 1 potato?

7.14 If the mean length of hospitalization of 140 patients was 11.4 days and the standard deviation of patient days is assumed to be 2.5 days, what is the 99% confidence interval for average length of stay?

7.15 The mean diameter of 250 trees was found to be 3.67 inches (Problem 3.17). If the variance of diameters is known to be 0.7, what is the 98% confidence interval for mean diameter of trees?

7.16 Estimate the number of days between germination and the first pickable cucumbers using the following sample of cucumber plants.

DATE OF GERMINATION	FIRST FRUIT
May 1	June 17
May 4	June 18
May 8	June 21
May 5	June 16
May 12	June 28
May 18	July 3
May 11	June 25
May 9	June 26

What is the 95% confidence interval assuming $\sigma = 2$ days?

7.17 Light bulbs average 100 hours of life with a variance 36 hours. If 25 light bulbs are tested, then 90% of the time the mean should fall between what limits?

7.18 You are designing a key case for men and are interested in determining the keys most men carry. You sample 25 men and arrive at an average of 6.5 keys (with $s = 4.2$). What would you estimate is the true average number of keys carried by men? What is the 95% confidence interval for the true average? Is this confidence interval of any use in deciding how many keys to provide for (what does s tell you about the distribution)? (You should also note that s was used, therefore z is not actually the proper distribution to work with; later we will find that a value called a t-statistic will be more appropriate.)

7.19 If the monthly weight loss at a weight control group averages 4 pounds with a standard deviation of 2 pounds, what would you expect the 90% confidence interval for the yearly average loss to be? (*Hint*: Treat this as a sample of 12 months to obtain the limits, then multiply by 12.)

7.20 A discouraged train rider decided to keep a record of the times between trains. The user waited for 10 different trains and recorded an average waiting time of 27 minutes. (The standard deviation of 4.8 minutes was not significantly different from the σ of 4 minutes that the train company claims.) What is the 99% confidence interval for the mean waiting time? How does this compare to the company's claim of an average of 19 minutes?

CHAPTER EIGHT AN INTRODUCTION TO HYPOTHESIS TESTING

One of the main purposes of the gathering of data is to come to some kind of decision. The data are nice by themselves, but they are useless unless we are to make some evaluation of them. One of the principle kinds of decisions we make, or questions we answer, involves the action we take with a hypothesis.

8.1 WHAT IS A HYPOTHESIS?

Basically, a hypothesis is a speculation. It is a statement that you are interested in proving or disproving, declaring correct or incorrect. The hypothesis may deal with any kind of endeavor or question from "Blonds have more fun" to "Educable mentally retarded children progress faster in reading level when integrated in regular classrooms than when kept in special classes" or "The endurance limit of prestressed reinforced concrete is greater than nonprestressed reinforced concrete" or "The galaxy was formed by a 'big bang'."

Formulating a hypothesis is actually a first step. The gathering of data and the decision about it is the next important step. The final decision-making process should depend on the kind and amount of data that we have; ideally the reasoning will involve some statistical analysis of the data that have been obtained.

As a researcher or investigator we will propose a hypothesis, which is a formal question we intend to resolve. Next we would gather some data. Then we would use the data to decide if the hypothesis is true or false, but remember, we can only make *one* decision—true or false. Now, *how* did you arrive at your decision? Was it by "common sense" or because "it looked good" or "it was obvious?"

If so, perhaps your decision is correct. Fine. But what if it isn't correct? You really won't be able to tell if it is not correct. But can you state anything about the chance that you might be wrong? Can you tell me *what is the probability that you might be wrong*?

Here lies the important difference between answering a hypothesis with "common sense" as opposed to the use of statistical methods. In either case *we may be correct in our decision* (the key word is *may*). However, using the principles of probability and statistics we will have an opportunity to disclose, and even limit, *our chance of making certain kinds of errors*. This is an extremely vital factor of which to be aware.

8.2 KINDS OF ERRORS: DECISION PROBLEMS

Let us look at what these errors I am alluding to are (there will be no data yet).

EXAMPLE 8.1
Consider the following situation you may face in traveling a particular road.

There is a small hill on a one lane road. You cannot see over the hill, nor could you stop quickly enough if there were an obstacle just on the other side (a stuck car, the end of the world). You therefore may formulate a hypothesis as you approach the top of the hill. (It may be so rapid you don't even realize you are doing it yourself.)

HYPOTHESIS There is no car stuck on the other side of the hill.

If I *accept* the hypothesis, then I will drive right on through at the legal 50 miles per hour. If I *reject* this hypothesis, I am saying that I believe that there is an obstacle ahead, and I will slow down considerably. But there may be a consequence of slowing down. If I slow down I may miss a meeting that could have led to a new important job. So there are alternatives to choose from and consequences to each alternative.

First, we should realize that *there may actually be a car on the other side of the hill* or there may *not* be a car stuck ahead. These two cases exist for the problem, but *we are not able to say what is correct* as we approach the hill. If we knew which was the correct case, there would be no difficulty and the problem would be trivial. A person flying in a helicopter watching us and the hill would be able to say which is correct. But we are not in the helicopter, we are in the car.

Since there are two possibilities for the actual case, let us treat them separately.

1. *If the hypothesis actually is true*: Then there would be no car stopped on the hill. The problem for us is *still* to decide if there is a car on the hill. We must make a decision. What two alternatives are available to us?

We could:

a. Accept the hypothesis—say there is no car ahead and ride on through.
b. Reject the hypothesis—say there is a car ahead and slow down.

In Figure 8.1 we examine the *consequences* of the decisions. When the hypothesis is true and there actually is no car ahead, then the two faces of the display show the correctness of our decision. The expression only describes our feeling with regard to the decision or the hypothesis.*

* The consequences of accepting the hypothesis, even though they result in ☺ may be such that we would not be pleased. However, with respect to the decision, we would be pleased. It is this satisfaction with the decision that the face is showing.

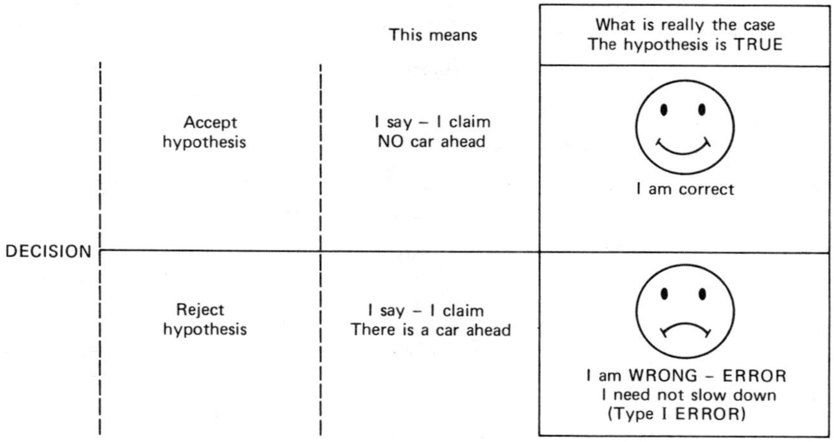

FIGURE 8.1 The two decisions to make with hypothesis that is TRUE. The hypothesis is—there is NO car ahead.

2. *If the hypothesis is actually false*: Then there is a car ahead, and we have the second case. But *our decision problem has not changed*. There still are only two alternatives—accept or reject the hypothesis. Figure 8.2 shows the effect of our choices—the faces are now reversed. This time the sad face indicates an accident ahead. As before, when we accept the hypothesis, we do not slow down. But, since this time the hypothesis is wrong and there is a car ahead, we misjudge with great consequences—an accident. (I emphasize, however, that the sad face is *not* because of the accident, but because we decided that the hypothesis was true when it was not true.)

What about the happy face? It made the correct decision as far as the *hypothesis* was concerned. But it missed the meeting. As for the indicated

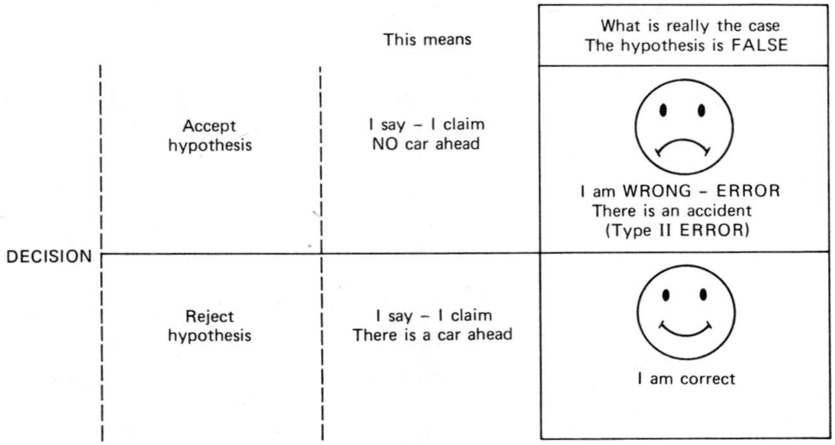

FIGURE 8.2 The two decisions to make with a hypothesis that is FALSE. The hypothesis is—there is NO car ahead.

face, again *it only concerns the hypothesis, not any indirect* (or direct) *consequences.*

We can now combine the two previous cases and draw up a decision box, which I will call our "four faces of consequences" (Figure 8.3). The two happy faces indicate correct judgments—correct decisions. The two sad faces we call error. Since the two types of errors will become very important, we have names for them.

Mistake Type I: The times *we reject* when we should not—we designate this occurrence by *Type I, alpha* (α) error, or *Error of the First Kind.**

Mistake Type II: The times *we accept* when we should not—we designate this occurrence by *Type II, beta* (β) error, or *Error of the Second Kind.**

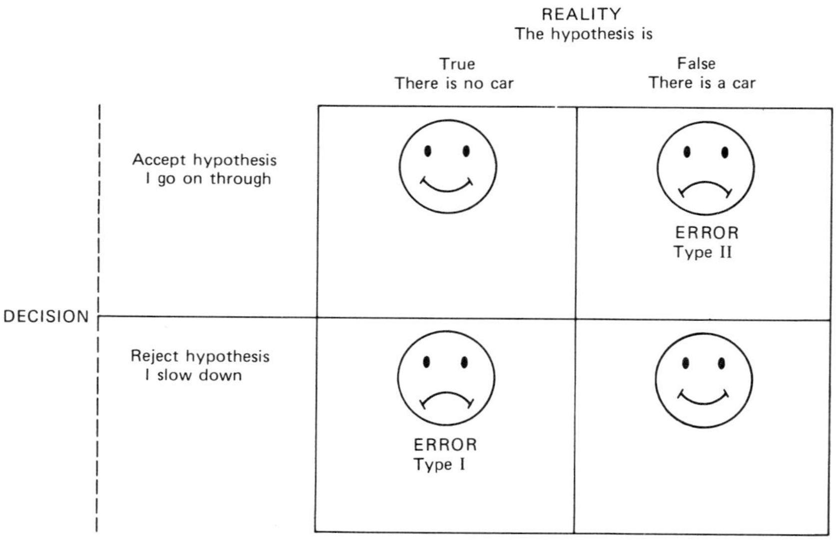

FIGURE 8.3 A combination of the two possibilities of reality. The "four faces of error" for our hypothesis, "There is NO car ahead."

(I suggest you go back over the notion of these two kinds of error several times. The idea is not a simple one, and I would recommend rereading this and the subsequent presentations as well.)

EXAMPLE 8.2
As another example, let us modify Example 8.1 as follows.

HYPOTHESIS There is no radar trap ahead.

* Usually in the formal statistical definitions the errors are not referred to as alpha or beta errors, but are called Type I and Type II errors. Alpha and beta are used to represent the probability of making these errors. In an introduction, however, it is useful to keep repeating the terms alpha and beta as we discuss the errors.

If I accept the hypothesis, I will continue traveling at a rate exceeding the speed limit. If I reject the hypothesis, then I slow down and I will miss my meeting, which may cost me the new job. We often call this hypothesis a "null" hypotheses.

Generally another term is introduced at this point. I have been saying that we may accept or reject the hypothesis. In the cases when we *reject* the hypothesis I have been just writing down what appears to be a logical statement. We usually call this statement an *alternate hypothesis*, and it is always given along with the hypothesis we are questioning. So in this case we could say that the alternate hypothesis is: There is a radar trap ahead. (Although this alternate is obvious, there will be future cases where there may be a number of alternatives with different consequences or chances of occurring.)

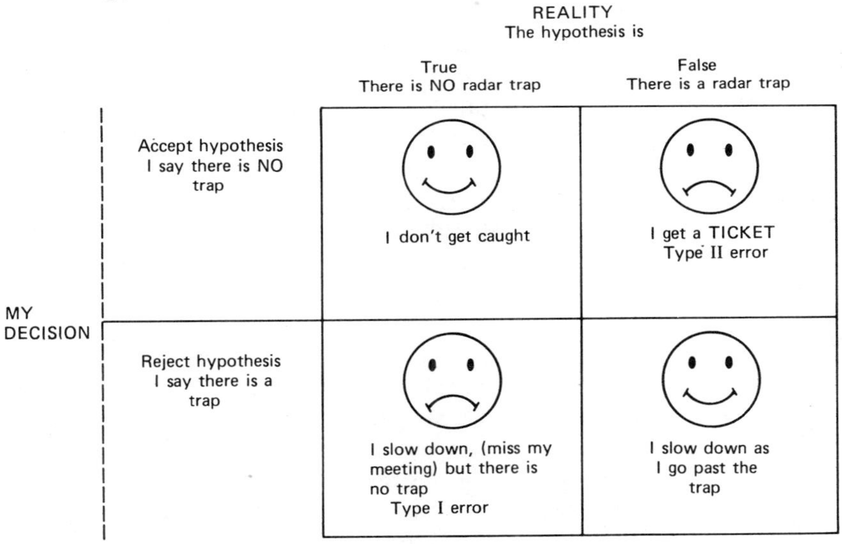

FIGURE 8.4 The "faces of error" for the hypothesis, "There is NO radar trap ahead."

Now let us look back at the consequences of accepting or rejecting the hypothesis. Figure 8.4 shows them graphically. Again, we can have two ways of making an error. This time, however, the relative weights or degrees of importance we may choose to give the errors will perhaps be reversed. Here the two possibilities of error are as before:

Type I (or alpha error): *Reject* when we should not reject; that is, I *say that* I expect a speed trap and I *slow down when there is not a trap.*

Type II (or beta error): *Accept* the hypothesis when we should not accept; that is, I say that *there is no radar trap* when there actually is one, with the result that *I get caught.*

Now, how much weight or emphasis might we place on the two alternative kinds of errors? This leads to the question of which mistake would cost more. Let us look at one possibility.

> A ticket is worth \$20; missing the meeting will result in a loss of a \$25,000-a-year job. (You might reason that getting a ticket would also entail missing the meeting, but I haven't told you of our progressive system here. The ticket is mailed to your home and you can pay by mail.)

Here, the Type I error, which was slowing down when we didn't have to, is a situation we really want to prevent. The Type II error we will be more readily willing to accept under these circumstances. From a purely monetary basis it is worth the risk of getting the \$20 loss to make the meeting and get the job.

We must look at the weights we place on the two possible types of errors that we can make. First, we should realize that *we can never eliminate both errors at the same time*, unless we know something special, like the radar unit is being repaired, or the only patrolman in the area has the flu, in which case we need not bother with this discussion. If we are not privy to any confidential material, then at best we can eliminate only one error. But doing that means we will *always* choose one of the alternatives, and it hardly seems reasonable to call that a decision. (I realize I am forgetting the moral issue of breaking the law, but from within the academic confines, we can analyze the problem as one of economics only. This is a deliberate avoidance.)

Consequences are often extremely vague and may include some that we cannot directly analyze. Those that are simple are the ones to which we can assign a monetary value, or from which we may see direct consequence—good or bad. Thus, getting a ticket may also lead to losing a license, which may mean that we would not be able to take the job anyway. It might mean no dates or extra carfare or any number of things.

There often are nonmeasurable or nonequatible consequences of a decision such as a psychological trauma or perhaps future unpredictable circumstances (maybe the job wasn't what you wanted and you would have been better off missing the meeting after all). There might also be many other possible occurrences, such as a tire blowing out because of excessive speed, or a car suddenly stopping in front of you, all of which cannot be reasonably considered simultaneously. Most frequently, however, whenever we make a decision, we have already done some simplification of which we may not even be aware, eliminating many factors and making numerous assumptions.

After all of this we still have not made the decision; the problem of assigning actual weights to the possible decisions is what is still remaining. As you have already seen, in this example we wanted to limit the chance of making a Type I error, since we mainly were interested in making the meeting. In the previous example we wanted to eliminate the Type II error, since we did not relish the thought of a possible fatal accident. Often a problem may exist where the best decision must be a balance between the two errors. Under these circumstances, you will have to assign the weights yourself.

So far we have not used any statistics or statistical principles, since we have no data or numerical relationships to aid in our decision. We will soon attempt to find ways of describing the errors by numerical methods. We will be discussing the probability or the odds of making those errors. In the present problems we have not been able to do this, but our goal is to assign probabilities to the chance of making the two wrong decisions.

One very typical type of problem that readily illustrates the principle of testing a hypothesis would be acceptance sampling. Perhaps you are making or selling a particular type of item and I will be buying it in a large quantity. The items could be ball-point pens, transistors for radios, shotgun shells, cans of motor oil, beer, corn seeds, laboratory animals (frogs or mice), or penicillin tablets.

We both would like to believe that all the items are good, but we recognize there may be some bad ones. In addition, for example, neither of us would want to test 5000 pens to see if they write. I may not have the equipment to test 10,000 transistors rapidly. Also, exploding shotgun shells, opening up cans and testing motor oil or beer, planting the seeds to determine if they will germinate all require either destroying the original material or at least the containers. Thus, we must select a random sample, check the sampled items, and ask the question of whether the entire lot of items are good enough.

EXAMPLE 8.3

A hospital is to purchase several thousand bottles of penicillin powder. Before accepting the shipment, a sample of bottles are opened and mixed with water to determine if they dissolve properly.

HYPOTHESIS The shipment is good. There are very few bad penicillin bottles (although at this point we have not determined what percent or number of bad bottles is acceptable).

ALTERNATIVE HYPOTHESIS The shipment is not good. There are too many bad bottles.

We must decide on the number of bottles to be tested and the criterion for accepting or rejecting the hypothesis. We might test 10 bottles and plan our decision as follows.

1. If there are *none* or only *1* bad bottle among the 10 we will accept the hypothesis and say the shipment is acceptable.
2. If there are 2, 3, 4, 5, 6, 7, 8, 9, or 10 bad or defective bottles we will reject the hypothesis and accept the alternate hypothesis, saying the shipment is bad. By rejecting the hypothesis, we will reject or send back the entire shipment.

There is still the possibility of error.

Type I Error: We may have a good shipment but perhaps, unfortunately, we happen to select 2 bad bottles. The result would be that we reject the entire shipment, even though the overall quality of the shipment is good. This is sometimes called the producer's risk because the producer or manufacture will end up getting a large collection of good material returned.

It is also our alpha (α) error.

Type II Error: In some cases we will have a bad shipment but the sample we choose will have either all good bottles or only 1 bad one. As a result we will buy the entire shipment and we will be disappointed because it contains more defective bottles than we wanted. This is our risk, the consumer's risk, and it is the beta (β) error.

Figure 8.5 illustrates these two types of errors.

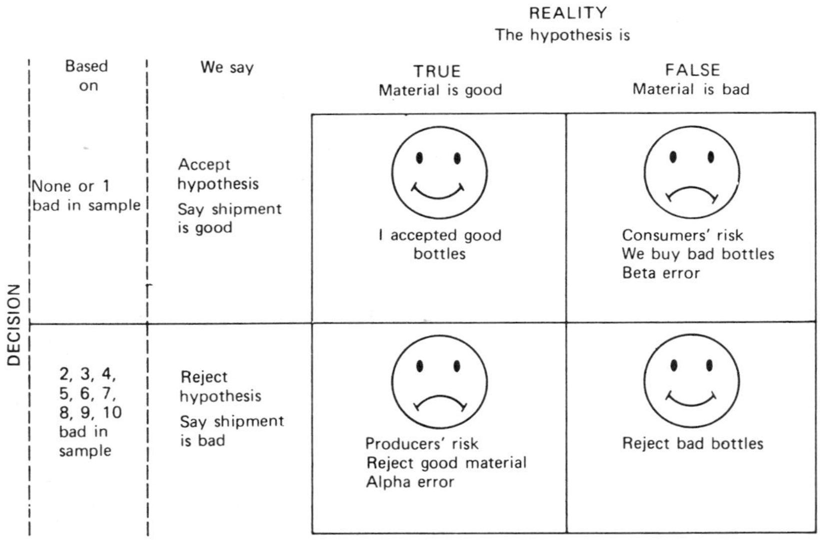

FIGURE 8.5 Decision box for Example 8.3, determining if we should accept a shipment of bottles of penicillin.

Now let us see how we may determine numerical values for the two types of errors. First we must define a little better just what we mean by a good or a bad shipment. For example, we might declare that good is when only about 5% of the shipment is defective (or 95% are good). If we further recognize that if the sample is randomly selected, and the bottles have the same independent chance of being defective, we have a binomial distribution. Therefore we are able to describe the probabilities of having different numbers of defectives in our sample.

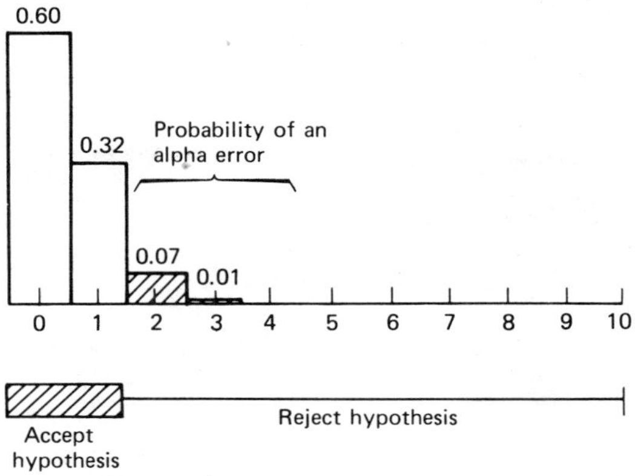

FIGURE 8.6 Distribution of number of defective bottles in a sample of 10. Probability of a defective bottle is 0.05. The acceptance and rejection regions are shown. The probability of an alpha error is the area in the rejection region and is about 0.08.

If you look at Figure 8.6 you will find a histogram for the distribution of number of defective bottles in samples of 10. Thus about 60% of the samples would have no defectives and 32% would have one defective bottle.

We have stated that if there were 0 or 1 defective item then we accept the hypothesis and the shipment. Below the values of 0 and 1 on the histogram, I have indicated that this is the acceptance region. Thus, in about 92% of the samples that we might take, we would accept the lot. *But*, in about 8% of the cases we would reject the hypothesis and the shipment. These situations correspond to the shaded section labeled *reject* on the histogram. These 8% of the samples result in an alpha error—rejecting when we shouldn't. Therefore the probability of making an alpha error is 0.08.

Next we come to the problem of beta error. This requires us to determine what we believe to be the alternate population. We actually can have a complete collection of such alternates; as a result, we would be able to arrive at a collection of different probabilities of beta errors, with each probability depending on a different alternate population.

For example, let us assume we have an alternate population with 25% defective and 75% good bottles of penicillin. What now happens if we are to take a sample of 10 bottles? The sample could contain 0 bad, 1 bad, and so on, up to 10 bad. Now if we were to take a large collection of samples from this alternate population, we would find that about 6% of the time we would have samples with 0 bad and about 19% of the samples would only have 1 bad. But what have we agreed to say if we get 0 or 1 defective? We accept the lot: Recall that we are not changing the acceptance or rejection criterion Figure 8.7. Therefore, over 25% of the samples that would come from this alternate population would fall in the accept region (6% + 19% = 25%). In these cases we will make a beta error—we will accept a shipment of material that overall contains about 25% bad material.

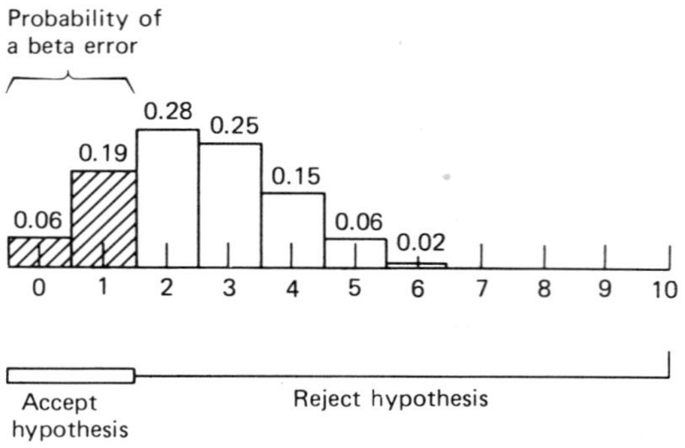

FIGURE 8.7 Finding the beta error for an alternate hypothesis probability of a defective bottle is 0.25 (population of 25% defective, 75% good). The shaded portion is the chance that the hypothesis of a good population will be accepted even though the product is poor. This results in a probability of a beta error of approximately 0.25 (or 25%).

Unfortunately, in most practical problems we are not dealing with a single alternate such as $p = 0.25$ but, instead, with the whole range of alternate *p*s. Each of these *p* values leads to a different probability of a beta error, since the location or shape of the alternate distribution would be different. As a result, the corresponding portion of the new distribution that is in the acceptance region will be different for each case.

Consider the possibility that the population is 10% bad and 90% good, or the probability of a defective bottle is 0.10. Now we find that the probability of obtaining a sample with no defectives is about 0.35 and for 1 defective it is 0.39. The probability of making a beta error is therefore 0.35

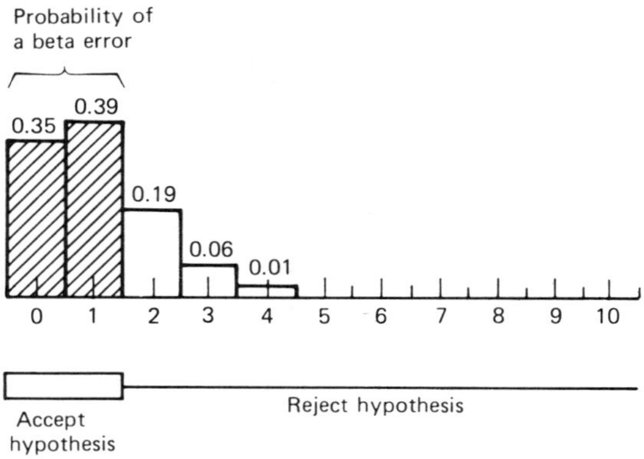

FIGURE 8.8 Beta error for another alternate hypothesis—probability of a defective bottle is 0.10 (10% bad, 90% good). The probability of a beta error is the shaded area and is approximately 0.74.

$+ 0.39$, or $\beta = 0.74$. Then about 74% of the time we will make this error. Figure 8.8 illustrates this second alternate population.*

In the next chapter we will come to hypotheses that will use the normal distribution as the model. There we will find that we will be able to manipulate more easily the location that separates the accept and reject regions and also will be able to tie this point ot more exact values of α (areas in the tail). However, we will not pursue the concept of beta error any further.

REVIEW

Hypothesis • Alternate Hypothesis • Error
Decision • Criterion • Acceptance • Rejection
Type I (Alpha) Error • Type II (Beta) Error
Probability of Type I Error = α = Alpha
Probability of Type II Error = β = Beta

PROBLEMS

SECTION 8.2

8.1 When a person is on trial, the jury formulates a hypothesis. What is the hypothesis that our judicial system operates under? What is the alternate hypothesis? What are the two errors that the jury can make? What are the consequences of these errors? How does society consider or compare the importance of these errors?

8.2 The President of the United States may sometimes be confronted with a statement that there is an impending nuclear attack. State the null hypothesis. What is the alternate hypothesis? What types of errors are possible? What are the possible consequences of each?

8.3 A test for cancer is administered. What are the two possible hypotheses? Which is preferred? What errors can be made? Which is the alpha error and which is the beta error? What might be the consequences of the errors?

8.4 When buying a stock, what are the possible hypotheses? Do you have a single alternative or are there many possible alternates? What kinds of errors can you make? Which would be labeled as an alpha error and which as a beta error? Can we assign any relative chance to these errors?

8.5 A biologist is studying the effect of a new fertilizer on corn plants and is hoping that the yield in bushels of corn per acre will increase substantially. What is the hypothesis that should be proposed? What is the alternate? Is the alternate to be concerned with a decrease in yield?

8.6 A stream is being monitored for the level of a particular pollutant. A large paper mill has been considered as the prime source of the pollutant. Federal law provides a large fine for polluting the waterway, and the paper mill has been installing equipment to reduce the

* These probabilities of various defectives are obtained from Table A.6 for $n = 10$, $p = 0.10$.

amount of pollution. From the standpoint of the federal inspector, what is the hypothesis? The alternate hypothesis? The alpha and beta errors? The consequences of the errors? What might be reasons for such errors?

8.7 When you buy a box of grass seeds, it states a percent germination (e.g., 90%). Would you expect that exactly 90% would germinate? How would you state the hypothesis? What alternate hypothesis would you suggest? Is it a one- or two-directional question? What would be a possible limit for a test value? How would you arrive at a statistic? (*Hint*: first think about planting a sample.) What are the possible decisions you can make? What are the errors?
* If you have covered the binomial distribution, then how might you apply that distribution to determining the test limit?

8.8 A broad and important portion of applied statistics is an area called *acceptance sampling*. Based on a sample, a manufacturer or a supplier of some item decides if the overall batch of items is acceptable to ship. Similarly, based on a sample, you as a purchaser decide if you will accept the product.

Ammunition is produced at an arsenal. When the arsenal is ready to ship some ammunition, a few of the items are fired. Why does the arsenal test only a sample?

a. Based on the number of acceptable firings, the arsenal decides to ship or not to ship. What are the alpha and beta errors and their consequences? (This ammunition cannot be repaired.)
b. If the ammunition is shipped, the receiving department of the Army tests some of the ammunition. Based on the sample firings, it decides to accept or reject them (send them back). What are the possible errors and the possible consequences? Is there really any difference?

8.9 A sportsperson who buys rifle shells must make a decision similar to the Army's decision in Problem 8.8. What are the possible errors the sportsperson can make? How does the seriousness of the possible errors compare to that of the Army's? Who has a more serious possible consequence as the result of a bad bullet?

8.10 A nail manufacturer states "less than 1% are bad nails." How would the nail manufacturer check to assure this statement is correct? Might an error be made? If he or she rejected the fact that "less than 1% are bad nails," what would this cause him or her to do? What might be the consequence of shipping a lot of nails that had more than 1% bad or defective nails?

8.11 A carpenter is about to buy 50 pounds of nails and examines a handful of 30, thinking: "If I find 2 bad ones I won't buy; if I see 1 or no bad ones I'll risk them." Determine the following.

a. The hypothesis.
b. The alternate hypothesis.

c. The limits.
d. The test statistic.
e. The possible decisions.
f. The chances and beta error.
g. The consequences of the possible errors.

8.12 How are the number of limits used related to the hypothesis? When would you only be concerned with one limit and when would you need two limits?

8.13 If we believe textbooks cost an average of \$13, with a standard deviation of \$2, what limits should the average of the cost of your 6 textbooks fall within for you not to reject the hypothesis that textbooks cost is no different than an average of \$13? Start by stating the hypothesis and the alternate hypothesis, then provide the test limits (use $\alpha = 0.10$).

8.14 A church official believes that the church members in Executive Township have an average income of \$25,000 with $\sigma =$ \$4000 (see Problem 5.10). What is the hypothesis? What is the alternate hypothesis? Fifteen families will be sampled. What are the 95% test limits? If the average income of the sample of 15 families is \$26,500, what is the decision?

8.15 The manufacturer of a stomach antacid (Antiseltzer) claims it relieves stomach discomfort in 5 minutes (with a standard deviation of 2 minutes). At a party 10 people volunteer to take part in an experiment and tell how long it takes for them to get relief. State the hypothesis and alternate hypothesis. Determine the test limit(s) for an α of 0.10. If the average time to get relief was 7 1/2 minutes, do you accept or reject the hypothesis?

8.16 When operating properly, a machine that fills soda cans dispenses exactly 16 ounces with a standard deviation of 0.9 ounces. If a sample of 5 cans is taken and α is set at 0.05, what are the test limits? Now if the machine goes out of adjustment and begins to dispense 17 ounces, what is the probability that you will *not* detect this change with the sample of 5? (This is the beta error for an alternate of 17 ounces.) What is the consequence to the company? If the machine begins to dispense 14.5 ounces, what is the chance you *will* detect this change? What is the consequence of not detecting this change?

*8.17 Refer to Problem 8.13. If the average cost of textbooks actually increases to \$15, what is the chance you will not reject the hypothesis?

*8.18 Refer to Problem 8.14. The church has tried to reach the lower income group in the community. If the church membership actually averages \$24,000, what is the chance of rejecting the original hypothesis? Find the chance of rejecting the hypothesis for true averages of 23,000, 22,000, 21,000, and 20,000. Plot these values. Which are they, β or $1 - \beta$? Why?

*8.19 If the stomach antacid (Problem 8.15) actually takes an average of 6 minutes to give relief, what is the chance that we will detect a significant difference? How about 7 minutes?

*8.20 The manufacturer knows its antacid really averages relief in 10 minutes, but it wants to advertise with a shorter time period. The company would like to state that if 4 people take Antiseltzer the average time for relief would be in a range of "*ave.*" ± 2 minutes (where "*ave.*" is the value the company wants to use in its advertisement and the ± 2 minutes are the 2 sigma or 0.05 limits). The company would like to pick the smallest possible value of "*ave.*" However, it realizes that a value too small will be rejected too often. It decides to pick a value of "*ave.*" that will be rejected only 25% of the time. What value should that be?

*8.21 Fire drills have shown a mean time for evacuation of 4 minutes 15 seconds, with standard deviation of 40 seconds. A new procedure for evacuating is instituted and 6 fire drills are run. If α is set at 0.10, what are the limits for declaring a significant change in time to evacuate has occurred? Now if the average time to evacuate has actually decreased to 3 1/2 minutes, what is the probability of *not* detecting the change (the β error)? What is the consequence? If the new procedure actually increases the time to 4 1/2 minutes, what is the chance we will end up with a sample that shows a significant decrease?

CHAPTER NINE HYPOTHESIS TESTS AND ESTIMATION UNDER AN ASSUMPTION OF NORMALITY

In the last chapter we discussed the concept and format of a statistical hypothesis test. In this chapter we will examine a collection of statistical tests that have one common feature; the source of the data will be considered to be normally distributed.

In some cases the actual population may not be normally distributed, yet the tests will be acceptable. This situation occurs because we are dealing with samples, and we have already observed (in Chapter 7) that the means of many kinds of samples closely approach normal distributions.

In fact, the whole overriding principle behind this chapter (and the next one) will involve the idea of calculating certain statistics (e.g., sample means) and comparing them to a theoretical distribution of these values. The comparison will be our test.

Basically, all of these tests will have a common format. It should be noted at this point that all of the tests in this section may be one-sided or two-sided tests. That is, we may concern ourselves with the possibility of a value being *only greater than* a fixed value or *only smaller than* a fixed value, or we may look *simultaneously* at either possible difference. This consideration affects our limits for particular levels of significance, and we will treat these as separate problems.

Since this chapter is intended to convey a feel for some of the scope of statistical analysis for relatively basic problems, we must first take an overview of the various tests and the ways they are interrelated.

9.1 AN OVERVIEW OF THE CHAPTER

You should realize that there are a number of different ways to classify the set of tests to be presented in this chapter, although we will only examine a few of the comparisons.

We will deal with several kinds of comparisons among *means* (Sections 9.2 to 9.5) and *variances* (Sections 9.6 to 9.7). The differences here I think are obvious. With both quantities, means, and variances, we will deal with single-sample (9.2, 9.3, 9.6) and *two*-sample (9.4, 9.5, 9.7) tests.

One-sample tests involve the comparison of a single-sample value against some hypothesized theoretical population value—such as the defective bottles discussed in the last chapter, the several populations given in Chapter Four, or the distribution of eggs from one supplier.

In this last case, perhaps we have found that the standard size of a large egg is known (by some previous measurements) to be normally distributed with a mean of 2.45 inches. We may now question if the eggs from a particular egg farm are smaller.

We will take a sample of eggs and measure them. Since this is a sample, by now you should expect that with a sample *we will undoubtedly find a difference* between the mean of these eggs and the population mean from which they came. Since we are really interested in the population of eggs from that farm, the real question involves our decision of whether this sample is unlikely to have come from the population with the mean of 2.45.

In Figure 9.1 we have a pictorial histogram that should give you some feel for the proportion of differently sized eggs in the population. Each egg in the diagram represents 2% of the population.

Now remember that you don't know what that distribution actually is. You may surmise it, but you don't know it. A sample like the one in Figure 9.1*b* might be selected from the population—at least we must say that we have little reason to *deny* that it might have come from the proposed population. The sample in Figure 9.1*c*, however, would cause me to be highly suspicious. It would be a very rare case for me to pick, *at random*, only the very small eggs.

Two sample tests deal with the question of whether or not two samples come from sources that are not different (with regard to the particular population value being tested).

Now if we obtain a sample of eggs from a new supplier we are going to ask a different question "are large eggs from the new supplier a different size than large eggs from the current supplier?"

Your approach is to take a sample of the eggs delivered by the new company and then compare these eggs to the ones we measured from our current supplier. We will not compare all the values but, instead, the averages of these two samples (here I am just discussing means, but obviously we can also compare variances). *We will undoubtedly find a difference in the average size of the two samples.*

We may find that the eggs from the current supplier have an average that is 0.05 inches smaller? The real question must be: *Could the two samples have come from the same population*? Is it unreasonable or improbable that two samples taken from a common, single, source would be as different as our samples?

In Figure 9.2 we have the case where two samples are taken from a common population. The samples differ. What we hope is that when we look at the difference between the sample means we can conclude that the

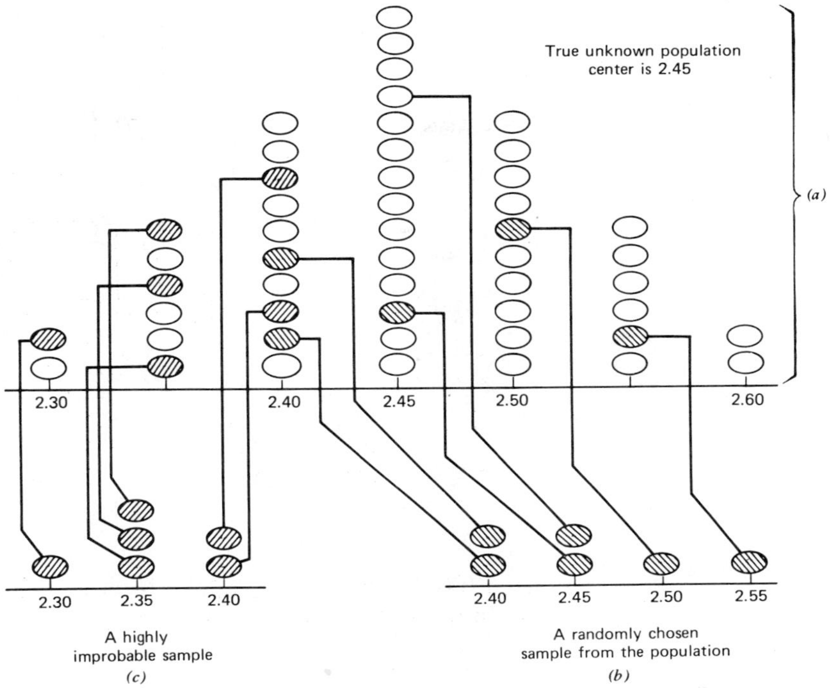

FIGURE 9.1 Taking a sample from a population of eggs that averages 2.45. (*a*) The true population (unknown to you). (*b*) A random sample from the population. Eggs are chosen from *all* sections of the population. (*c*) A very unlikely sample. All eggs are from the bottom section of the distribution. I would not expect such a sample to have come from the distribution in (*a*).

difference is not excessively large. We will declare that there is not a significantly large difference in our egg samples—that these are two typical samples from a single population. Thus, both companies supply eggs of equal size.

The alternate kind of decision that we can make is that there *is* a difference. This difference is *not* that of the *samples*, *but* of *the populations*. We want to determine if we really have two different population sources from which we are taking our samples. For example, Figure 9.3 shows two possible populations. Perhaps one is the source of the current company's eggs and the other is the source of the new supplier's eggs. The two distributions are different. However, the difficulty is that the samples might end up being very similar. The result would be that we would conclude there is no difference—an error on our part.

We will also consider one additional item in the comparison of two samples. In most cases we work with samples that are totally *independent*

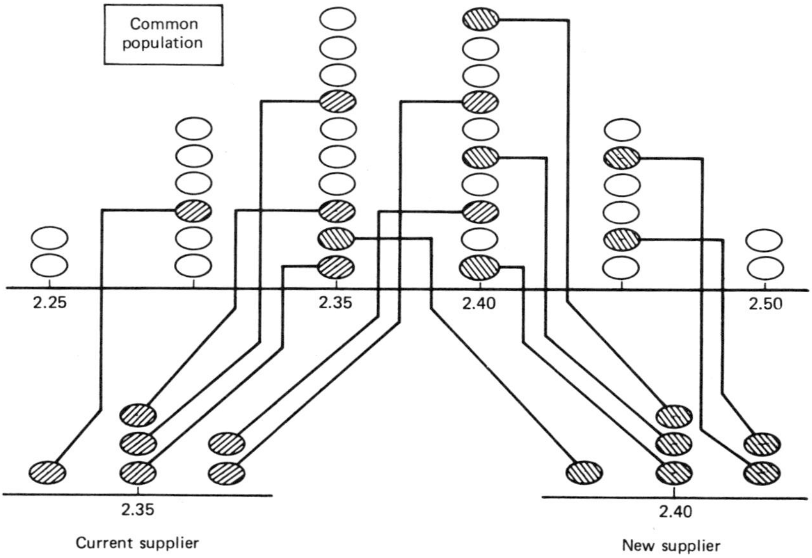

FIGURE 9.2 Two samples from a common population. The averages of the samples are different, but the source is the same.

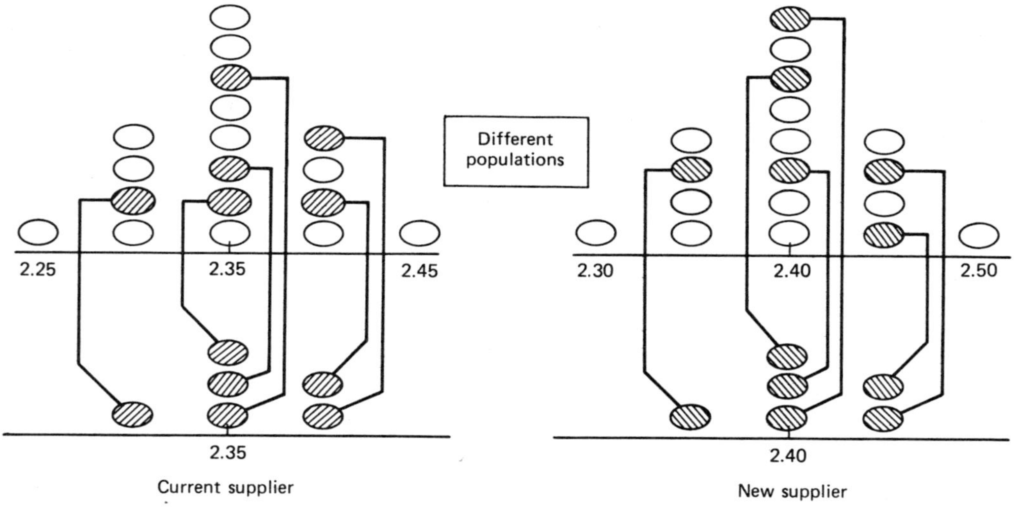

FIGURE 9.3 Two samples from two different populations. The averages here are different, but observe that you *might* possibly have chosen two samples that would be identical, even though the populations were different (e.g., a sample consisting of 2.40, 2.40, 2.40, 2.35, 2.35, 2.35).

of each other. In some cases, however, we have *related* samples. These generally occur when we have have some connection between the individuals of each sample. Here are a few examples of related cases.

Before and after tests, or test-retests, such as: A driving test that measures reflexes is taken both before and after a drug (or alcohol) is administered. Absorbency of towels is measured before and after washing. A group of adults is given a test of attitudes, shown a film, and then retested to see if a change in attitude has occurred. A metal bar is tested for hardness, heat treated, and then retested to see if the hardness has changed.

Matched conditions, such as: Pieces of wood are painted with two different paints to determine which wears best. Two brands of tires are tested on a car at the same time. Differences between sexes may be tested by considering husband and wife, or brother and sister. A control can be chosen with attributes as similar as possible to the subject or individual being tested (this may be important if other factors such as time may also influence the experiment, we assume the control is similarly affected in every way except for the test).

In this text we will only examine the *means* of related samples (not the variances of related samples.)

Another major consideration also applies only to the comparison of means. However, the question here concerns itself with how we obtained the estimate of the variance that must be used in conducting a test for means. In our previous discussions of sampling distributions of means and confidence intervals of means, we always required a value of spread in our calculations. The main question that must be answered is how that value of spread will be determined.

1. It is *given or known* from prior information.
2. It is *unknown* but we will use the *sample variance to estimate* the population variance.

TABLE 9.1 Statistical Tests in This Chapter.

		ONE-SAMPLE TESTS	TWO-SAMPLE TESTS	
			INDEPENDENT	RELATED
Means	Large sample or variance known	z-test Section 9.2	z for difference in means Section 9.4	
	Small sample and variance unknown	t-test Section 9.3	t for difference in means Section 9.4	Paired t-test Section 9.5
Variance		χ^2-test Section 9.6	F-test Section 9.7	

However, the only true difference is in the name of the test statistic and the table we use to find the limit of the test statistic. The general format will be the same.

The distinction between these two methods is often also made on the basis of the sample size. If the sample size is large (say 50 to 100), even if the variance is unknown we will be able to treat the estimate as though it were the true variance. Hence we use the term *large sample* tests for the first case and *small sample* tests for the second case.

Table 9.1 provides a summary of the various tests in this chapter and includes the name and section in which it is discussed. We should also note that the format for the sections will be: hypothesis, alternate hypothesis, limit(s), test statistic, and decision (and confidence interval where applicable).

9.2 COMPARING THE MEAN OF A SAMPLE TO A THEORETICAL OR HYPOTHESIZED MEAN—VARIANCE KNOWN

When we discussed the central limit theorem and confidence intervals in Chapter Seven, we arrived at the observation that the means of samples were normally distributed. The standard deviation of the means, the standard error, was $\sigma/\sqrt{n}$ where σ was the true standard deviation of the population and n is the size of the samples. We now also have a hypothesized population mean, which we label as μ. Given this information, we can describe the distribution of samples, which is a normal distribution.

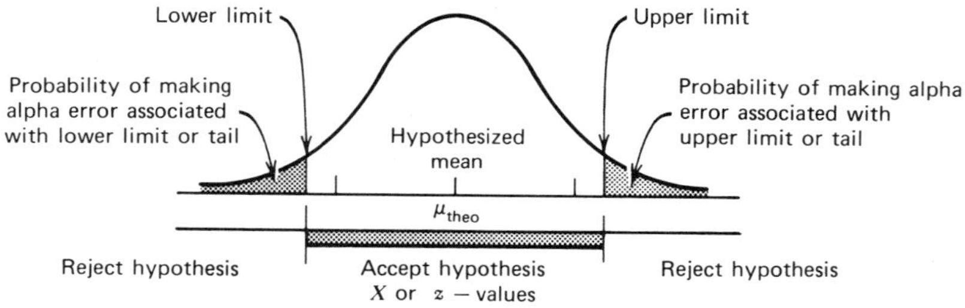

FIGURE 9.4 . Interpreting the limits and the errors for a distribution of sample means.

We will use a standardized normal distribution and therefore we will be able to determine easily the limits for which specific percentages would be in the tails (see Table A.1b). The percentage in the tails of this normal distribution is our probability of making an alpha error (also called the alpha level)—see Figure 9.4. Next we calculate the z-value for the sample values that fall in the area of the curve that is rare, the rejection region (the region in which our decision is to reject the hypothesis and accept the alternate hypothesis).

EXAMPLE 9.1

The manager of quality control for a supermarket chain was to check the meat scale in a supermarket. A sample of five 2-pound steaks were weighed on a special scale and the weights were found to be

1.96, 2.02, 1.92, 1.94, 1.97

If this type of scale is *known* to have an error in measurement that is equal to a σ of 0.04 pound, is there a significant difference between these values and a true mean of 2.00? Consider an α of 0.05.

HYPOTHESIS The population mean of the readings from this scale, μ, is not significantly different from the hypothesized mean, called μ_{theo}. This is also stated as $\mu = \mu_{\text{theo}}$ or $\mu = 2$.

ALTERNATE HYPOTHESIS The sample comes from a population whose mean, μ, is *greater than or less than* μ_{theo}: $\mu \neq \mu_{\text{theo}}(\mu < \mu_{\text{theo}}$ or $\mu > \mu_{\text{theo}})$.*

GIVEN $n = 5$, sample mean $= \overline{X} = 1.962$, and $\mu_{\text{theo}} = 2.00$.

LIMITS

$$\text{Upper:} \quad z_{1-\alpha/2} = z_{0.975} = 1.96$$

$$\text{Lower:} \quad z_{\alpha/2} = z_{0.025} = -1.96$$

It is important that you keep in mind the difference between the $\overline{X}$ of 1.962 and the limit of 1.96 (or 1.960). These are completely different kinds of quantities and their similarity is only coincidental.

TEST STATISTIC

$$z = \frac{\overline{X} - \mu_{\text{theo}}}{\sigma/\sqrt{n}} = \frac{1.962 - 2.00}{0.04/\sqrt{5}} = \frac{-0.038}{0.04/2.236} = -2.124$$

DECISION The test statistic, z, is outside the lower limit, therefore *we reject the hypothesis* that this scale accurately gives an average weight of 2.00 pounds.

The consequence of rejecting this hypothesis and accepting the alternate hypothesis, that the scale is not reading correctly, might lead to different results, depending on the type of difference that occurred. If the readings are low and are then followed up by an inspector from the city (e.g., from the Weights and Measures Department), a large fine might result. If the readings are too high there could be a loss of profit with which the store might not be pleased. Refer to Figure 9.5 for a graphic illustration of this example.

* The symbol $\neq$ means "does *not* equal," $<$ means less than $(3 < 4)$, and $>$ means greater than $(6 > 2)$.

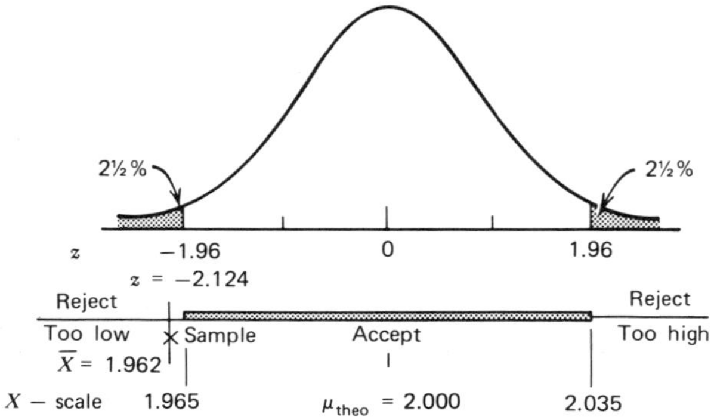

FIGURE 9.5 A two-sided z-test for a hypothesized mean of 2.000 and an alpha of 0.05. The sample mean from Example 9.1 is shown to fall in the *rejection region*. Therefore we reject the hypothesis that on the average the scale accurately gives the weight of 2-pound steaks.

We can also determine the confidence interval for the mean of the population of weights.

$$\bar{X} + z_{\alpha/2}\frac{\sigma}{\sqrt{n}} < \text{true mean} < \bar{X} + z_{1-\alpha/2}\frac{\sigma}{\sqrt{n}}$$

$$1.962 + (-1.96)\frac{0.04}{\sqrt{5}} < \mu < 1.962 + (1.96)\frac{0.04}{\sqrt{5}}$$

$$1.962 - 0.035 < \mu < 1.962 + 0.035$$

$$1.927 < \mu < 1.997 \tag{9.1}$$

Note that this interval does not actually include the hypothesized mean, although the upper limit is very close to the hypothetical mean of 2.00.

In this example we have looked at the hypothesis from both directions. We approached the question of whether the weights were *either* greater than or less than the hypothesized mean weight. This type of test is called a *two-sided* or *two-tailed* test.

Not in all cases, however, would we be interested or concerned about possible differences in *both* directions (increases and decreases). For example, an inspector from the Consumer Protection Agency might do the checking. In this case the inspector probably does not care if the store is generous in giving more meat, but is very concerned about underweight packages. Thus the test would now just have a rejection region for the lower values. All the area associated with the probability of an alpha error would be in the one lower tail. Thus we would have a *one-sided* or *one-tailed* test (see Figure 9.6*a*).

If, however, we could test the weights from the point of view of an unscrupulous store manager or butcher, we now find the only concern is that the weights are not too large and therefore only the upper tail is of

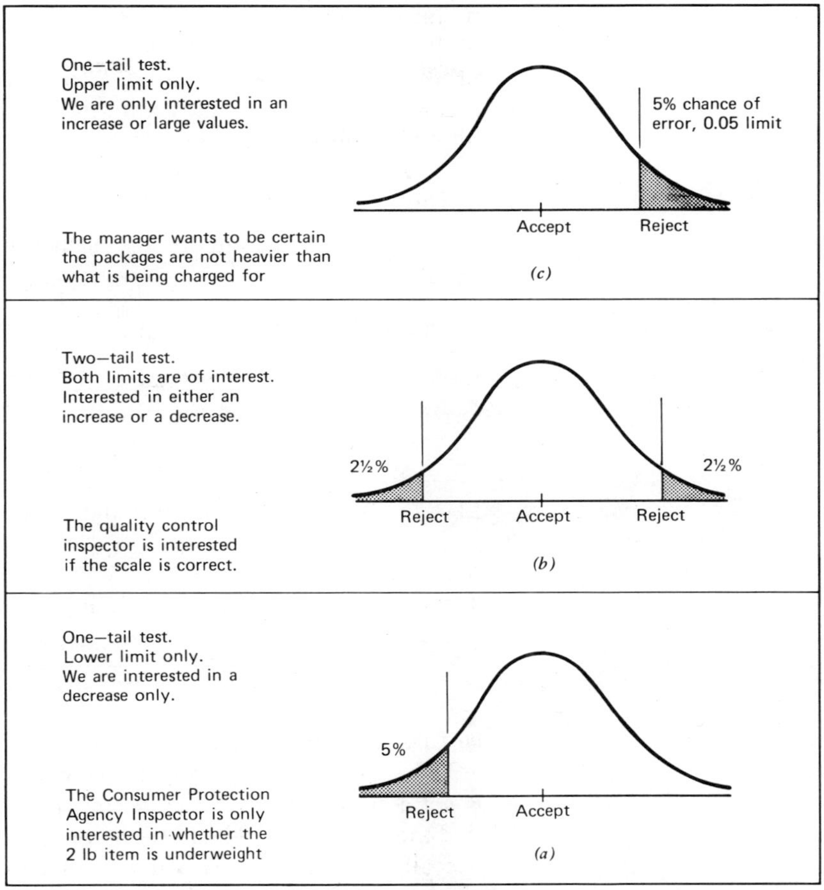

FIGURE 9.6 Comparing one- and two-tailed tests and the limits for accepting or rejecting a hypothesis. All cases are totals of 5% chance of error. (*a*) and (*c*) are one-tail tests, and (*b*) is a two-tailed test.

interest. Figure 9.6*c* illustrates this case. For comparison, Figure 9.6*b* shows our quality control person's view and the two-tail test. Observe that when we have a two-tail or two-sided test we find limits, so that half the error is in each tail.

A one-sided test

We will now examine a one-sided test. In this case we will only compute a single limit, and the alternate hypothesis determines which limit that would be. We look at an example in which we will only be concerned about a change to a smaller population.

EXAMPLE 9.2

A pollution control device for automobiles was being tested for long-term use. It has been known from previous tests that a particular type of pollutant is present in automobile exhaust with an average of 90 parts per million and a standard deviation of 25. Ten tests with the device had produced an average exhaust level of 75 parts per million. The company is not ready to begin marketing the device until it is 99% sure that an improvement in lowering the emission level is evident. (Therefore we allow a 1% chance of Type I error, or $\alpha = 0.01$.)

This is a one-sided test and σ is known.

HYPOTHESIS The population mean, μ is 90.

$$(\mu = \mu_{\text{theo}} = 90)$$

ALTERNATIVE The population mean is *less than* 90.

$$(\mu < \mu_{\text{theo}} \quad \text{where} \quad \mu_{\text{theo}} = 90)$$

GIVEN

$$n = 10, \quad \sigma = 25, \quad \bar{X} = 75, \quad \alpha = 0.01.$$

LIMIT We only use one limit and the full α is associated with this limit.

$$z_\alpha = z_{0.01} = -2.33$$

TEST STATISTIC There is *no* change in how we compute this value.

$$z = \frac{\bar{X} - \mu_{\text{theo}}}{\sigma/\sqrt{n}} = \frac{75 - 90}{25/\sqrt{10}} = \frac{-15}{25/3.162} = -1.90.$$

DECISION Since the test statistic of -1.90 is larger than the limit of -2.33, it is in the accept region (see Figure 9.7), and we declare that there is *not* a significant decrease in emission level. We accept the hypothesis that the mean is still 90 and the company will keep working on the device. (Recall, though, there is still a chance of making a Type II error. The alternate hypothesis may be true, but we haven't been able to recognize it or to reject the hypothesis at the level of significance we desired.)

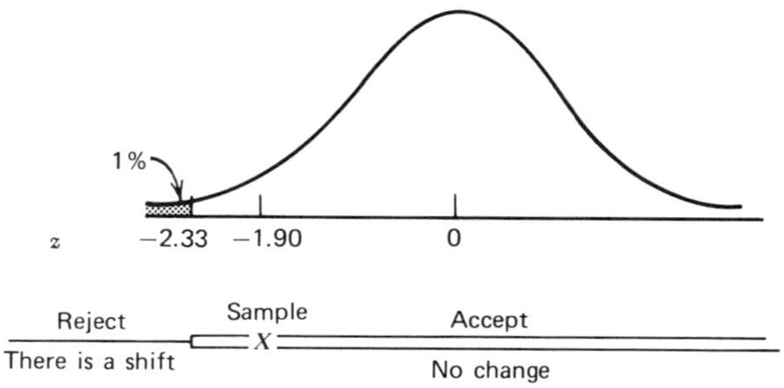

FIGURE 9.7 A one-sided z-test for an alpha of 0.01. This is for testing a hypothesis of no change in population mean to a smaller value. The sample mean from Example 9.2 is shown to fall in the accept region. We therefore state that we do not have evidence to reject the hypothesis of no decrease in emission level.

You should keep in mind that the only major difference in setting up a one-sided test where you would be concerned with an *increase* in the mean would be that the limit would be positive ($z = +2.33$ in this case) and the rejection region would be in the upper area of the curve.

9.3 COMPARING THE MEAN OF A SAMPLE TO A THEORETICAL POPULATION MEAN—VARIANCE UNKNOWN: *t*-TEST

When we worked with a normal distribution we *assumed* that we knew what the population variance or standard deviation was—and we assumed it was known without error (it does not vary). Now we are going to assume that we do not have knowledge of the population standard deviation. Instead, we are going to use the standard deviation *of the sample* to estimate the population standard deviation. Because we are now working with an estimation, we need to recognize that this will create inexactness in whatever other comparisons we work with.

Furthermore, the smaller the sample, the poorer our estimate of variance will be. If the sample is large enough the estimate will approach the true variance, and we should end up with a normal distribution.

The distribution called the *t-distribution* is a symmetric distribution that resembles the normal distribution and can be shown to be associated with the distribution of sample means. Figure 9.8 shows several *t*-distributions. Observe that the *t*-distributions are distinguished by the value called degrees of freedom, *df*, which generally will be $n - 1$ (where n is the number of values in the sample). The *t*-distribution has a larger spread than the normal distribution but, as the value called degrees of freedom increases, the *t*-distribution becomes closer and closer to the size of the normal distribution.

Since this distribution and several other distributions that we will discuss later require us to work with the quantity "degrees of freedom," it seems appropriate at this time to make a few comments about the term.

Basically, it is a value that tells how many of the quantities may be independently chosen. For example, if I am allowed to pick any 5 numbers, then there is *no* restriction on *any* of the numbers. Thus, I might choose 1, 3, 5, 2, and 6. Since all of them have the freedom to vary, we say that the

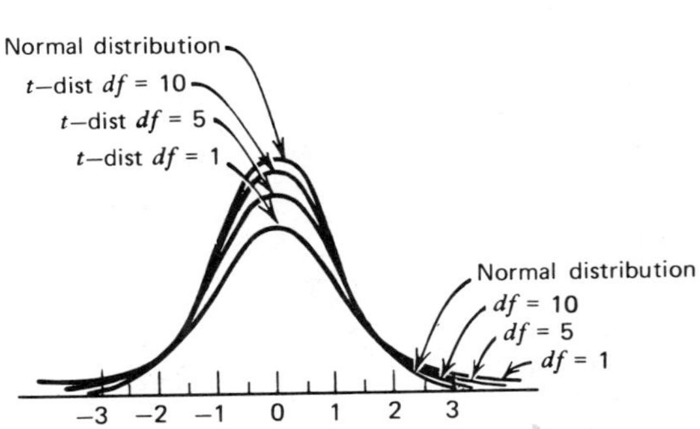

FIGURE 9.8 Several *t*-distributions and their relationship to the normal distribution. The quantity *df* is degrees of freedom.

total number of values that may freely vary, the degree of freedom, is 5 (one for each number I was able to freely pick).

If we add a restriction such as the sum of the numbers is to be 18, I can no longer freely pick *all* the individuals. If I were to pick the first 4 values in the same manner as before (1, 3, 5, 2), in order to obtain a sum of 18, the 5th value *must* be 7. So we now only have 4 values that may be freely chosen, and we state that the degrees of freedom are 4.

In the case of calculating sample variances (which is involved in discussing the t-values) we end up with $n - 1$ degrees of freedom. There are several ways of showing this but perhaps the easiest to think of is the fact that the sample variance can be expressed as the sum of the squares of the differences between the individual values and the sample mean. Thus there are n differences. But there is one restriction; the sum of these differences is *zero*.* Therefore, after you obtain $n - 1$ of the differences, the last difference cannot be freely chosen. Thus, we have $n - 1$ degrees of freedom.

As with the normal distribution, we will work with a table of values (Appendix Table A.2). Since we have such a large variety of possible t-distributions (one for *each* value of degrees of freedom), the table contains only selected critical values. In addition, just as we "standardized" the many possible normal distributions, we do a similar operation with the t-distribution. All t-distributions in the table have a center of 0 and all values along the X-axis are considered as t-values (just as we had called them z-values for the normal distribution). Thus, for 4 degrees of freedom, we have Figure 9.9. For example, the $t_{0.975}$ says we are looking at the t-value below which 97.5% of the possible cases occur.

Similarly, 2.5% are above this point (called $t = 2.776$) and we say that there is a probability of 0.025 for a value to be in this tail. In Appendix Table A.2 you will find this value under the "level of significance for one-tailed tests" of 0.025. Two-tailed cases are those cases where we are interested in the area between $\pm t$. In these cases we use the line labeled *level of significance for two-tailed test* to determine what are the limits. Thus, if $t = 2.776$, 2.5% of the values are above $+t$ and 2.5% are below $-t$ (see Figure 9.9), or a total of 5% are outside the limits and 95% are inside the limits.

The test statistic we calculate is called a t-statistic and is found as

$$t = \frac{\overline{X} - \mu_{\text{theo}}}{s/\sqrt{n}} \tag{9.2}$$

Note that the *only* difference between this and the z-statistic is the use of s in place of σ and the t instead of z.

As an example, let's redo Example 9.1.

* If, for example, you go back to Example 3.2 on pages 62 and 63, and add up the $f_i(X_i - \overline{X})$s, you will find that they sum to zero (roundoff error considered). See if you can demonstrate why it must be true in all cases.

df		$t_{0.975}$	
		Level of significance-one tail	
		0.025	
4		2.776	

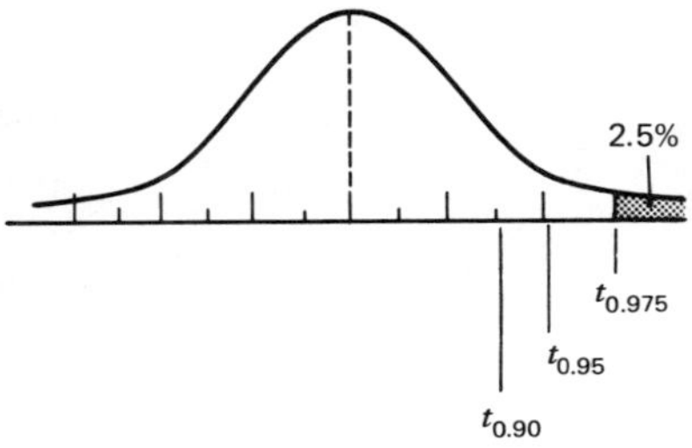

FIGURE 9.9 *t*-distribution with 4 degrees of freedom. Finding the *t*-value for which 2.5% are above (in one tail). This is the same point that 97.5% are below and therefore is labeled as $t_{0.975}$.

EXAMPLE 9.1 (Cont.)
This time we are to consider the σ-value as unknown. We still have $\overline{X} = 1.962$, $\mu_{\text{theo}} = 2.00$, and $\alpha = 0.05$. The hypothesis and alternate hypothesis are unchanged.

We calculate s from the sample and find that $s = 0.038$. Degrees of freedom $= df = n - 1 = 4$.

LIMITS We again have 2 limits (found under $df = 4$); see Figure 9.10.

$$\text{Upper:} \quad t_{1-\alpha/2} = t_{0.975} = 2.776$$

$$\text{Lower:} \quad t_{\alpha/2} = t_{0.025} = -2.776$$

TEST STATISTIC

$$t = \frac{\overline{X} - \mu_{\text{theo}}}{s/\sqrt{n}} = \frac{1.962 - 2.00}{0.038/\sqrt{5}} = \frac{-0.038}{0.038/2.24} = -2.24$$

DECISION The *t*-value is between the two limits. Therefore, we accept the hypothesis that there is no difference between the sample mean and a theoretical mean of 2.00.

Thus our decision is different than before (we will shortly have one explanation of why).

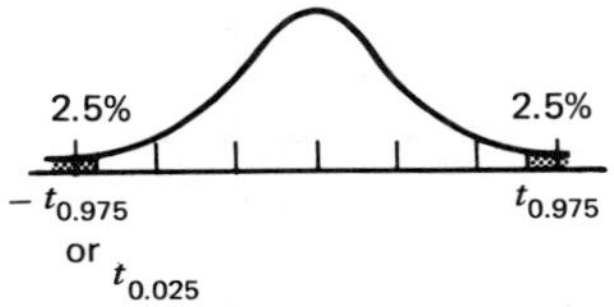

FIGURE 9.10 *t*-distribution showing how two-tailed limits are determined. Here the total area that is outside the limits is 5%.

CONFIDENCE INTERVAL

$$\bar{X} + t_{\alpha/2}\frac{s}{\sqrt{n}} < \mu < \bar{X} + t_{1-\alpha/2}\frac{s}{\sqrt{n}}$$

$$1.962 - 2.776\frac{0.038}{\sqrt{5}} < \mu < 1.962 + 2.776\frac{0.038}{\sqrt{5}}$$

$$1.962 - 0.047 < \mu < 1.962 + 0.047$$

$$1.915 < \mu < 2.009$$

Observe that even though the standard deviation is smaller than the previously given theoretical standard deviation (it was stated as being 0.04), the confidence interval and the test limits are much larger than in the z-test. This is because with small sample sizes t is much larger than z for a given α level. Note, however, as the sample size increases, the t-values and z-values become more and more alike (mathematically we say that the t-values approach the z-values) and s becomes more like σ, so that the tests become essentially identical.

Let us now examine a one-sided test.

EXAMPLE 9.3

The job placement department at a large college issued a statement that the starting salaries for graduates of a particular program average \$10,000 or more per year.

Before entering the program, you take a random sample of 7 students who recently received job offers. The salaries (in dollars) are: 9500, 10,000, 9700, 9600, 10,100, 9800, and 9200. You are willing to take no more than a 1 in 20 risk (5%) that the placement departments assertion is correct before committing yourself to the program. Therefore, we set the alpha level at 0.05. Since we are only concerned with whether the salaries are lower, this is a one-sided test.

HYPOTHESIS The salaries offered are not different than \$10,000.

$$(\mu = \mu_{\text{theo}} = 10{,}000)$$

ALTERNATIVE HYPOTHESIS The salaries are lower, the true population mean is less than \$10,000 ($\mu < \mu_{\text{theo}}$).

GIVEN $n = 7$, degrees of freedom $= df = n - 1 = 6$, $\bar{X} = 9700$, $s = 306$, $\alpha = 0.05$

LIMIT There is only one limit, a lower limit.

$$\text{Lower } t_{\alpha/2} = t_{0.05} = -1.943$$

TEST STATISTIC

$$t = \frac{X - \mu_{\text{theo}}}{s/\sqrt{n}} = \frac{9700 - 10{,}000}{306/\sqrt{7}} = \frac{-300}{306/2.646} = -2.59.$$

DECISION Since this test statistic (-2.59) is below the limit (-1.943), we reject the hypothesis of no change and accept the alternate. We declare that the salaries actually offered are less than the \$10,000 amount claimed by the placement office (see Figure 9.11).

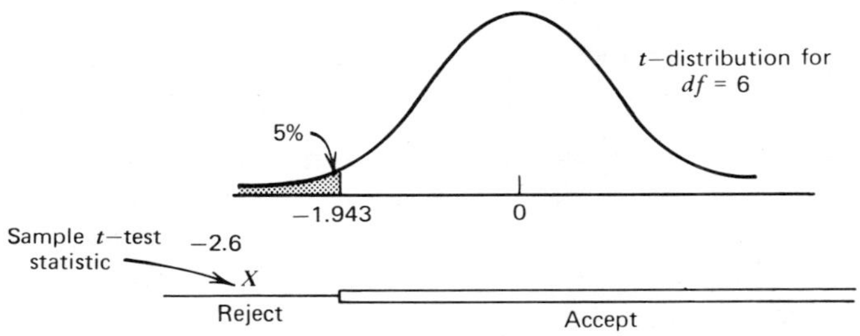

FIGURE 9.11 Acceptance and rejection region of t-distribution for Example 9.3. The sample mean of -2.6 falls in the rejection region.

9.4 DIFFERENCE IN THE MEANS OF TWO INDEPENDENT SAMPLES

One of the most common types of problems encountered concerns the question of whether there are differences in the means of two populations. This section will present several tests for cases when two independent samples are drawn from distributions that are relatively normal (or else are moderately large samples). Actually, one basic test is being employed here. The only distinctions between the several variations to be discussed have to do with what we use for standard deviation. In particular we need to distinguish between situations of the following types.

1. We know or do not know what the standard deviation is.
2. We consider the standard deviations of the two populations to be equal or not equal.

The first question establishes the limit value and the second provides us with the test statistic.

1. If we *know the standard deviation* the limit is a *z-value* and we have a *z-test*. If we do *not* know the population standard deviation, we must estimate it from the samples. Then if the sample size is small, we use a t-limit and call this a t-test.
2. If the standard deviation is to be the same for both samples the z-test is somewhat simplified, as we only have one value of σ. For a t-test we should first check to see if the variations are essentially the same (by

using an F-test—see Section 9.7). If the test of variances shows no differences,* we usually go ahead and average them together into what we call a pooled variance—a single estimate of σ. If they are different we have some complications in the calculations.

Here we have a collection of similar tests, all following a common format. The main differences are in the limits and the test statistics for each case.

HYPOTHESIS There is no difference between the population means (μ_1, mean of the first population $= \mu_2$, mean of the second population).

ALTERNATIVE HYPOTHESIS There is a difference between the population means.

1. If this is a two-sided test, we say $\mu_1 < \mu_2$ or $\mu_1 > \mu_2$.
2. If this is a one-sided test, we say $\mu_1 > \mu_2$, and we choose sample 1 to be the sample that we believe might come from the population with the larger mean.

Significance level, α, is to be chosen. $\bar{X}_1$ and $\bar{X}_2$ are the means of the first and the second samples, respectively.

TEST STATISTIC Table 9.2 provides a general listing of test statistics and the appropriate limits.

DECISION As before, these depend on whether we have a one- or two-tailed test.

1. *Two-tailed or two-sided test.* If our test statistic is between the two limits we accept the hypothesis of no difference between the means. If the test statistic is outside either limit we accept the alternate hypothesis.
2. *One-tailed or one-sided test.* If our test statistic is less than the limit we accept the hypothesis of no difference. If the test statistic is more than the limit we accept the alternate hypothesis that there is a difference.

EXAMPLE 9.4

Students were randomly selected and placed into two classes. Two different teaching techniques, one traditional and one experimental, were then tried with the hope that the students in the experimental group would progress faster.

A standardized test was given to the two classes of students to determine if there was a difference between the groups. The test has a known standard deviation of 25.

* It should be noted that the reason the test of variances will show no differences *might* be just that we do not have a large enough sample. In fact, if the sample size is large enough, we should *expect* to obtain a significant difference. In such cases we also need to consider whether the difference is great enough to us to show concern.

TABLE 9.2 Table of Test Statistics and Limits for a Variety of Different Parametric Tests for the Difference of Two Means

WHAT IS KNOWN	TEST STATISTIC		ONE- OR TWO-SIDED TEST	LIMIT(S)
Standard deviations known	Standard deviations equal	$z = \dfrac{\bar{X}_1 - \bar{X}_2}{\sigma\sqrt{\dfrac{1}{n_1} + \dfrac{1}{n_2}}}$	Two	Upper $= z_{1-\alpha/2}$ Lower $= z_{\alpha/2}$
	Standard deviations *not* equal	$z = \dfrac{\bar{X}_1 - \bar{X}_2}{\sqrt{\dfrac{\sigma_1^2}{n_1} + \dfrac{\sigma_2^2}{n_2}}}$	One $(\bar{X}_1 > \bar{X}_2)$	$z_{1-\alpha}$
Standard deviations *not* known but assumed equal		$t = \dfrac{\bar{X}_1 - \bar{X}_2}{s_p\sqrt{\dfrac{1}{n_1} + \dfrac{1}{n_2}}}$	Two	Upper $= t_{1-\alpha/2}$ Lower $= t_{\alpha/2}$ $df = n_1 + n_2 - 2$
	where	$s_p^2 = \dfrac{(n_1 - 1)s_1^2 + (n_2 - 1)s_2^2}{n_1 + n_2 - 2}$	One	$t_{1-\alpha}$ $df = n_1 + n_2 - 2$

In this text we will not consider the case of standard deviations not known and not equal.

n_1 = size of sample 1, n_2 = size of sample 2
$\bar{X}_1$ = mean of sample 1, $\bar{X}_2$ = mean of sample 2
σ_1 = standard deviation of population 1 σ_2 = standard deviation of population 2
s_1 = standard deviation of sample 1 s_2 = standard deviation of sample 2
σ = common population standard deviation

The traditional group had 22 students and a mean score of 127 was recorded. The experimental group had 18 students and scored an average of 136. If there is a significant increase we will attempt to use the experimental method on a broader scale. An α level of 0.10 is chosen.

HYPOTHESIS There is no difference between techniques. ($\mu_{\text{experimental}} = \mu_{\text{traditional}}$).

ALTERNATIVE HYPOTHESIS There is a difference, the experimental group does better ($\mu_{\text{experimental}} > \mu_{\text{traditional}}$).

LIMIT Since the standard deviation is known, this is a z-test. It is also a one-tailed test, therefore

$$\text{Limit} = z_{1-\alpha} = z_{0.90} = 1.64$$

TEST STATISTIC Since the standard deviations are equal for both tests, we use the formula

$$z = \frac{\overline{X}_1 - \overline{X}_2}{\sigma\sqrt{\dfrac{1}{n_1} + \dfrac{1}{n_2}}} = \frac{136 - 127}{25\sqrt{\dfrac{1}{18} + \dfrac{1}{22}}}$$

$$= \frac{9}{25\sqrt{0.10}} = 1.13$$

DECISION Since the test statistic is smaller than the limit, we conclude that there is no significant increase (we do not reject the hypothesis). The experimental method will be discontinued or more data should be collected before making the decision. It must be emphasized again that accepting this hypothesis does *not* constitute a proof that it is true.

EXAMPLE 9.5

In deciding what type of paving surface to use on a highway one consideration is the stopping distance for cars. A number of tests were conducted on two types of surfaces by measuring the distance it took for cars traveling at 50 miles per hour to come to a complete stop. The results in feet were as follows.

ASPHALT A				HARD SURFACE P			
127	108	136	119	129	118	144	117
121	126	142	134	119	129	151	130
$\overline{X}_1 = 126.6, s_1^2 = 116.6$				$\overline{X}_2 = 129.6, s_2^2 = 153.1$			

n_1 and n_2 are equal to 8 and we have pooled or averaged the variances to obtain

$$s_p^2 = \frac{(8-1)116.6 + (8-1)(153.1)}{8+8-2} = \frac{816.2 + 1071.7}{14} = \frac{1887.9}{14} = 134.8$$

The limit to be used depends on whether σ is known and if this is one- or two-sided test. See Table 9.2. You should also realize that the general case of the test statistic is

$$\frac{\overline{X}_1 - \overline{X}_2}{\text{Standard deviation value}}$$

The standard deviation value, the standard deviation of the difference, is also called the standard error of the difference. The particular cases are found in Table 9.2.

HYPOTHESIS There is no difference in stopping distance between the pavement surfaces.

ALTERNATIVE HYPOTHESIS There is a difference between these pavement surfaces. This is a two-sided test ($\mu_1 < \mu_2$ or $\mu_1 > \mu_2$).
Significance level is chosen as 0.05.

LIMITS Standard deviations are considered unknown, therefore this is a t-test. We will consider that the variances are essentially equal. Therefore degree of freedom $df = n_1 + n_2 - 2 = 8 + 8 - 2 = 14$.

$$\text{Upper limit} \quad t_{1-\alpha/2} = t_{0.975} = 2.145$$
$$\text{Lower limit} \quad t_{\alpha/2} = t_{0.025} = -2.145$$

TEST STATISTIC

$$s_p = \sqrt{134.8} = 11.6$$

$$t = \frac{\bar{X}_1 - \bar{X}_2}{s_p\sqrt{\dfrac{1}{n_1} + \dfrac{1}{n_2}}} = \frac{126.6 - 129.6}{11.6\sqrt{\dfrac{1}{8} + \dfrac{1}{8}}} = \frac{-3.0}{11.6\sqrt{\dfrac{2}{8}}} = -0.517$$

DECISION Since the test statistic is *between* the limits, we accept the hypothesis that there is *no* difference between the pavements.

9.5 A STATISTICAL TEST FOR COMPARING TWO RELATED SAMPLES: PAIRED *t*-TEST

If we gave a reading test to a group of 9th graders we would find that the distribution of reading grades would have a large variance. If we then selected another group of 9th graders at a later time and gave them a similar test, we would find that the variance of their reading scores would also be large. In addition, because of these very broad distributions, we would probably be unable to detect any significant change unless the samples were very large. Even then, we might question the practical significance of such a difference.

In Table 9.3 we have a pair of samples of the reading scores of 10 students given at two different times. A t-value for the difference in two means has been calculated and is not significant. In Figure 9.12 we see the histograms for the two samples. You may detect what appears to be a difference, but remember that these are *small* samples and samples are expected to differ.

Now let us consider what might happen if we were to match up the students at the two times or test the same individual at the two times in the 9th grades. You might expect that there should be a very high relationship

TABLE 9.3 Reading Grade Levels of a Group of 9th-Grade Students. Tests Taken at Two Different Times

Beginning of the year	9.4	10.3	8.4	6.8	7.8	9.8	9.2	11.2	9.4	9.0
End of the year	9.3	10.6	8.8	7.0	7.7	10.0	9.8	11.7	9.7	9.0
Difference	−0.1	0.3	0.4	0.2	−0.1	0.2	0.6	0.5	0.3	0.0

$\bar{X}_{\text{beg}} = 9.13$. $\bar{X}_{\text{end}} = 9.36$, $s_p = 1.30$, $t = 0.396$

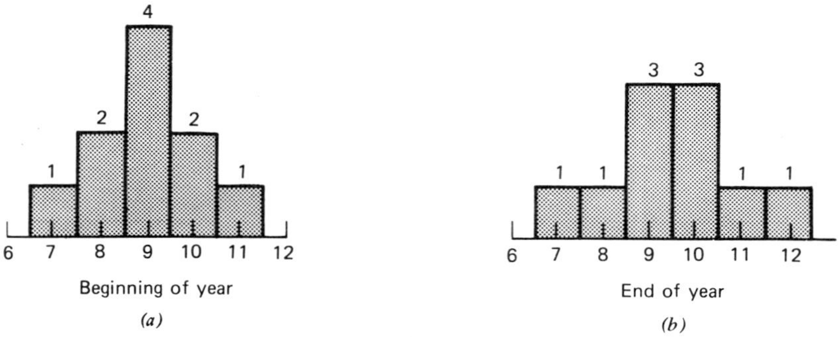

FIGURE 9.12 Histograms for reading grade test scores at the beginning of the ninth grade and at the end of the ninth grade.

between reading levels at these two times. A student who was a poor reader in the beginning of the 9th grade is probably still a poor reader at the end of the 9th grade, and a good reader will probably stay good relative to the overall class. We no longer have independent samples. Look at Table 9.3 again; this time compare the score for each student the first time with the score below it at the later time. Is there a pattern? Twice as many students show an increase versus those decreasing. Is the pattern significant?

We will look at the change for each student. In this case we can:

1. Calculate actual differences for *each* of the pairs or related values.
2. Consider the values (the differences) to be approximately normally distributed.

Under these circumstances we can calculate the mean and standard deviation of the *differences* and then apply a t-test. Our hypothesis is that if there is no *real overall* difference, then the average of the differences should be zero. So we will conduct a t-test to determine if there is a significant difference between the mean of these differences and a hypothetical mean of zero ($\mu = 0$).

The *alternate hypothesis* is that the mean difference is *not* zero ($\mu \neq 0$).

For a one-sided test choose the *order* of subtraction so that the expected difference will be positive.

For a two-sided test we consider both differences.

We will consider a one-sided test with a significance level of $\alpha = 0.025$.

The mean of the differences

$$\bar{d} = \frac{2.3}{10} = 0.23$$

and the standard deviation s_d for the 10 differences is 0.241. Also, n = number of pairs = 10.

LIMIT $df = n - 1 = 9$ (one-sided test).*

$$t_{1-\alpha} = t_{0.975} = 2.262$$

TEST STATISTIC

$$t = \frac{\bar{d}}{s/\sqrt{n}} = \frac{0.23}{0.241/\sqrt{10}} = \frac{0.231\sqrt{10}}{0.241} = \frac{0.23(3.16)}{0.241} = 3.02$$

DECISION Since $t = 3.02$ exceeds the limit of 2.262, we accept the alternate hypothesis that there *is* a difference (see Figure 9.13).

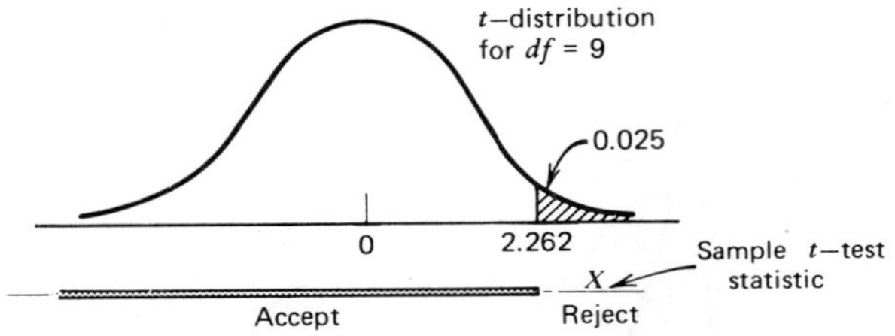

FIGURE 9.13 Acceptance and rejection region for the paired t-test example.

9.6 COMPARING A VARIANCE TO A THEORETICAL VARIANCE

In these next two sections we will examine variances, first by comparing a sample value to some theoretical value and then by comparing variances from two samples to determine if there are differences in the population variances. In both of these cases we will encounter new and very different test statistics. The statistic we will work with in this section is called the chi-square value and is symbolized as χ^2.

The χ^2-distribution is a distribution we encounter when we deal with collections of values that involve adding or summing up squares. Variances of samples, which after all require you to add a collection of squared quantities, have distributions that are related to χ^2-distributions.

If we take each one of a collection of sample variances, divide them by the known (or theoretical) population variance, and multiply these quotients by $n - 1$ (where n is the sample size), we will obtain a χ^2-distribution. Thus

$$\frac{s^2}{\sigma^2}(n - 1) = \frac{s^2}{\sigma^2}(df)$$

would have the same distribution as a χ^2-distribution with $n - 1$ degrees of freedom.

Since the larger the sample the closer the sample variance and the true variance would become, we would expect that the χ^2-distribution would change with sample size. Again we relate sample size and degrees of freedom. However, since there are other applications of the χ^2-distribution where degrees of freedom are not connected to sample size, the degrees of

* In a two-sided test you need 2 t-values $-t_{1-\alpha/2}$ and $t_{\alpha/2} = -t_{1-\alpha/2}$.

freedom notation is even more important with the χ^2-distribution than it was with the t-distribution.

The χ^2-distribution is not symmetrical and all the values are positive. Again we need to know the degrees of freedom, since for different degrees of freedom we have different curves. The smaller the number of degrees of freedom the more skewed the distribution is (see Figure 9.14). Appendix Table A.3 gives selected critical values of χ^2 for several different numbers of degrees of freedom. Note that the χ^2-values are the quantities recorded on the X-axis, and the values in the table are *areas below that value.* (See Table A.3 for an explanation on its use.)

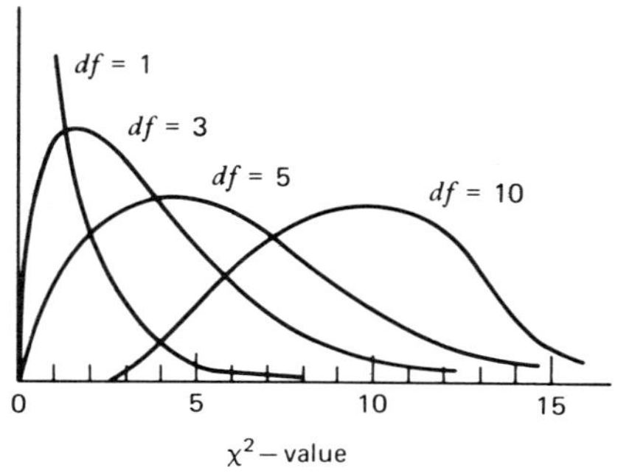

FIGURE 9.14 Some χ^2-distributions and the relationship of degrees of freedom (df).

It must also be pointed out that this test *assumes that the population is normally distributed.* If the population is not normal this test may make incorrect decisions much more frequently than the alpha level leads us to expect. Some statistical tests react more than others to differences in the kind of distribution we have. For example, the t-test for means is relatively insensitive to different distribution shapes.

EXAMPLE 9.6

Gravel that is used on a road should have a certain amount of variation, yet not too large a variation. Most of the stone to be used is about 3/4 inch diameter with some finer and some coarser stone included. The theoretical standard deviation of gravel from the Crushed Stone Corp. quarry is 0.20 inches. Thus, $\sigma^2_{\text{theo}} = (0.20)^2 = 0.04$. A sample of 25 stones had a variance of 0.10. Is there a significant difference in the variability of this sample and the typical stones?

GIVEN Sample size $n = 25$, $df = n - 1 = 24$, $\sigma^2_{\text{theo}} = 0.04$, $s^2 = 0.10$. A significant difference generally implies at least an α of 0.05 (a 5% chance of error or a 1 in 20 chance of a Type I error).

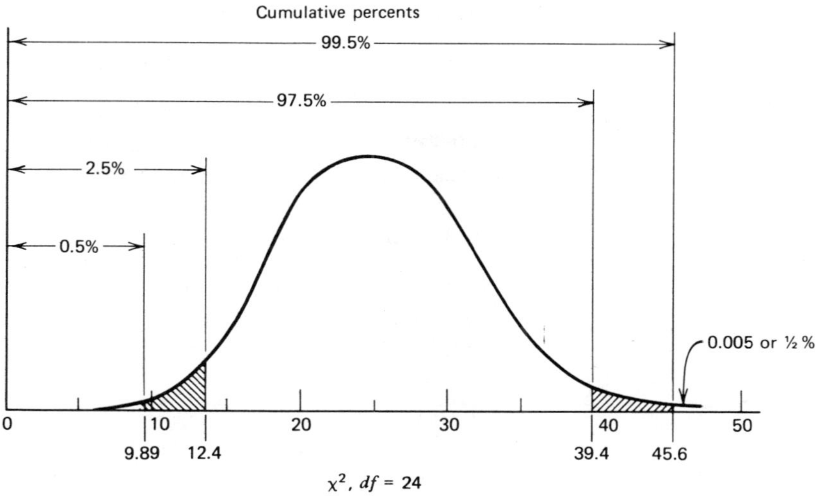

FIGURE 9.15 Graph of the χ^2 (chi-square) distribution with 24 *df*. Four critical values are indicated. Note that between 9.89 and 45.6 lies 99% of the distribution and between 12.4 and 39.4 lies 95% of the distribution.

LIMITS

$$\text{Upper:} \quad \chi^2_{1-\alpha/2} = \chi^2_{1-0.05/2} = \chi^2_{0.975} = 39.4$$
$$\text{Lower:} \quad \chi^2_{\alpha/2} = \chi^2_{0.025} = 12.4$$

These values are from Table A.3. (See Figure 9.15 for a graph of the χ^2-distribution and Figure 9.16 for an illustration of obtaining these values from the table.

df	Probability								
		0.005		0.025		0.975		0.995	
24		9.89		12.4		39.4		45.6	

FIGURE 9.16 Finding the critical values of χ^2 from Table A.3. Several critical probability levels are shown for degrees of freedom equal to 24. (Generally, $df = n - 1$.)

TEST STATISTIC

$$(n-1)\frac{s^2}{\sigma^2_{\text{theo}}} = (24)\frac{0.10}{0.04} = 24(2.5) = 60$$

DECISION Since the test statistic is greater than the upper limit (60 is greater than 39.4), we declare that there is a significant difference in variability between this gravel and the theoretical variance of gravel from Crushed Stone Corp.

CONFIDENCE INTERVAL What is the 99% confidence interval for the true population *for this gravel*? Based on this sample, the 99% confidence interval for the variance, σ^2, is between the following two limits.

$$(n-1)\frac{s^2}{\chi^2_{1-\alpha/2}} \quad \text{and} \quad (n-1)\frac{s^2}{\chi^2_{\alpha/2}}$$

where $\alpha = 0.01$.

The limits are then

$$(24)\frac{0.10}{\chi^2_{0.995}} \quad \text{and} \quad (24)\frac{0.10}{\chi^2_{0.005}}$$

or

$$(24)\frac{0.10}{45.6} = 0.053 \quad \text{and} \quad (24)\frac{0.10}{9.89} = 0.243$$

so that

$$0.053 < \sigma^2 < 0.243$$

(Observe that this interval does not include the theoretical σ^2 of 0.04.)

EXAMPLE 9.7

It has been found that not speeding but, instead, the variability of speeds among drivers is the main cause of accidents. A very slow driver may cause an accident as frequently as a very fast driver. A study showed an optimum standard deviation of speeds on a major highway to be 3 miles per hour. The variance would be 9. A sample of the speeds of 16 cars had an s^2 of 14. At the 5% level of significance, is this a greater degree of variability than optimum?

GIVEN Sample size $n = 16$, $df = n - 1 = 15$, $\sigma^2_{\text{theo}} = 9$, $\alpha = 0.05$.

LIMIT

$$\chi^2_{1-\alpha} = \chi^2_{0.95} = 25.0$$

TEST STATISTIC

$$(n-1)\frac{s^2}{\sigma^2_{\text{theo}}} = (15)\frac{(14)}{9} = 23.3$$

DECISION The test statistic is *not* greater than the limit of 25.0, therefore we conclude that there is no significant difference between the variability of this sample and our theoretical variance of 9.

On the basis of this sample data we cannot conclude that the actual variability is greater than the desired optimum.

9.7 DIFFERENCE IN VARIANCES: *F*-TEST

Many times we take two samples and want to know if the populations they came from have the same spread. In addition, a number of statistical tests that are used to study other characteristics of data, such as the mean, make the assumption that the two samples came from populations whose spreads are the same. If there is any suspicion that the spreads may not be the same, you should compare the spreads of the two samples. Here we will make one assumption, that the samples *both came from normal distributions.* If this is true, and if the spreads *are* the same, then the ratio of the variances of the two samples will fall into an F-distribution. We therefore call this the F-test. Tables of critical values for various F-distributions exist. For example, see Table A.4 and the accompanying instructions for its use.

We take two independent samples and compute the sample variances. If we label the two samples 1 and 2:

s_1^2 = variance of the first sample
n_1 = size or number of values in the first sample
σ_1^2 = variance of the population that the first sample came from

and

s_2^2 = variance of the second sample
n_2 = size of the second sample
σ_2^2 = variance of the population that the second sample came from

We will assume that s_1^2 and s_2^2 will undoubtedly be different for any given set of samples, but σ_1^2 and σ_2^2 could be the same. The underlying question is *whether the population variances are the same.*

We will look only at a two-tailed test, but I will comment about the difference between one- and two-tailed tests.

HYPOTHESIS There is no difference between the population variances of the two samples ($\sigma_1^2 = \sigma_2^2$).

ALTERNATE HYPOTHESIS There is a difference. One population variance is larger than the other ($\sigma_1^2 < \sigma_2^2$ or $\sigma_1^2 > \sigma_2^2$).

Both sample sizes are known, α is to be provided, σ_1^2 and σ_2^2 are *not* specified in any way. s_1^2 and s_2^2 are computed from the data.

TEST STATISTIC

$$F = \frac{\text{Larger } s^2}{\text{Smaller } s^2} = \frac{s_1^2}{s_2^2}$$

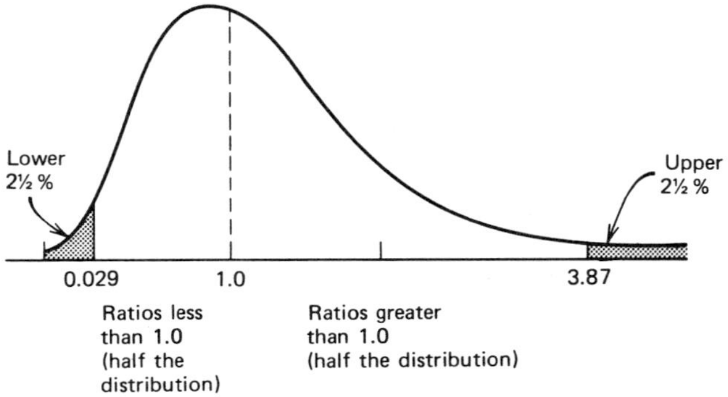

FIGURE 9.17 A typical F-distribution. F (12, 9).

By stating beforehand that we will place the larger s^2 in the numerator, we guarantee that the value we calculated is greater than 1.0 and we only need to concern ourselves with one limit—the upper limit. Figure 9.17 shows a typical F-distribution.

LIMIT

$F(f_1, f_2)$ (From Appendix Table A.4* listed as two-tailed test. You select the α level.)

where

$$f_1 = df \text{ in numerator} = n_1 - 1$$

and

$$f_2 = df \text{ in denominator} = n_2 - 1$$

DECISION If the calculated F is smaller than the limit, accept the hypothesis of no difference between the variances. If the calculated F is larger than $F(f_1, f_2)$ we reject the hypothesis and accept the alternate hypothesis. We declare that σ_1^2 does not equal σ_2^2.

EXAMPLE 9.8

The quality of paper used for printing is very dependent on the variation in thickness. Two different brands of paper are to be compared. Ten readings of paper taken from Quality Paper Co. had $s^2 = 0.0022$ inches; 13 readings taken on paper from Superior Paper Co. had $s^2 = 0.0031$ inches.

HYPOTHESIS There is no difference in variability between paper from Quality Paper Co. and Superior Paper Co. ($\sigma_1^2 = \sigma_2^2$).

* See Appendix Table A.4 for an explanation on the method of obtaining values from the F-tables.

ALTERNATE HYPOTHESIS There is a difference in variability between the brands of paper.

We will use a 5% level of significance and a two-tailed test.

TEST STATISTIC

$$F = \frac{\text{Larger } s^2}{\text{Smaller } s^2} = \frac{0.0031}{0.0022} = 1.41$$

Now we can identify which is s_1^2 and which is s_2^2, since s_1^2 is to be associated with the larger value of variance. Therefore

$$\begin{aligned} s_1^2 &= 0.0031 & s_2^2 &= 0.0022 \\ f_1 &= n_1 - 1 & f_2 &= n_2 - 1 \\ &= 13 - 1 & &= 10 - 1 \\ &= 12 & &= 9 \end{aligned}$$

LIMIT The 5% F-limit is

$$F(f_1, f_2) = F(12, 9) = 3.87$$

This is really an $F_{0.975}$. We have defined the F-ratio in the test statistic as larger variance/smaller variance. Therefore, all ratios that have been in the lower 2 1/2% area, if the test statistic permitted it, are now included in the upper 2 1/2% area. Thus the single area, the upper one, is now equivalent to the two-tail area and reflects a 5% significance level.

DECISION Since the test statistic, 1.41, is less than the limit, 3.87, we *accept* the hypothesis that there is no significant difference in variability between the two kinds of paper.

When you want to perform one-tailed F-tests, the only difference in the above procedure is that you select which sample goes on top of the test statistic prior to performing the calculation. You select the variance of the sample that you expect should be the larger one to go in the numerator. Then you select the limit with the appropriate significance level from the table labeled One-Tailed Test.

REVIEW

Statistical test • One-Sided Test • Two-Sided Test

Limits—Upper • Lower • Both

One Sample Two Sample Means Variances

Test for Means—Variance Given or Known: z-test

Variance Unknown: t-test

Difference in Means: Independent •

Related • Paired t-test

In general:

One Sample $\quad$ Statistic $= \dfrac{\text{Sample mean} - \text{Theoretical mean}}{\text{Standard error of the mean}}$

$$z = \frac{\overline{X} - \mu}{\sigma/\sqrt{n}} \qquad t = \frac{\overline{X} - \mu}{s/\sqrt{n}}$$

Two Sample $\quad$ Statistic $= \dfrac{\overline{X}_1 - \overline{X}_2}{\text{Standard error of difference}}$

Degrees of freedom • χ^2

Test for Variance • χ^2 test • Difference in Variances • F-test

PROBLEMS

SECTION 9.2

9.1 A class of 25 6th graders are given a standardized reading test. The national norm on the test is a mean of 100 with standard deviation of 10. The class averaged 108. Is this class unusual? Remember that the limits would have been decided on before the test was given and the evaluators are interested in knowing if the class is good or bad. (Use a 5% level.)

9.2 A special automobile tire has been developed that is believed to reduce the distance it takes for your car to stop. On a special surface the old tires have required an average of 110 feet (with $\sigma = 15$ feet) to stop from 55 miles per hour. Ten sets of new tires were tested and they averaged 98 feet. You want at least 99% assurance they are significantly better before marketing will take place. Are they?

9.3 A newspaper has been selling an average of 150,000 copies every weekday (with $\sigma = 20{,}000$). As a gimmick to increase circulation the paper ran a contest. Circulation for the first week then was

153,000; 158,000; 161,000; 160,000; 167,000; 161,000.

Has the contest been a success?

* If the cost of the contest was $30,000 and the paper sells for 15 cents, has it been a success monetarily?

9.4 An airline claims its departures average only 8 minutes delay with a standard deviation of 4 minutes. One day while waiting for a flight I checked 9 flights and found they averaged 12 minutes delay. Is the claim correct? (Test at $\alpha = 0.05$.)

9.5 A common question that is asked is how large a sample must I take. For example, the weights and measures inspector in Example 9.1

must decide beforehand how many readings to take. Also required will be the approximate value of σ, α, and how small a confidence interval or difference is permissible. This last value is essentially the maximum amount the inspector is willing to accept as an error in the location of the true population. The formula for sample size, n, is then

$$n = \left(\frac{z_{1-\alpha/2}\,\sigma}{\text{Error}}\right)^2 \tag{9.3}$$

where *error* is the maximum difference he or she will accept.

Thus if the inspector desired to be off in the estimate of true weight no more than 0.02 pounds, error $= 0.02$. Since $\alpha = 0.05$, $z_{1-\alpha/2} = z_{0.975} = 1.96$; σ was 0.04. Then

$$n = \left(\frac{(1.96)(0.04)}{0.02}\right)^2 = (3.92)^2 = 15.37$$

So the inspector must take a sample of 16.

a. How large a sample must be taken to be off by no more than 0.01?
b. How large a sample must be taken to have 99% assurance of being off by no more than 0.02 pounds?

9.6 Referring to Problem 9.2, how large a sample of tires must be used if we want 99% assurance of detecting a decrease of 5 feet? (Note that you should be using $z_{1-\alpha}$ not $z_{1-\alpha/2}$.)

9.7 Referring to Problem 9.3, if the newspaper raises its price to 20 cents, how long will it take to be 99% sure that a decrease in circulation of 5000 has or has not taken place? How long will it take to detect a 10,000 decrease?

9.8 A chemist has known that 95% of the time a certain polymer reacts between 18 and 24 seconds after adding the standard ingredients. After adding a new catalyst, reactions averaged 19.5 seconds for 6 tests. Has the catalyst changed the average reaction time? (Use the 0.05 level.)

SECTION 9.3

9.9 For the following samples, test if the means are significantly different from zero. Test at the 0.05 and 0.01 levels and also state the level at which they would be significant.

a. 16, 20, 5, −2
b. 0.00, 0.01, 0.02, −0.02, 0.03, 0.05.
c. 18, 25, −9, 35, 2, 17.
d. 50, −25, 0, 30, −20, 40, 25, 15, −20, 30.
e. 0, 1, 1, 2, 1, 3, −1, 3, 2.

9.10 For the following samples test if the means are significantly different than the given theoretical mean. Test at the 0.05 and 0.01 levels and state the level at which they would be significant.

a. $\mu = 1000$: 1000, 1500, 1400, 1200, 900.
b. $\mu = 0.05$: 0.02, 0.03, 0.01, 0.02.
c. $\mu = 100$: 105, 107, 106, 104, 98, 103, 104, 107, 101.
d. $\mu = 15.5$: 16, 15, 16, 15, 15, 14, 15, 15, 14, 14, 14, 15, 16, 15, 15, 16.
e. $\mu = 125$: 124, 125, 123, 125, 122, 123, 126, 125, 125, 124.
f. $\mu = 50$: 55, 48, 51, 48, 50, 52, 57, 53, 47.

9.11 The number of hours before each of 10 light bulbs burned out was

1900, 2000, 1875, 2000, 1950, 1800, 1850, 1950, 2100, 2075

How do the results of these tests compare to the statement on the package that claims they burn "at least 2000 hours."

9.12 The same light bulbs (discussed in Problem 9.11) are supposed to yield 900 lumens of light. They were measured as 875, 880, 890, 920, 880, 895, 900, 905, 895, and 900. At what level are they considered significantly different than stated?

9.13 An olive jar contains the statement "net weight 12 ounces—average of 50 olives." Ten jars were picked and the numbers of olives were counted. The jars contained the following number of olives.

43, 43, 47, 44, 48, 51, 46, 48, 47, 53

Do you agree that the olives could be averaging 50? Within what limits do you expect the true mean to be? (Use 95% limits.)

9.14 Refer to the airline statement in Problem 9.4. Typically, they would state that flights averaged only 8 minutes delay time. What would the decision be if the sample of 9 flights had a variance of 21?

9.15 A large orchard has averaged a yield of 140 pounds of apples per tree. A new fertilizer is tested to try to increase the yield. Twelve trees are selected and the following yields in pounds were recorded.

155, 145, 140, 165, 150, 145, 145, 160, 150, 140, 135, 155

Is there a significant increase? (Use a 0.10 level.)

9.16 A neighbor claims it should only take 14 minutes to get to school. Does the time it takes me (Example 1.2) indicate otherwise? (I don't talk back to this neighbor unless I am 99% sure.) We previously found the mean time of 20 trips was 15.5 minutes and the standard deviation was 0.85 (Problem 3.36).

SECTION 9.4

9.17 If the average time between arrivals of express buses is not significantly longer than for local buses, it will pay for me to wait for an express. The standard deviation between arrivals of buses is 4 minutes. I checked a sample of buses and found that for 6 locals the average time between buses is 14 minutes and for 8 express buses it was 18 minutes. Should I wait for the express? (Test at the 5% level.)

9.18 One Sunday two fathers had an argument over whose teenager talked more on the phone. So they both started keeping a record of how many hours their children talked on the phone each night. The following Saturday they compared notes (times in hours).

Mr. C's child:	1.4	2.1	1.7	1.9	2.3
Mr. K's child:	1.3	1.6	1.5	0.8	1.8

Is there a significant difference? (Test at the 10% level.)

9.19 A new form of sole for shoes was being tested for wear. This was an artificial sole that was to be compared to leather soles. After 2 months of use, two sets of shoes were measured to determine the amount of wear in thousandths of an inch.

Artificial	12,	13,	12,	9,	13,	12,	14,	10,	11,	8,	13,	11
Leather	16,	13,	13,	14,	14,	15,	12,	16,	13,	12,	15,	16

Is there a significant difference (improvement)? (Test at the 5% level.)

9.20 A telephone company did some investigation of the effect of reducing the rate for long distance calls. A record of the number of long distance calls for the 15 minute interval from 6 to 6:15 P.M. was kept for 2 working weeks before and after the change.

BEFORE		AFTER	
47	47	61	50
52	53	60	54
57	56	59	63
48	51	71	65
49	50	49	61

Was there a significant change in the number of long distance calls? (Test at the 1% level.)

9.21 An investigation of employees who were ill in a company revealed the following number of days absent for those who had the flu vaccine and those who did not have the flu vaccine.

Had vaccine	3, 2, 1, 1, 2, 3 1/2, 4, 3, 2, 2, 3, 4, 5, 2, 2 1/2, 2, 1, 1/2, 1, 2, 1, 1, 3, 2, 1, 3, 2 1/2, 1 1/2, 3, 1
Did not have vaccine	2, 2, 1, 2, 3, 3, 5, 4, 2, 4, 3, 3, 2 1/2, 3 1/2, 5, 7, 3, 2, 1, 2, 4, 6, 3, 2, 3.

Is there a difference in the average number of days ill for these two groups of people?

9.22 Ten garments are washed in "X" detergent and 10 other garments are washed in "X" detergent without phosphates. The whiteness was rated on a scale of 1 to 100 as follows (100 was the whitest).

"X" detergent:	37, 48, 49, 61, 55, 71, 59, 78, 67, 66
"X" without phosphate:	44, 33, 27, 36, 27, 68, 49, 51, 60, 63

Is there a significant difference in the whiteness? (Test at the 5% level.)

9.23 It is now revealed that two people did the rating of whiteness (in Problem 9.22). The first 5 readings in each group were by Mr. A. and the second 5 readings were by Mrs. T. Is there a difference in the way these people rate the whiteness? How might this affect the results of the previous comparison? How might you redo Problem 9.22 to account for any difference in rater? (*Hint*: Think in terms of two tests.)

9.24 Refer to Problem 7.16. All the seeds were planted on April 18. If the first 4 plants were in fertilized soil and the second 4 were in unfertilized soil, is there a difference in the time to germinate for seeds in fertilized and unfertilized soil? (Test at the 5% level.)

SECTION 9.5

9.25 A heating oil company has advertised that by installing a special device on the furnace your heating bills will be reduced substantially. Before I was going to have the device installed, I questioned 7 people who had it installed. Their heating bills in dollars the year before and the year after installation were as follows.

Before	225	195	235	165	260	245	230
After	200	205	215	170	250	255	225

Is there a significant decrease in cost?

9.26 The pulse rates of 12 people were recorded before and after taking a new drug.

Before	68	71	84	93	67	74	82	77	71	83	62	66
After	71	70	81	97	73	80	90	76	80	79	80	67

Is there a significant increase in pulse rate? (Test at the 10% level.)

9.27 Two hair sprays were being compared—"The End" and Brand "X." The company that makes "The End" was comparing it to Brand "X" to determine if its brand had longer holding power. Ten women had half their head sprayed with "The End" and the other half with Brand "X." A strong fan blew in their faces and the number of minutes for their hair to become windblown was recorded as follows.

The End	8	6	9	5	7	4	8	8	9	10
Brand "X"	10	5	7	6	4	4	7	7	7	8

Is "The End" a better hair spray? (Test at the 10% level.)

9.28 A botanist was interested in knowing if there was a difference in the time fruits matured on different parts of a plant, and recorded the day of the first fruit on the top and on the bottom for 15 plants. All the fruits came out during July.

Top	3	6	7	5	8	9	10	10	7	8	6	9	10	12	4
Bottom	7	9	5	8	8	10	11	12	6	9	7	13	8	13	8

Is there a significant difference in the time to mature (at the 5% level)?

9.29 Refer to Problem 9.19. If each of the values of artificial soles is listed above the corresponding value for leather sole on the second shoe for pairs of shoes (in other words one sole goes on one shoe, perhaps the left, and the other type of sole goes on the other shoe, the right), then we have matched pairs of shoes to compare and the wear is more readily compared (if people are to do the testing). Use a paired *t* test to determine if there is a significant difference between the two types of soles. (Use $\alpha = 0.05$.)

9.30 Most meats lose weight when they are cooked. A restaurant owner carefully weighed 10 1-pound steaks before and after cooking. The owner is willing to accept an average of a 10% loss in weight without complaining to the meat supplier. The weights were as follows.

Before	1.10	1.08	1.03	1.09	0.93	0.99	1.01	1.07	1.01	0.94
After	0.92	0.95	0.91	0.96	0.85	0.88	0.89	0.93	0.96	0.87

Is there a difference in weight significantly greater than 10%? (*Hint*: First decrease the before weights by the 10%, then compare them.)

SECTION 9.6

9.31 From the tables find:

a. $\chi^2_{0.95}$ for 10 *df*.
b. $\chi^2_{0.05}$ for 5 *df*.
c. χ^2 for which 10% of the distribution values are between ($df = 12$).
d. χ^2-values for which 80% of the distribution values are beyond ($df = 25$).
e. 10% upper limit of χ^2 for 200 *df*.

9.32 I have 4 light bulbs in a ceiling light fixture that is inconvenient to reach. I have decided that if the variance of the life of light bulbs is small it pays to change all 4 bulbs when 1 bulb burns out. I have decided that if σ is less than 50 hours I will change all the bulbs at one time. I tested a group of 12 light bulbs to determine the variation among them. The variance of these light bulbs was 1600. Should I change all light bulbs? (First decide if this is a one- or two-sided test. Then test at the 5% level.) What are the possible errors and the consequence of these errors?

9.33 An oven that is regulated must not vary much for consistent baking. A particular food requires the variance of the temperature not to be more than 30. The following eight readings were taken in the oven:

$$350, 370, 360, 355, 360, 365, 355, 365 \qquad (\bar{X} = 360)$$

Is the variation in the oven excessive? (Test at the 5% level.)

9.34 A machinery manufacturer claims that a new machine will reduce the variation in the size of parts. The company that is interested in purchasing the machine has the manufacturer make 10 parts that usually have a size of 1.000 inches with variance of 0.000005. The variance in the size of the 10 parts was 0.000002. Is this a significant reduction in variation? (Test at 1% level, since the machine is very expensive.)

9.35 The precision of an instrument is related to the variance of repeated readings. A new type of laser beam transit is claimed to repeat its distance measurements so that 95% of the time it is within ± 1.2 inch (over a distance up to several miles). (These are 2σ limits.)

Two points were chosen approximately a mile apart and 16 measurements were taken. The variance of these measurements was found to be 2.43. Do you challenge the manufacturer's claim? What would you give as 99% confidence intervals that the variance of the readings are?

9.36 A psychologist has been attempting to determine if marijuana affects response time to a stimulus. The variance of previous tests of response time for a volunteer has been 0.5 seconds. The volunteer then smoked a cigarette and tried the test 5 times. The times in seconds were

$$2.6, 4.3, 2.2, 4.4, 5.5$$

Has the variation in response increased? (Use $\alpha = 0.05$.)

9.37 With certain types of plants harvesting is made easier if all the fruits or vegetables on a tree or in a field mature at the same time. A certain type of corn has matured in 45 days with a standard deviation of 3 days. Twenty-five plants of a new strain of corn plant were grown. They matured in 48 days with a standard deviation of 1 day. Is there a significant change in the variation? (Test at the 10% level.)

9.38 If the width of floor tiles varies too greatly from tile to tile the pattern does not match when the tiles are set down on the floor. A tile manufacturer asserts that the 12-inch tiles have a standard deviation of only 0.05 inches. I measured the tiles in one carton of 24 tiles and found the variance of the tiles was 0.01. Is this significant (with $\alpha = 0.05$)? What interval would the true variance probably be within (95% *CI*)? Under the worst condition (largest variance), try to picture the possible effect such a variation would have if you were setting down tiles of this size. Would they match poorly?

SECTION 9.7

9.39 From the tables find:

a. $F_{0.95}(3, 5)$.
b. F-value for which 15% of the distribution values are above, *df* in numerator = 5, *df* in denominator = 20.
c. F-value for which 15% of the distribution values are above, *df* in numerator = 20, *df* in denominator = 5.
d. 1% upper critical value for $f_1 = 30$, $f_2 = 20$.
e. $F_{0.01}(6, 10)$.
f. $F_{0.01}(1, 5)$.
g. A pair of F-values that enclose 95% of the values with $f_1 = 3$, $f_2 = 15$.

9.40 A researcher was interested in knowing if there is a difference in the variability of the response for a person taking marijuana versus a person drinking alcohol. Ten similar subjects smoked a test cigarette and 10 others drank alcohol. The variance on the test of those smoking marijuana was 175 and the variance among the drinkers was 150. Is there a difference in variability? (Test at the 5% level.)

9.41 The variation of speeds of cars on a highway is considered at least as important as the actual speeds (see Example 9.8). Two superhighways were compared. Thirty vehicles were clocked on an interstate highway and 25 cars were clocked on a toll superhighway (automobiles only). The variance of speeds was 400 for the interstate highway and 40 for the toll road. Is there a significant difference in variation? Comment on the problem of the sampling. (Was it random?)

9.42 The variation in voltage at the wall socket can affect the performance of many electrical appliances. An inspector checked the voltage in a house and in a factory and took 15 readings in the house and 12 in the factory. The resulting variances were

$$s^2_{\text{house}} = 83 \qquad s^2_{\text{factory}} = 18$$

Is there a significant difference in the variation of voltage for these two locations?

9.43 Two paving contractors were asked to give estimates to pave identical driveways. The estimates were as follows.

	BEST PAVING CO.	PAVE-IT CO.
Driveway A	450	525
Driveway B	375	500
Driveway C	675	475

Do the two contractors differ in the amount of variation in their quotes? (Test at the 10% level.)

9.44 The degree of variation in the readings of an instrument is very important. A micrometer and a vernier caliper were compared by measuring the same part six times with each instrument. The variance among the six readings with the micrometer was 0.00000004 and for the six readings with the vernier caliper is was 0.0000016. Are the readings with one instrument more variable than with the other? (Test at the 5% level.)

9.45 An aspirin manufacturer that does considerable advertising claims its aspirin are more uniform than the very inexpensive brand. Samples from several bottles of both manufacturers were checked for the amount of aspirin in the tablets.

MUCH ADVERTISED BRAND	INEXPENSIVE BRAND
$n = 30$	$n = 20$
$s^2 = 0.0009$	$s^2 = 0.0016$
$\bar{X} = 5.01$	$\bar{X} = 4.98$

Is there a significant difference in the brands? Is this a one- or a two-sided test? (Test at the 5% level.)

9.46 Refer to Table 7.1. For each successive pair of samples of 4 compute the ratio of the variances (s^2/s^2). Thus the first ratio is 4.33/2.67. You can calculate 49 ratios. Plot these and compare the 90% and 95% percentages of this distribution to the theoretical 90% and 95% limits of an F-distribution (3, 3 df).

CHAPTER TEN HYPOTHESIS TESTING AND ESTIMATION UNDER ASSUMPTIONS OF BINOMIAL DISTRIBUTION

The binomial distribution acts as one of the most commonly encountered models for data. If, for example, I am interested in knowing the proportion of letters needing only 1 postage stamp, I do not have to weigh and record carefully the weights of each letter. I only need to determine if the letter is less than 1 ounce; this measurement will be much easier to perform.

I can just note if a student is late or not late—I don't record the time arrived in class. I can either measure the actual gap in a spark plug (to the nearest 0.02 millimeter) or check if it is in the correct range using a go no-go gage. A person can either say yes on a poll, or can use a much more complicated technique of identifying a numerical value on some scale of degree of approval.

It should be evident that it is usually easier to obtain binomial-type data than to take measurements. Sometimes, the only kind of data is binomial (there is no such thing as a woman being half pregnant).

We will now examine some one- and two-sample tests that involve hypotheses concerning the population proportion (percent or probability). We will also obtain confidence interval estimates for proportions. Incidentally, this is the only value to estimate (as opposed to the mean *and* variance of the normal distribution).

10.1 COMPARING THE SAMPLE PROPORTION TO A THEORETICAL VALUE

Whenever we obtain a sample from a binomial type source, we obtain a single value. It may either be written as an X-value (the number of times one of the possibilities occurred out of a total sample of size n) or it may be computed as a proportion (the value $p = X/n$). In either case we may freely convert from one form to the other. The types of questions we shall attempt to answer are: "Is there a significant difference between this sample proportion and some hypothetical, theoretical proportion?" or "Within what interval would we expect the true population to lie (confidence interval for proportion)?"

Before considering these questions, we will have to recognize the role that the size of the sample plays in making a decision.

With small samples we should be working with the exact binomial distribution and with large samples we can use the normal approximation to the binomial (when $n \cdot p$ and $n \cdot p(1 - p)$ are both at least about 10).

Throughout this chapter p will be the true population proportion, p_o will stand for the hypothetical population proportion (used in hypothesis testing), and $\hat{p}$ will be the sample proportion that is used as an *estimate* of the population proportion (and is generally X/n).

Using the exact binomial test (small sample size)

Basically, Chapter Six concerned itself with the question of describing the complete set of probabilities for all of the possibilities in a given binomial distribution. It provided a technique for establishing a theoretical distribution to use as a guide to compare to a particular sample. We are now considering a technique used to decide if the sample could be expected to have come from such a population. The procedure we will follow here will simply involve examining the binomial distribution tables to decide if a sample falls in a tail, as an extreme value. If our *one sample* falls within the *extreme* 5% or 1% of the given theoretical distribution, we declare that the sample comes from a different population. Essentially, Example 8.3 was a binomial test, since we hypothesized that p (or, more properly, p_o) (page 176) was 0.05 and we determined a limit and the probability of making an alpha error associated with having a value fall outside that limit. We then made a decision based on the sample. We will now consider another example. Observe that we will accumulate all of the probabilities in the tail until we reach the total probability we want to be in that particular tail (for our limit). The X-values associated with the tail will constitute our rejection values, or those cases that will lead us to declare the hypothesis to be false.

The *test statistic* will be our actual number of cases, which we will then compare to the limits.

EXAMPLE 10.1

We are going to examine a total of 12 flashlight batteries that were randomly selected from several packages of batteries. We find that 4 of the batteries are defective. Let us further assume that the probability of a defective battery, or the proportion of defective batteries, has been 0.10. Is there now a change in p from the 0.10? (This is essentially a continuation of Example 6.1.)

Our procedure is as follows.

HYPOTHESIS The probability of a defective battery is 0.10 ($p_o = 0.10$).

ALTERNATE HYPOTHESIS The probability of a defective battery is greater than 0.10.

LEVEL OF SIGNIFICANCE We will choose an alpha level of 0.05 or we will limit ourselves to a 5% chance of a Type I error.

LIMIT Examining Table A.8 for $n = 12$ and $p = 0.10$, we find the probability of 7, 8, 9, 10, 11, or 12 defectives is 0.0000, 6 is 0.0005, 5 is 0.0038, and 4 is 0.0213. Adding these together gives us the probability of 4 or more defectives as

$$0.0213 + 0.0038 + 0.0005 + 0.0000 = 0.0256$$

or a 2.56% chance of error.

If we include 3 defectives this tail increases to a probability of

$$0.0852 + 0.0256 = 0.1108$$

or 11.08% chance of error. Since we desire a 5% limit, we must take the next smaller probability, which is the 2.56%. We declare the limit to be 3 defectives (any more will be beyond this limit and are included in the rejection area in the tail). Figure 10.1 shows the histogram for this probability distribution and indicates the rejection region.

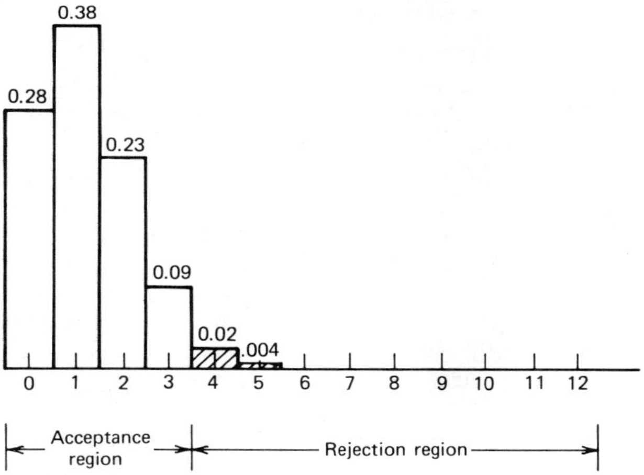

FIGURE 10.1 Distribution of possible samples for binomial test when $n = 12$ and $p = 0.10$.

TEST STATISTIC Our statistic is merely the number of defectives we observed. We found a total of 4 defectives.

DECISION Since the test statistic, 4, is greater than the limit 3, we declare that the hypothesis is not true. We state that the proportion of defective batteries is *not* 0.10, it is probably *greater* than 0.10.

CONFIDENCE INTERVAL It would be desirable to place a confidence limit or interval around the actual proportion that we have observed. In this example the observed proportion, $\hat{p}$, was $X/n = 4/12 = 0.33$.

Unfortunately, with small samples, confidence intervals are so large that they provide little aid in decisions. In addition, to obtain meaningful limits on the intervals when working from the tables requires a much more extensive collection of tables, as well as a slightly different technique of working with the table. We will defer the discussion of confidence intervals until the next section.

Large samples: The normal approximation

With larger samples we will use the normal distribution as an approximation to the binomial. This will greatly simplify and at the same time broaden the types of statements we can make. I should note that we will limit our discussions to hypothesis and estimation of *proportions*. If you require estimations regarding X-values or means, then you merely multiply the corresponding proportions, or p-values, by n.

As before, we hypothesize that the population has a particular known value of p associated with the source of our sample. We will then compare the proportion of the sample to this p-value, although we will be converting to a z-value. The procedure will be demonstrated in the following example.

EXAMPLE 10.2

A museum purchased a \$1 million painting, expecting that it would attract 75% of its visitors. To verify this estimate, 50 people were chosen randomly as they entered the museum and followed until they left. Of these 50, it was found that 27 looked at the painting. Does this contradict the original estimate?

HYPOTHESIS The proportion of all visitors looking at the painting, p_o, is 0.75.

ALTERNATE HYPOTHESIS The proportion of all visitors is significantly different than 0.75 (larger or smaller).

We will use an α of 0.05 (5% chance of error). Sample size $n = 50$.

LIMITS We are to use a z-statistic.

$$\begin{aligned} &\text{Upper} \quad z_{1-\alpha/2} = z_{1-0.05/2} = z_{0.975} = 1.96 \\ &\text{Lower} \quad z_{\alpha/2} \quad\; = z_{0.05/2} \quad\; = z_{0.025} = -1.96 \end{aligned} \tag{10.1}$$

TEST STATISTIC We are now dealing with estimates and hypotheses about p. The variance for the distribution of sample p-values of $\hat{p}$ from a population with the hypothesized parameter of p_o is*

$$\sigma^2 = p_o(1 - p_o) \tag{10.2}$$

or

$$\sigma = \sqrt{p_o(1 - p_o)} \tag{10.3}$$

* Note that the only change in the formulas 10.2 to 10.4 when a true population p is used is to replace p_o with p.

This differs from our previous work with the binomial distribution (Equation 6.5, page 136) because at that time we dealt with the distribution of X-values and now we are concerned with proportions. The X-values fall in a range of 0 to n, while proportions are to be between 0 and 1.0.

Next we will need the standard error (for proportions), and this will be $\sigma_p = \sigma/\sqrt{n}$. Thus:

$$\text{Standard error} = \sigma_p = \frac{\sigma}{\sqrt{n}} = \frac{\sqrt{p_o(1 - p_o)}}{\sqrt{n}} = \sqrt{\frac{p_o(1 - p_o)}{n}} \tag{10.4}$$

Since the sample value is $p = X/n$, the z-statistic will be†

$$z = \frac{p - p_o}{\sqrt{p_o(1 - p_o)/n}}. \tag{10.5}$$

Since $p = X/n = 27/50 = 0.54$, we have

$$z = \frac{p - p_o}{\sqrt{p_o(1 - p_o)/n}} = \frac{0.54 - 0.75}{\sqrt{(0.75)(0.25)/50}} = \frac{0.54 - 0.75}{\sqrt{0.00375}}$$

$$= \frac{-0.21}{0.0612} = -3.43$$

DECISION If the value of z obtained in Equation 10.5 is between the two limits, we accept the hypothesis that there is *no* significant difference between the sample proportion and our hypothesized p_o.

Since the test statistic of -3.43 is outside the test limits, we *reject* the hypothesis that $p_o = 0.75$. We declare that p is significantly different from 0.75. It appears to be considerably less.

CONFIDENCE INTERVAL We can draw a confidence interval around the particular *sample* proportion. This provides us with a range of possible values of the population proportion from which we would assume our sample may reasonably have been taken. Also, we are really not sure what p is and we are estimating it with $\hat{p}$.

$$\hat{p} + z_{\alpha/2}\sqrt{\frac{\hat{p}(1 - \hat{p})}{n}} < p < \hat{p} + z_{1-\alpha/2}\sqrt{\frac{\hat{p}(1 - \hat{p})}{n}} \tag{10.6}$$

Observe that we are now regarding X/n as an estimate of the true population p and therefore we call it $\hat{p}$, or

$$\hat{p} = \frac{X}{n} = \frac{27}{50} = 0.54$$

† The p here is the sample proportion but is *not* being interpreted as an *estimate* of true p. That comes later.

and

$$\hat{p} + z_{\alpha/2}\sqrt{\frac{\hat{p}(1-\hat{p})}{n}} < p < \hat{p} + z_{1-\alpha/2}\sqrt{\frac{\hat{p}(1-\hat{p})}{n}}$$

$$0.54 - 1.96\sqrt{\frac{(0.54)(0.46)}{50}} < p < 0.54 + 1.96\sqrt{\frac{(0.54)(0.46)}{50}}$$

$$0.54 - 1.96\sqrt{0.005} < p < 0.54 + 1.96\sqrt{0.005}$$

$$0.54 - 1.96(0.071) < p < 0.54 + 1.96(0.071)$$

$$0.54 - 0.14 < p < 0.54 + 0.14$$

$$0.40 < p < 0.68$$

Thus we give 19 to 1 odds that the true proportion of people looking at the painting is between 0.40 and 0.68 (or between 40% and 68% of the visitors).

One additional important consideration should be made regarding confidence intervals. It is often relatively easy to obtain yes-no answers or take go no-go measurements for a large group of individuals. However, our "common sense" is not always accurate for predicting what to expect as a valid confidence interval.

Most people are usually surprised to hear that with a poll of only 1000 people from a population as large as the United States that we should claim that 95% confidence interval would cover a mere ± 3 percentage points. In other words, when a Gallup poll of 1000 persons is shown to have a result of 46% favoring an action of the President, we claim that we are reasonably assured that for the entire population approximately

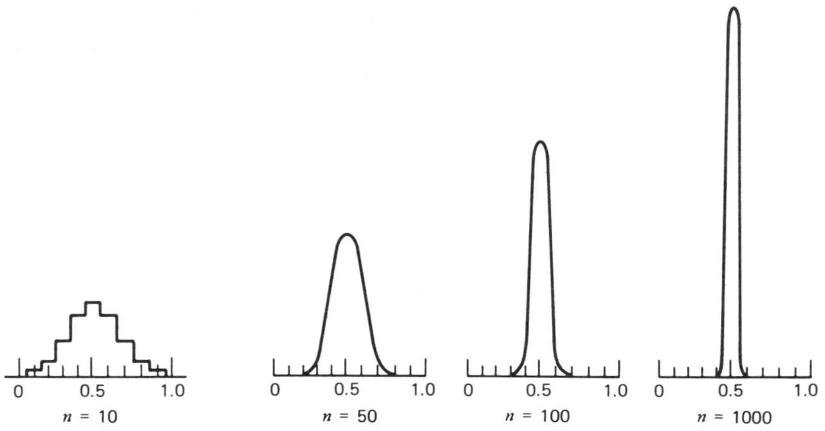

FIGURE 10.2 Distributions of sample ps (or $\hat{p}$s) from binomial distributions with population $p = 0.50$. Four different sample sizes are shown.

43 to 49% of the people favor the action (the confidence interval for p being from 0.43 to 0.49).

In Figure 10.2 we have a collection of four distributions of sample proportions all taken from a population with $p = 0.5$. Observe how the range of sample ps (or $\hat{p}$s) shrinks as the sample size increases. When $n = 1000$, the distribution is extremely narrow.

10.2 DIFFERENCE IN PROPORTIONS TEST: INDEPENDENT SAMPLES

Quite often we have two independent pairs of proportions to compare (two samples from binomial distributions). In Chapter Six we had a collection of examples of types of measurements that could be reported as proportions. We now will look at how some of these examples could be extended to cases involving the questions of differences or changes in proportions.

EXAMPLE 6.1 (Cont.)

a. Is there a difference in the proportion of heads between pennies and nickels?
b. A poll is taken before and after a major speech. We question if there has been a change in the percent of the people in favor of the candidate.
c. Is the proportion of defective batteries different for alkaline versus standard batteries?
d. Is there a difference in the proportion of time two different traffic lights are red?
e. Is there a significant difference in the proportion of male and female customers who buy something?
f. Is the percentage of bills paid on time different for families than for businesses?

In all these cases we are to take two samples and determine two proportions, which we will then designate as $X_1/n_1 = \hat{p}_1$ and $X_2/n_2 = \hat{p}_2$ (where the subscripts 1 and 2 designate the values for the two samples).

HYPOTHESIS There is no difference between the population proportions for the two samples. That is, p_1, population proportion for first sample, $= p_2$, population proportion for the second sample.

We will also make one important restriction—that both sample sizes are large (all $n \cdot p$s and $n \cdot (1 - p)$s are larger than about 5). This will permit us to use a normal approximation and a z-statistic. Keep in mind that if the two samples came from the same population, we would intuitively *expect* that the difference between the sample proportions, $\hat{p}_1 - \hat{p}_2$, would be zero. Obviously, for any particular set of samples we would not necessarily obtain identical $\hat{p}$s, and the difference would not be 0. For example, if the population had a true p of 0.5 and we took two samples of 10 each, we might obtain $\hat{p}_1 = 0.4$ and $\hat{p}_2 = 0.6$. Then $\hat{p}_1 - \hat{p}_2 = -0.2$. However, if we were to repeat this process over and over, the average difference would be 0. Furthermore, when the sample sizes are large enough, these differences would have a z-distribution, as in Equation 10.7.

$$z = \frac{\hat{p}_1 - \hat{p}_2}{\sqrt{\hat{p}(1 - \hat{p})(1/n_1 + 1/n_2)}} \tag{10.7}$$

where

$$\hat{p}_1 = X_1/n_1, \qquad \hat{p}_2 = X_2/n_2$$

$\hat{p}$ = estimated population proportion based on the combined samples

$$= \frac{X_1 + X_2}{n_1 + n_2} \tag{10.8}$$

EXAMPLE 10.3

The flu vaccine was given to 125 employees; 75 other employees did not receive it. Of the employees who lost days because of illness, 30 had received the vaccine and 25 did not. Thus $\hat{p}_1 = X_1/n_1 = 30/125 = 0.24$ (the proportion of those that received the vaccine and became ill), and $\hat{p}_2 = X_2/n_2 = 25/75 = 0.33$ (the proportion of those that did not receive the vaccine but became ill).

HYPOTHESIS The proportion of people who became ill after receiving the vaccine is not different from the proportion of people who became ill but did not receive the vaccine.

ALTERNATE HYPOTHESIS There is a difference in proportions.

We will use a 5% level of significance ($\alpha = 0.05$).

LIMITS

Upper $z_{1-\alpha/2} = z_{0.975} = 1.96$

Lower $z_{\alpha/2} = z_{0.025} = -1.96$

TEST STATISTIC First we calculate the overall average $\hat{p}$.

$$\hat{p} = \frac{X_1 + X_2}{n_1 + n_2} = \frac{30 + 25}{125 + 75} = \frac{55}{200} = 0.275$$

Then

$$z = \frac{\hat{p}_1 - \hat{p}_2}{\sqrt{\hat{p}(1 - \hat{p})(1/n_1 + 1/n_2)}} = \frac{0.24 - 0.33}{\sqrt{(0.275)(1 - 0.275)(1/125 + 1/75)}}$$

$$= \frac{-0.09}{\sqrt{(0.275)(0.725)(0.008 + 0.013)}}$$

$$= \frac{-0.09}{\sqrt{(0.275)(0.725)(0.021)}}$$

$$= \frac{-0.09}{\sqrt{0.0042}} = \frac{-0.09}{0.065} = -1.38$$

DECISION Since -1.38 is not larger than 1.96 nor smaller than -1.96, we accept the hypothesis that there is no difference. We decide that our chance of becoming ill is essentially the same whether or not we receive the vaccine.

CONFIDENCE INTERVAL Since for large n_1 and n_2 the standard error for the difference in two proportions is

$$\sqrt{\frac{\hat{p}_1(1-\hat{p}_1)}{n_1}+\frac{\hat{p}_2(1-\hat{p}_2)}{n_2}}$$

and is also normally distributed, then the confidence interval is

$$\hat{p}_1-\hat{p}_2+z_{\alpha/2}\sqrt{\frac{\hat{p}_1(1-\hat{p}_1)}{n_1}+\frac{\hat{p}_2(1-\hat{p}_2)}{n_2}}<p_1-p_2$$

$$<\hat{p}_1-\hat{p}_2+z_{1-\alpha/2}\sqrt{\frac{\hat{p}_1(1-\hat{p}_1)}{n_1}+\frac{\hat{p}_2(1-\hat{p}_2)}{n_2}} \qquad (10.9)$$

$$0.24-0.33+(-1.96)\sqrt{\frac{(0.24)(0.76)}{125}+\frac{(0.33)(0.67)}{75}}<p_1-p_2$$

$$<0.24-0.33+1.96\sqrt{\frac{(0.24)(0.76)}{125}+\frac{(0.33)(0.67)}{75}}$$

$$-0.09-1.96\sqrt{0.00146+0.00295}<p_1-p_2$$

$$<-0.09+1.96\sqrt{0.00146+0.00295}$$

$$-0.09-1.96\sqrt{0.00441}<p_1-p_2<-0.09+1.96\sqrt{0.00441}$$

$$-0.09-0.13<p_1-p_2<-0.09+0.13$$

$$-0.22<p_1-p_2<0.04$$

Thus we say the difference in proportions (taking vaccine versus not taking vaccine) can be anywhere from $+0.04$ (a 4% increase in illness) to -0.22 (a 22% decrease in illness). Since this straddles 0%, there is a possibility of *no* change.

10.3 THE SIGN TEST—A TEST FOR RELATED DATA

In Section 9.1 we discussed a number of different kinds of situations where the data can be viewed as related. (If you did not study that section, I would suggest that you read it now.) In all those cases we had two sets of values, but they involved two measurements on the same (or comparable) individuals. Sometimes the only thing we can really say after a test or examination of some characteristic, unless we were to use a very elaborate method, is that there *is* a difference or a change. As long as we can identify the direction of change in each case, such as improvement in reflexes, increase in absorbency, positive change in attitude, decrease in hardness, and so forth, *we can apply the sign test*. We can use it when we have numerical differences (I finished the race in 1 minute 20 seconds, you finished in

1 minute 12 seconds) or when we have categorical values for difference (you won). When we deal with these changes ,we will call one of the possible changes a *plus*, + (possibly increase, won, or improvement) and the other a *minus*, − (decrease, lose, or fail). In other situations, such as Example 10.4, we will use letters to designate the differences.

In all cases we assume that the chance of having a plus is the same as that of a minus. Thus, the probability of either sign or letter is $\frac{1}{2}$. If this is true, the possible samples of plus and minus signs will be a binomial distribution with $p = 0.50$ and $n =$ number of differences or pairs of values.

Since we are really working with a binomial test, we must consider two cases: *small samples* using the table and *large samples* using the normal approximation. The only difference, however, is in the determination of the limit. The rest of the test procedure is the same.

HYPOTHESIS There is no difference in population averages.

ALTERNATE HYPOTHESIS There is a difference in population average. (The sign test actually tests to determine if the medians are different.) We may perform a one- or two-tailed test. As always, this influences our limit ar ˀ requires us to decide if we are interested in +, −, or both.

We must choose α. With small samples it can be approximate. When using the sign test, it is usually better to calculate the test statistic first. If there is no change, in some cases, we call the pair tied. Ties will influence the computation of the limit. For each pair of values enter a +, −, or 0, depending on whether there is a + change or difference, a − change, or no change (or an A, B, or 0, if these are the categories). The total number of plusses *or* minusses constitutes our test statistic. (However, the test statistic will depend on whether we have a one- or two-sided test. If the test is to be *one*-sided, the sign we are to add up must be decided on first.)

LIMIT The total number of plusses and minusses (which is also the *total number of pairs* minus *the ties*) must be computed. This is n, the sample size.

If n is 20 *or less:* Use the binomial tables, Appendix Table A.8 with $p = 0.5$. For a one-tailed test add up the probability in one tail until you reach α. For a two-tailed test add up the probability in one tail until you reach $\alpha/2$; then you can subtract the X from n to find the second limit.

If n is more than 20: Use the normal approximation. To find the X-limits we would choose the appropriate equation(s) below:

One tailed

$$X = \frac{n}{2} + z_{(1-\alpha)}\tfrac{1}{2}\sqrt{n}$$

Two tailed

$$\text{Upper } X = \frac{n}{2} + z_{1-\alpha/2}\,\tfrac{1}{2}\sqrt{n} \tag{10.10}$$

$$\text{Lower } X = \frac{n}{2} + z_{\alpha/2}\,\tfrac{1}{2}\sqrt{n}$$

(Remember that $p = 0.5$.)

DECISION If the total plusses (or minusses) are within the limits accept the hypothesis of no difference. If the plusses (or minusses) are outside the limits accept the alternate hypothesis that there is a difference.

EXAMPLE 10.4

Ten persons were selected to taste two brands of tea to decide which they preferred. They were to identify which was better by writing the letter **B** for better and W for worse. In one case the person felt there was no difference and put down a 0 for both categories. In Table 10.1 the results are listed.

TABLE 10.1 Results of 10 Tea Tasters in Comparing Two Brands of Tea. B Stands for Better, W for worse, and 0 for no Difference

Taster	1	2	3	4	5	6	7	8	9	10
Brand A	W	B	B	W	W	B	0	B	B	B
Brand B	B	W	W	B	B	W	0	W	W	W

Totals For Brand A: 6 Bs, 3 Ws and 1 no change.

Note: (We don't worry about totals for Brand B because the test statistic and decision will only need the results from one brand. In addition, the total Bs and Ws are just reversed from Brand A.

Since we have no prior anticipation of which will be better, we will consider it a two-tail test and use an alpha level of 0.05. (Note that if Brand A was much less expensive we might ask the question "Is Brand B significantly better?" Then we would have a one-tail test.)

HYPOTHESIS There is no difference between brands of tea.

ALTERNATE HYPOTHESIS There is a difference, one brand is better than the other (not specifying which).

LIMIT The sum of the Bs and Ws was 9 (6 + 3); then $n = 9$. Consulting Table A.7 for $n = 9$ and $p = 0.50$ we have: $X = 9$, $p = 0.0020$; $X = 8$, $p = 0.0176$; and the sum is 0.0196. We are looking for 2.5% in each tail. If we add the $X = 7$, $p = 0.0703$ we now have a sum of 0.0899, or about 9%. Therefore we stop at $X = 8$ (about 2%). Similarly, we find that 0 and 1 would constitute the other rejection region. The acceptance region would be from $X = 2$ to $X = 7$.

TEST STATISTIC We will use the number of Bs for Brand A. This is 6.

DECISION Since the test statistic is *not* in the tail or critical region, we accept the hypothesis that there is no difference between the two brands of tea.

REVIEW Population Proportion—p • Hypothesized Population Proportion—p_o
Estimate of population proportion—$\hat{p} = X/n$

$\sigma_p = \sqrt{\frac{p(1-p)}{n}}$ • Difference in Proportions (Independent Samples)

Sign Test for Related Data.

PROBLEMS

SECTION 10.1

10.1 The management of a large department store has felt that if the number of shoplifters is more than 20% of the customers who enter the store, extreme measures will be necessary. Since the cost of investing in security equipment is very large, a risk of less than 1% is considered necessary.

a. A group of 150 people were randomly chosen as they entered the store and followed or observed until they left. Of these 150 people 41 were seen to have taken something without paying. Is there sufficient reason for the management to consider the purchase of security equipment?
b. What would you estimate the proportion of shoplifters to range within? (Assume 1% limits.)
c. If the number of customers averages about 6000, what would you estimate the number of shoplifters to range within?

10.2 During a hot June, 12 false fire alarms were reported out of a total of 20 "reported" fires. If the expected proportion of false alarms has always been 30%, is this an unusual month?

10.3 A dishonest cashier believed that only 10% of the customers count their change. The cashier then proceeded to shortchange every customer and, by the end of the day, saw 200 people and 28 complained of receiving incorrect change. How good is the estimation of the proportion of people who count their change? (Use an $\alpha = 0.05$.)

10.4 A prospective father decided to estimate how many cigars he would need when the new baby was to be born. He polled 20 of his friends and co-workers, asking them if they accept cigars, and he found only 15 "takers." If there are a total of 250 men to offer cigars to, how many cigars should he plan on giving out (with 95% assurance of having enough)? (*Hint:* Use the sample to estimate the population proportion and then go to the population.)

10.5 The recreational department of a small township is attempting to determine if the town should invest in two new tennis courts. The town council will approve them only if there is reason to expect them to be used 85% of the nights (weather permitting). They decide to observe each night at the old courts to see if there are people waiting for others to finish. After checking on 50 nights, they find people waiting on 40 of the nights. Is there reason to believe that there would be people waiting significantly less than 85% of the time (at the 0.10 level)?

10.6 A new pizza parlor believes that if it gives away free pizzas the first day, 75% of the original customers will return in 2 weeks. On the first day 385 pizzas were served, with each person filling out a card with his or her name and telephone number. Two weeks later the pizza owner called 30 of the people and found that 23 had already purchased another pizza. Is the owner's assertion correct? (Test at the 10% level.)

10.7 A printer feels that a printing machine is operating satisfactorily when less than 2% of the pages are poorly printed. The printer examines 100 sheets every hour, and, after a week (40 hours), finds a total of 98 bad sheets among those checked. Should the machine be repaired? What do you estimate the total number of bad sheets to be if 600 pages are printed an hour?

10.8 The president of a large company was persuaded to purchase a $100,000 automatic machine. The chief engineer claimed it would operate 60% of the time. The president randomly selected 15 times to observe if the machine was in operation. Seven of these times it was not being run. Should the president complain to the engineer (the president wants to be wrong less than 5% of the time)?

*10.9 The president now asks how many times must the machine be sampled to make certain the sample values will be off no more than 3%. The formula to use is

$$n = p(1 - p)(z_\alpha/\text{error})^2 \qquad (10.11)$$

where *error* is the maximum amount the president wants to be off (it is $\frac{1}{2}$ of the confidence interval). If α is 0.05 (95% assurance)

$$n = p(1 - p)(1.96/\text{error})^2$$

If p is expected to be 0.60 and the error is to be 0.03 (3%), how large a sample must be taken?

*10.10 If you have no idea what p should be, then the largest possible value of Equation 10.11 would be when

$$p(1 - p) = \tfrac{1}{2}(1 - \tfrac{1}{2}) = \tfrac{1}{4}$$

Thus, you can always be safe and use the formula as

$$n = \tfrac{1}{4}(z_\alpha/\text{error})^2$$

Redo Problem 10.9. How different is the necessary sample size?

*10.11 The department store owner in Problem 10.1 wanted to be 99% certain that the estimate of the number of shoplifters was good within ± 2%. If the owner doesn't want to make a preliminary guess of p, how large a sample should be taken? (Note that z_α is not 1.96 but 2.58 because of the 99% significance level.)

*10.12 If the department store manager feels that p is about 0.20, then how large a sample should be taken? How does this sample size compare to the one in Problem 10.11?

*10.13 Derive Equation 10.11. Start with Equation 10.5, where X/n is the extreme and $(X/n) - p$ would be the error.

SECTION 10.2

10.14 A grocery store kept a record of the sales of its own brands of coffee (in pounds). It found the following results for sale days and nonsale days (the sales were on other brands).

	SALE DAYS	NONSALE DAYS
Total coffee	185	153
Its brand	27	51

Is there a difference in the proportion of sales at the different times? (Test at the 10% level.)

10.15 A department store had found that 41 out of 150 customers had been shoplifting. A closed circuit TV system was installed with warning notices posted. After a month another study was conducted. Another 150 people were followed and 28 were observed taking unpaid merchandise. Was there any significant decrease in the amount of theft? (Justify at the 1% level.)

10.16 A student stood at the entrance to a large discount shopping center and polled people entering the store. He recorded yes, no, no opinion, and refused to answer. On a day when he dressed up in a suit he had 45 refusals out of a total of 140 persons asked. The second time he had a beard and dirty clothes. This time 60 out of 135 persons refused to answer. Is there a significant difference in proportions?

10.17 A group of students protested to the Dean that two teachers were giving very different numbers of failure grades. Teacher A failed 28 out of 100 students, while Teacher B failed 43 out of 125 students. What would the Dean's response be if an investigation would be continued only if there was less than a 1% chance that the grades could be essentially from the same population?

10.18 After the telephone company instituted a reduced rate for long distance calls after 6 P.M., a study was conducted to see if the proportion of long distance calls changed. Before the change took place, 200 calls were checked and 37 were found to be long distance calls. After the change, another 200 calls were checked; this time 42 were long distance calls. Was there a significant change? (Test at the 5% level.)

SECTION 10.3

10.19 Ten children were given the choice of playing with either of two toy cars. The cars were identical except that one was red and the other was yellow. If 8 picked the red car, is this sufficient reason to state there is statistically significant preference for red? (Use a sign test.)

10.20 A series of chemical reactions was observed to decide if a particular chemical acted as a catalyst to speed up the reaction. Twelve pairs of tests were conducted and the reactions with the "catalyst" were found to be faster in 10 of the pairs of tests. Do you conclude that this "catalyst" does speed up the reaction?

10.21 The number of wrong answers to a set of simple arithmetic problems were recorded for 15 subjects in the morning and in the evening. The results were:

Morning	3	2	8	6	3	5	3	2	1	2	4	3	2	1	1
Evening	7	1	10	10	2	1	8	6	0	2	3	1	6	2	0

Use the sign test to determine if there is a difference in the number or errors made in the morning and evening.

10.22 A wear test for two different highway paints was conducted by painting stripes on a road. Pairs of lines were drawn. After several months, the sets of stripes were compared and the paint sample in better condition was noted. The two paints were listed as A and B. The results of 20 samples were:

A, A, A, A, B, A, A, B, A, A, Neither, B, A, A, B, A, B, B, A, B

Use a sign test to determine if one paint is superior.

10.23 Following a rainstorm, the change in temperature was noted. After 50 such rainstorms it was found that there was an increase in temperature 8 times, no change 5 times, and a decrease the remaining times. Is there evidence to state that there is a statistically significant change in the temperature following rainstorms?

CHAPTER ELEVEN SOME NONPARAMETRIC METHODS

In the last two chapters you were introduced to some statistical tests. In Chapter Nine we discussed a variety of tests that all required assumptions of normality. In addition, what we actually were doing involved decisions regarding population *parameters* (these are, for example, the population means or variances, and are really numerical values that identify particular distributions). However, many statistical tests have been developed that do not depend on the shape of the distribution or deal with the parameters of the underlying population. We have already considered one such test in the last chapter, the sign test. These tests fall in the large classification called nonparametric or distribution-free tests. We shall only look at a few such tests here, but they should give you some understanding of the variety and use of these statistical methods.

11.1 OVERVIEW

In this chapter we will look at three different nonparametric statistical tests or methods. The major part of the chapter will be devoted to the χ^2 (chi-square) test. This test can be used to determine if categorical data shows dependency or if two classifications are independent. It can also be used to make comparisons between theoretical populations and actual data when categories are used.

A second method, called the Mann-Whitney U-test, will compare two independent distributions to determine if they are different.

The third test will consider a *single set of values* and will attempt to provide a way of deciding if one of the measurement in a set is an *extreme* value that we might either reasonably decide doesn't belong with the rest of the values or provides an indication that perhaps some other population or source is present.

11.2 THE χ^2-TEST

Quite frequently we are easily able to identify simultaneously each individual according to two *separate* classifications. In an opinion poll we may find reported results that are broken down into designations such as region of country or age group, and we may ask if the responses are independent of the subdivisions mentioned.

We may ask if the use of different methods of birth control are different for families of perhaps different religious beliefs or income levels.

A study may be made of the defects in new cars, identifying both the model and the kind of defect, possibly to determine if there might be differences in the frequency of certain defects for different models.

Obviously, many other examples could be listed, but we will now look at how we attempt to determine if such a dependency among classifications exists.

First we should use a simple example to illustrate the concept and then we will broaden the scope of the categories. I should emphasize at this point that the chi-square test is generally not considered very good for comparisons with only two possiblities for each of the characteristics. There are correction factors (one called the Yates correction) that make the results more realistic. However, a two-by-two arrangement is excellent for illustrating the technique.

EXAMPLE 11.1

There is a continual argument by the "older" generation that long hair and "hippie" style are related to smoking marijuana—that is, long-haired "hippies" smoke marijuana, clean-cut students don't. A survey of 75 students provided the following two pairs of breakdowns.

Long hair	25	Smoke marijuana	33
Clean-cut	50	Don't smoke marijuana	42

This listing is not a satisfactory arrangement for observing a possible relationship between marijuana and hair length. The question of whether two things that are being measured are independent is really a question of relationship. Thus, we must have pairs of observations for each individual. In this case we have the arrangement in Table 11.1. Here we find, for example, that of the 33 students who smoke marijuana, 13 have long hair and 20 are clean-cut. Of the 50 who are classified as clean-cut, 20 smoke marijuana and 30 do not, and so forth.

The problem is that just looking at Table 11.1 does not really give you an indication of whether or not there is any relationship between these characteristics. We need to compare the values in the table against some

TABLE 11.1 Breakdown and Relation Between Marijuana Smoking and Life Style (as Expressed by Hair Length)

STYLE	MARIJUANA		TOTAL FOR HAIR
	SMOKE	DON'T SMOKE	
Long hair	13	12	25
Clean-cut	20	30	50
Total for marijuana	33	42	Overall total 75

kind of reference. We will now attempt to determine what values to expect in *each* box.

We start off by agreeing on the totals. They are the values we listed before and are listed along the sides of the chart in Table 11.1 (see also Table 11.2a). Then

If the two characteristics are independent, each subgroup will be divided up into the same proportions as the totals are divided.

Thus, we find that the total of 75 students are split into two groups according to hair length with 25/75 or 1/3 having long hair and the other 50/75 or 2/3 labeled as clean-cut. We then reason that *if there is no relation* between hair length and smoking marijuana then each of the two categories

TABLE 11.2 Arriving at the Set of Theoretical Frequencies Assuming Independence

	MARIJUANA SMOKE	MARIJUANA DON'T SMOKE	
Long hair			25
Clean-cut			50
	33	42	75

a. Identify the totals for each row and each column.

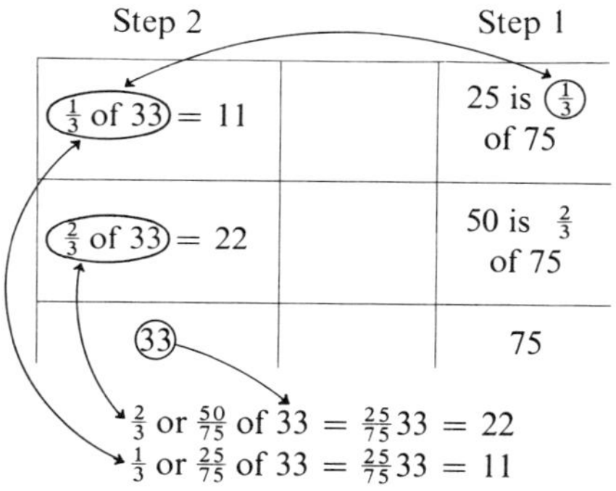

b. Find the proper proportions for one column (or row) by calculating the proportion of the row total and multiplying the proportion times the total in the particular column you are splitting.

11	14	25
22	28	50

c. Final theoretical values can be found by subtraction. The 14 is found as 25 − 11, and 50 − 22 = 28

Also $\frac{33}{75}$ of 25		25
		50
33	42	75

d. We can also find the same 11 in the upper left corner as 33/75ths of the 25 in the row total.

under marijuana will *also* be subdivided 1/3 long hair and 2/3 clean-cut. Therefore, we expect 25/75 of 33 or 11 of the 33 students who smoke to have long hair and 50/75 of 33 or 22 students who smoke to be clean-cut (see Table 11.2b).

In a similar manner we can subdivide the 42 students who don't smoke into two categories of $(25/75) \cdot 42 = 14$ *with long hair* and $(50/75) \cdot 42 = 28$ *clean-cut*. We finally arrive at Table 11.2c.

Two additional important observations should also be made

1. It does not matter which order we use to arrive at the theoretical values. We can also calculate the upper left corner as $(33/75) \cdot 25 = 11$. Here we are finding the proportion of long-hair students who also smoke marijuana. The net result, however, is the same.
2. In this problem, no matter what value you start with in the table, once you have found *one* value, all of the others are established. Once you have found the 11, you can immediately find the 22 and the 14 by subtraction. After you find either the 22 or the 14, you can also find the 28 by subtraction. So only one value can be freely varied, or there is only 1 degree of freedom.*

We have now found a collection of *theoretical values* to complement the actual observations. In Table 11.3 we have both sets of values entered in the same areas (the theoretical or expected values in the small boxes). Next we will compare the differences by computing a value called χ^2 (chi-square) which is found as

$$\chi^2 = \sum \frac{(\text{Actual frequencies} - \text{Theoretical frequencies})^2}{\text{Theoretical frequencies}}$$

The sum here is to include the calculations for *all* of the boxes.

TABLE 11.3 Final Table for Determining if there is a Relation Between Marijuana and Hair Type. The Vaues in the Small Boxes are the Theoretical Values (Assuming Independence)

STYLE	MARIJUANA		TOTAL
	SMOKE	DON'T SMOKE	
Long hair	[11] 13	[14] 12	25
Clean-cut	[22] 20	[28] 30	50
Total	33	42	75

* Discussions of degrees of freedom occurred in several places including pages 193–4 and 203–4.

To illustrate the computations we will perform them separately from the table. The first step is to compute the difference between the actual and theoretical frequencies for each case.

	actual frequency	−	theoretical frequency		
Long hair—smoke	13	−	11	=	2
Long hair—don't smoke	12	−	14	=	−2
Clean-cut—smoke	20	−	22	=	−2
Clean-cut—don't smoke	30	−	28	=	2

Next we square each of these differences. This will get rid of the minus signs and will emphasize large differences.

	(actual frequency − theoretical frequency)^2	
Long hair—smoke	$(2)^2$	$=4$
Long hair—don't smoke	$(-2)^2$	$=4$
Clean-cut—smoke	$(-2)^2$	$=4$
Clean-cut—don't smoke	$(2)^2$	$=4$

We must recognize an important characteristic of the data. A difference of 2 out of an expected 11 is a much larger relative difference than a difference of 2 out of 28 (the first is a difference of 18%; the second is only 7% different). This becomes even more critical when the differences are not all the same value (as will be the case when we enlarge the size of the table). We adjust the values by dividing by the theoretical values.

Since we are comparing squared differences, we will divide each square difference by its theoretical or expected frequency. Thus, for each difference we find

$$\frac{(\text{actual frequency} - \text{theoretical frequency})^2}{(\text{theoretical frequency})} \tag{11.1}$$

or

Long hair—smoke	$\frac{4}{11} = 0.36$	Clean-cut—smoke	$\frac{4}{22} = 0.18$
Long hair—don't smoke	$\frac{4}{14} = 0.29$	Clean-cut—don't smoke	$\frac{4}{28} = 0.14$

We sum up these values to obtain the χ^2-value.

$$\chi^2 = 0.36 + 0.29 + 0.18 + 0.14 = 0.97$$

Finally, we compare this to the critical values of the χ^2 in Table A.3. We choose values for a one-sided test, since we are only interested in if the χ^2-value is very large. As we noted earlier, there is only 1 degree of freedom. We will consider a 5% level of significance. Therefore, we find the 95% χ^2-limit to be 3.84. As before, if our computed statistic is less than the limit, we would state that there is no significant difference, or these two general categories are independent of each other. The computed statistic is much less than the limit and is therefore *not* significant. We state there is no relation between smoking marijuana and hair style; they are independent (or the data is not sufficient to show a relationship). Now let us look at an example of a larger table of possibilities.

EXAMPLE 11.2

Two hundred people were polled for their opinion of the economic situation and the President's overall job performance. The complete breakdown for the responses are given in Table 11.4.

TABLE 11.4 Comparison of Attitudes of 200 People to President's Overall Job Performance and the Economic Situation

ECONOMIC SITUATION	PRESIDENT'S PERFORMANCE				TOTAL
	VERY GOOD	O.K.	POOR	NO OPINION	
Good	31	9	4	6	50
Poor	10	21	60	9	100
No opinion	9	16	10	15	50
Total	50	46	74	30	200

Observe the similarity between this table and Table 11.1 for the previous example. The only difference is its size. As in the previous example, we must find a set of theoretical or expected frequencies given the assumption that the responses on the two questions are unrelated. If they are unrelated, when we look at the responses to the question on economic situation, we observe the distribution of these responses (1/4 good, 1/2 poor, and 1/4 no opinion). We then assume that the 50 very good responses to the *second* question *will be distributed in the same* manner (1/4 good, 1/2 poor, and 1/4 no opinion). Thus we would expect 12.5 good, 25 poor, and 12.5 no opinion responses. Table 11.5 includes all of the

TABLE 11.5 Calculating the Expected Frequencies (Theoretical) for Poll on Attitude on President's Performance and Economic Situation

ECONOMIC SITUATION	PRESIDENT'S PERFORMANCE				
	VERY GOOD	O.K.	POOR	NO OPINION	
Good	12.5	11.5	18.5	7.5	
Poor	25.0	[23.0]	37.0	15.0	[100]
No opinion	12.5	11.5	18.5	7.5	
		[46]			[200]

As one example of the calculations the box under row "Poor" and column "O.K." is emphasized. It is found by

$$\frac{\text{(Row total for row of box)(Column total for column of box)}}{\text{Grand total}} = \frac{(100)(46)}{200} = 23.0$$

expected responses. Note that in general the expected frequency for any given box is

$$\frac{\text{(Row total for the row of that box)(Column total for column of that box)}}{\text{(Grand total)}} \qquad (11.2)$$

We compute the χ^2 as before, and Table 11.6 provides the computations.

TABLE 11.6 Calculations to Determine the χ^2-Value for Independence of Attitudes on Presidential Performance and Economic Situation

ACTUAL	EXPECTED	(ACTUAL − EXPECTED)	(ACTUAL − EXPECTED)^2	$\frac{(\text{ACTUAL} - \text{EXPECTED})^2}{\text{EXPECTED}}$
31	12.5	18.5	342.25	27.3
9	11.5	− 2.5	6.25	0.5
4	18.5	−14.5	210.25	11.4
6	7.5	− 1.5	2.25	0.3
10	25.0	−15.0	225.00	9.0
21	23.0	− 2.0	4.00	0.2
60	37.0	23.0	529.00	14.3
9	15.0	− 6.0	36.00	2.4
9	12.5	− 3.5	12.25	1.0
16	11.5	4.5	20.25	1.8
10	18.5	− 8.5	72.25	3.9
15	7.5	7.5	56.25	7.5
				$\sum = 79.6$

$\chi^2_{0.99} = 16.8$
$df = 6.$

If you prefer, you could also do all the calculations in the boxes themselves. The total is found to be 79.6. This is to be compared to a χ^2-value in the Appendix Table A.3. The only question remaining is degree of freedom.

Degrees of freedom

A single formula is available to calculate the number of degrees of freedom for any table of the form we have been using (commonly called a contingency table).

$$\begin{aligned}\text{Degrees of freedom} &= (\text{Number of rows} - 1)(\text{Number of columns} - 1)\\ &= (r - 1)(c - 1) \qquad (11.3)\end{aligned}$$

where r = the number of rows and c = the number of columns.

For this example $r = 3$, and $c = 4$. Therefore

$$(r - 1)(c - 1) = (3 - 1)(4 - 1) = (2)(3) = 6 \text{ degrees of freedom}$$

Another way of arriving at the number of degrees of freedom is to find how many numbers you can freely choose to place in the body of the table (remember that the outside values are already established). For example, starting in the upper left corner you can put any of a variety of values in each of the first 3 columns of the top row. After you pick 3 values, the 4th value can be only one number. You can then do the same thing in the second row, now having a total of 6 values that you can choose. But that is where it ends. In the third row there is no longer any choice—try it. Thus we have 6 degrees of freedom.

Now that we have determined the number of degrees of freedom, we can find the χ^2-limit. Looking in Appendix Table A.3, we find under $df = 6$ that the 1% limit, $\chi^2_{0.99}$, is 16.8. Our χ^2 of 79.6 is much larger than this limit and therefore we say that there is much less than 1% chance of error in declaring that these two questions have responses that are *not* independent. We conclude that the responses *are dependent*—the attitude toward presidential performance and the economic situation are related to each other in some way.

χ^2 as a goodness-of-fit Test

Quite often we believe a single set of data should follow the pattern of a certain type of distribution. We might believe the measurements of some characteristic to be normally distributed or occurrences of some type to fall in a binomial distribution. A geneticist might believe the distribution of certain hair type follows, for example, a 9:4:4:1 proportionality. An engineer might want to compare the wearout of bearings to some special mathematical distribution. Whatever the problem, we can ultimately end up with:

1. A histogram (or bar chart) with the distribution of actual frequencies (or relative frequencies).
2. A corresponding theoretical distribution with theoretical frequencies (or theoretical relative frequencies).

We should therefore be able to use the χ^2-statistic to compare these two distributions.

EXAMPLE 11.3

An insurance company is interested in knowing if the manufacturers of the cars owned by students on a particular college campus are distributed differently than the general sales of automobiles (Table 11.7).

The question, then, is how well does this sample compare to a theoretical distribution. *One* such possible distribution might be the sales of new cars in the United States for a given year. In Table 11.8 I have provided the sales for 1976 (including the percentages and proportions of sales).

We also need to compute the expected or theoretical frequencies for the 300 cars we sampled. This is obtained by multiplying 300 by the proportion of sales, and is found in the last column of Table 11.8 (e.g., we expect the number of GM cars to be (0.4754)(300) = 142.6).

TABLE 11.7 Manufacturer of Car Operated by a Group of Students (Number of Students = n = 300)

MANUFACTURER	SAMPLE
General Motors	131
Ford	75
Chrysler	29
American Motors	13
Imports	52

The procedure for calculating χ^2 remains the same. We compute the difference between the actual and theoretical frequencies, square each difference, and divide each squared value by its corresponding theoretical frequency. These calculations are summarized in Table 11.9. We then add up the values in the last column to obtain the χ^2-value:

$$\chi^2 = 0.94 + 0.96 + 2.43 + 4.24 + 1.34 = 9.91$$

Finally, we compare this to the critical value of the χ^2 in Table A.3 (under one-sided test). In this type of problem the *degrees of freedom* are equal to *categories minus* 1. Since there were 5 categories, the test limit would be χ^2 with $5 - 1 = 4$ degrees of freedom. (Note that once 4 of the frequencies are chosen: 131, 75, 29, and 13, the last must be 52; therefore you only have the freedom to vary 4 of the frequencies.) The 95% value, the critical value that has a 5% chance or error, is 9.49. Our quantity exceeds this, and we therefore state that there is a significant difference between the sample distribution and the theoretical population of cars. Apparently, the students have different buying preferences than the overall population (or some other reason may be proposed, for example, see Problem 11.13).

TABLE 11.8 Passenger Car Sales for 1976. This is to Serve as a Theoretical Population to Compare to the Actual Sample of Student Cars

MANUFACTURER	ACTUAL SALES	PERCENT OF SALES	PROPORTION OF SALES	THEORETICAL FREQUENCY FOR 300
General Motors	4,800,716	47.54	0.4754	142.6
Ford	2,256,177	22.34	0.2234	67.0
Chrysler	1,301,940	12.89	0.1289	38.7
American Motors	247,640	2.45	0.0245	7.4
Imports	1,491,600	14.77	0.1477	44.3

TABLE 11.9 Computations of χ^2-Value to Compare Car Distributions

	1	2	3	4	5
CATEGORY	ACTUAL FREQUENCY F_a	THEORETICAL FREQUENCY F_t	$F_a - F_t$	$(F_a - F_t)^2$	$(F_a - F_t)^2/F_t$
General Motors	131	142.6	−11.6	134.56	0.94
Ford	75	67.0	8.0	64.00	0.96
Chrysler	29	38.7	−9.7	94.09	2.43
American	13	7.4	5.6	31.36	4.24
Imports	52	44.3	7.7	59.29	1.34
Sum	300	300.0	0		9.91

11.3 MANN-WHITNEY *U*-TEST

This is a nonparametric test that will permit us to compare two distributions to determine if they are different. It is sensitive to differences in mean, variance, or shape, although we may not be able to identify which of these differences is the cause for a significant result in the test.

The requirements of the test are that the data must be at least ordered and that the population values actually cover a continuous range or scale. The measurements that we observe may be discrete, such as "I like the cafeteria food, I hate the cafeteria food" or your grade is "A, B, C, D, or F." In both of these cases there are obviously many differences within the categories so that we could hypothetically rank the individual observations even further.

When the data is merely categorical or is ordered, we should not use tests involving an assumption of a normal distribution, as for example t, z, χ^2/df tests. In addition, even when the data are measured on an interval scale, there are occasions when we do not believe that the population is normally distributed. In such cases, the Mann-Whitney test is quite useful.

If we take two samples from the same population and rank the values in each of the samples we would expect that considerable overlapping between the samples should take place. If we lump all of the values into one ordered sequence we would expect a random pattern to occur. On the other hand, if the two samples came from different populations, we would expect the two samples to tend toward different ends of the overall group. A statistic called U can be used to gauge the degree of randomness between the two samples.

We are to look at two samples, one of size n_1 and the other of size n_2. We then combine all of the values by placing them in a single sequence in increasing order. In addition, we label or identify in some way the group the values belong to. We then calculate for each and every value of *one* of the samples the number of values of the other sample that is below it. The overall sum is U. The following example should explain this procedure.

EXAMPLE 11.4

Two brands of flashbulb batteries are tested for the number of flashes they can trigger before they are used up. Six "Long-Life" batteries and 8 "Brand X" batteries are tested and the number of flashes are recorded in Table 11.10. We can assume that the true population includes the possibility of less than a full flash, which often is the case with a partly dead battery. It is also possible for the distribution to be quite skewed.

TABLE 11.10 Number of Flashes with Different Batteries Before Being Used Up

"LONG-LIFE"			"BRAND X"			
64	71	68	69	78	85	80
84	93	76	97	81	104	98
	$n_1 = 6$			$n_2 = 8$		

First we rank all of the fourteen batteries in a single sequence. Table 11.11 shows the ranking. Next to each value is the symbol representing the type of battery associated with the number of flashes. LL stands for "Long-Life" and X stands for "Brand X."

Next we are to calculate the value of U. First we select one of the types of values, for example the LLs. For *each* LL count how many Xs come before the particular LL. Thus we observe that there are *no* Xs before the first LL and we calculate a 0 for the first LL. In a similar fashion, there are 0 Xs before the second LL. For the third LL there is 1 X before it, and we count

TABLE 11.11 Mann–Whitney Test to Determine if there is a Difference in the Number of Flashes Different Brands of Batteries can Trigger

NUMBER OF FLASHES	BRAND OF BATTERY	RANK	NUMBER OF Xs BELOW THE LLs	NUMBER OF LLs BELOW THE Xs
104	X	14	–	6
97	X	13	–	6
95	X	12	–	6
93	LL	11	5	–
85	X	10	–	5
84	LL	9	4	–
81	X	8	–	4
80	X	7	–	4
78	X	6	–	4
76	LL	5	1	–
71	LL	4	1	–
69	X	3	–	2
68	LL	2	0	–
64	LL	1	0	–
			Total, U 11	37

1 for this LL. The next LL also has 1 X before it. The fifth LL has a *total* of 4 Xs before it and we assign a 4 for this LL. The final LL has 5Xs preceding it and therefore is assigned a 5. We then *add up all of the numbers* associated with the LLs and we obtain an 11. This is the value of U.

Appendix Table A.12 provides a set of critical values for U. Let us consider making our decision as a two-tailed test with $\alpha = 0.05$ (5% level). Since $n_1 = 6$ and $n_2 = 8$, we find the minimum critical value of U to be 8. The value we obtained is larger (11) and is therefore *not* significant.

Although we have satisfactorily completed this example, go back to Table 11.11 and look at the line labeled "number of LLs below the Xs." Here you can see that if you had selected the other sample values for comparison, you would have obtained a much larger value of U, 37 in this case. We label this larger value of U as U' (called U prime). There is a relationship between these two values of U.

$$U = n_1 n_2 - U' \tag{11.4}$$

Thus, when we know that one value of U is 11, we can calculate the other as

$$U = n_1 n_2 - U' = 6 \cdot 8 - 11 = 48 - 11 = 37$$

which agrees with the value we found by adding up the values. This is a convenient and simple way of finding the smaller of the two possible values of U or may serve as a check if you have computed both U-values.

Another way of computing U is to work directly with the ranks. This is particularly useful if the n are large. We choose one of the groups and add up the ranks for each of the values. We call this R_1 if we are working with the n_1-values and R_2 if we are working with the n_2-values. Then U is

$$U = n_1 n_2 + \frac{n_1(n_1 + 1)}{2} - R_1$$

or (11.5)

$$U = n_1 n_2 + \frac{n_2(n_2 + 1)}{2} - R_2$$

If we choose LL then from Table 11.11 we find that the ranks are 1, 2, 4, 5, 9, and 11. Therefore

$$R_1 = 1 + 2 + 4 + 5 + 9 + 11 = 32$$

Since $n_1 = 6$ and $n_2 = 8$

$$U = n_1 n_2 + \frac{n_1(n_1 + 1)}{2} - R_1$$

$$= 6 \cdot 8 + \frac{6(6 + 1)}{2} - 32$$

$$= 48 + \frac{6 \cdot 7}{2} - 32$$

$$= 48 + 21 - 32 = 37$$

To find the smaller value of U, we would then substitute into Equation 11.4, and we would obtain $U' = 11$.

TIES Sometimes we obtain several measurements with the same value. Perhaps several batteries had the same number of flashes. If the ties were with batteries of the same brand, samples from the same group, there is no effect on U. However, if the ties are with values from both groups, then there is a problem. The recommended policy is to use the average rank for all of the tied values and then proceed as usual.

For example, if the long-life battery that had 84 flashes instead had 85 flashes, then there would be a "tie" at the 9th and 10th ranks. We would consider both as being at rank 9.5 and instead of 4 Xs before this LL, we would consider there to be 4 1/2 Xs. Similarly, there would be 4 1/2 LLs below this X. The remaining values would be the same and the total Us would be 11.5 and 36.5 instead of 11 and 37. Unless there is a very large proportion of ties, the effect of having ties is negligible and any correction is generally unnecessary.

WHEN n IS GREATER THAN 20 Appendix Table A.12 stops with $n = 20$. When either n is larger than 20 (or even somewhat less), the normal distribution can be used to find limits. You calculate the value of U in the same way as before. Then, knowing n_1 and n_2, we consider that U should be distributed with mean

$$\mu = \frac{n_1 n_2}{2}$$

and standard deviation

$$\sigma = \sqrt{\frac{n_1 n_2 (n_1 + n_2 + 1)}{12}}$$

Then the test statistic would be

$$z = \frac{U - \mu}{\sigma} = \frac{U - n_1 n_2/2}{\sqrt{\dfrac{n_1 n_2 (n_1 + n_2 + 1)}{12}}} \tag{11.6}$$

For a two-tailed test and significance level of alpha, the limits would be

Upper $z_{1-\alpha/2}$
Lower $z_{\alpha/2}$

(Problems 11.18 and 11.19 are examples that require this technique).

11.4 A TEST FOR EXTREME VALUES

Sometimes one value of a set of data appears to be exceptionally large or small. Frequently this may be the result of a gross error in measurement, such as reading your weight on a scale as 132 instead of 137 because you misinterpreted the various lines on the scale itself. Body temperature may be read as 102.3 degrees instead of 102.6 degrees (since every line stands for 0.2 and not 0.1 degrees as might be expected). An instrument or item to be tested may be dropped, damaged, or bumped and cause an erroneous reading. A distraction during a test may result in a poor performance. In these and many other ways we find that a sample may easily contain values that would be classed as blunders, errors, or extremes. Such an extreme could be particularly serious in making a decision based on a small sample (less than about 20). We will look at one test that provides an indication of whether an end value (the largest or smallest value) of a distribution is unusual.

In this test the data must be placed in an ordered arrangement—in an increasing or decreasing sequence. We then label the values of X_1, X_2, $X_3, \ldots$ all the way to X_n for the last or nth value in the sample (of size n). The range of this set of data will be $X_n - X_1$, the difference between the extremes. Now consider two possibilities.

1. If there are *no unusual values*, then the difference between the two smallest values, $X_2 - X_1$, will be small compared to the total range. Figure 11.1*b* shows how these differences will appear on an evenly distributed sample (the total range and the $X_2 - X_1$). These two differences are very different in size.
2. If an *unusual (extreme) value is present*, then the total range will consist mostly of the distance between that single value and the bulk of the distribution. In this case the two ranges (the total $X_n - X_1$ and $X_2 - X_1$) will begin to approach each other (see Figure 11.1*a*).

A table of critical values of the ratio of certain ranges is provided in Appendix Table A.11. You will find some cases when different points are used. This generally occurs when n begins to get large. Figure 11.1*c* indicates in part why this is true. You should also recognize that the same table can be used for either extreme; if you are comparing values on the large end you just use the second set of equations.

EXAMPLE 1.2 (Cont.)
There are a total of 20 trips recorded in Table 1.2. If they are placed in increasing order they are 13.6, 14.1, 14.3, 14.5, . . . , 16.4, 16.4, 16.5, 16.5, 16.7. Is the 13.6 an extreme?

Consulting Appendix Table A.11, we find that for $n = 20$ we are to compute the statistic called r_{22}, which is

$$r_{22} = \frac{X_3 - X_1}{X_{n-2} - X_1} \tag{11.7}$$

Furthermore, there is less than 10% chance that r_{22} would be greater than 0.401, a 5% chance of greater than 0.450, and a 1% chance of greater than 0.535. For our data: $X_1 = 13.6$, $X_3 = 14.3$, $X_n = X_{20} = 16.7$, so

$$X_{n-2} = X_{20-2} = X_{18} = 16.5$$

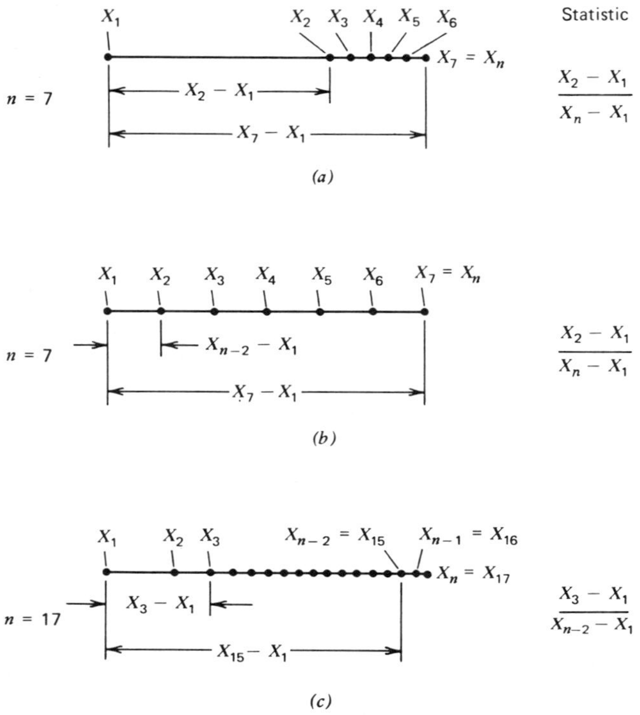

FIGURE 11.1 Considering whether one of the individuals is an extreme. Essentially, two ranges are being compared. (*a*) The n value is very small. The distance from the extreme under consideration (X_1) to its nearest neighbor (X_2) is compared to the total range ($X_n - X_1$). The closer this is to 1.00. the more unusual the value. (*b*) n is the same as in (*a*), but this time there seems to be no unusual value. Observe how small $X_2 - X_1$ is in comparison to $X_7 - X_1$. (*c*) The sample size is somewhat larger ($n = 17$). Instead of dealing with the nearest neighbor and the total range, slight changes are made in the distances to compensate for the possibility of several errors.

Therefore

$$r_{22} = \frac{X_3 - X_1}{X_{18} - X_1} = \frac{14.3 - 13.6}{16.5 - 13.6} = \frac{0.7}{2.9} = 0.241$$

This is smaller than the critical value for the chosen levels, so we declare it is *not* unusual.

EXAMPLE 11.5

The amount of tread on the original tires on my car were measured after 30,000 miles of driving. The depth of tread on the tires was

0.095, 0.102, 0.098, 0.215, 0.108 inches

Is the 0.215 an extreme value? If we order these values with the extreme as X_5 we have the following.

X_1	X_2	X_3	X_4	X_5
0.095	0.098	0.102	0.108	0.215

For $n = 5$ the statistic to be calculated (see Table A.11) is

$$r_{10} = \frac{X_2 - X_1}{X_n - X_1}$$

or

$$r_{10} = \frac{X_n - X_{n-1}}{X_n - X_1} \tag{11.8}$$

where $X_n - X_{n-1}$ is the difference between our extreme and its nearest value. For our problem this is

$$r_{10} = \frac{X_5 - X_4}{X_5 - X_1} = \frac{0.215 - 0.108}{0.215 - 0.095} = \frac{0.107}{0.120} = 0.892$$

Appendix Table A.11 lists the 1% critical limit of 0.780. Since 0.892 exceeds 0.780, we declare that the 0.215 is a very significantly extreme value. It happens to correspond to my spare tire, which essentially was never used.

REVIEW

Parameter • Independence • Categorical
Actual Frequencies • Theoretical Frequencies

Chi-Square Test • Expected Frequency $= \dfrac{\text{(Row total)(Column total)}}{\text{Grand total}}$

Degrees of Freedom $= df =$ (Rows − 1)(Columns − 1)

$$\chi^2 = \sum \frac{(\text{Actual frequency} - \text{Theoretical frequency})^2}{\text{Theoretical frequency}}$$

Mann-Whitney U-Statistic $U = n_1 n_2 + \dfrac{n_1(n_1 + 1)}{2}$ • Rank

$U' = n_1 n_2 - U$ • Extreme Value

PROBLEMS

SECTION 11.2

11.1 In examining 150 grades given by a teacher it was found that this teacher taught two courses—a basic one and an advanced course. The grade distributions for the two courses were as follows.

	A	*B*	*C*	*D*	*F*
Basic	10	20	38	12	10
Advanced	6	12	20	12	10

Does the teacher appear to give different grades for the two courses? (Test at the 5% level.)

11.2 A survey was conducted by a heating contractor to determine preference in method of heating a house. Five hundred homeowners were asked to declare how well they liked the method of heating they were using. The following results were obtained.

	OIL	GAS	ELECTRICITY
Very good	56	79	15
Satisfactory	65	105	30
Adequate	51	46	3
Poor	28	20	2

Do these results indicate that there are different types of preferences for the different heating methods? (Use the 10% level.)

11.3 A sociologist surveyed a group of students graduating from high school and asked them if their parents had gone to college and if they were planning to go to college. The results were as follows.

STUDENT PLANNING TO ATTEND	ONLY FATHER ATTENDED	ONLY MOTHER ATTENDED	BOTH ATTENDED	NEITHER ATTENDED
Yes	36	6	18	10
No	27	8	5	50
Maybe	37	11	52	40

Is there a relationship between parent and child?

11.4 A study conducted to determine if the weather on a given day is independent of the weather on the previous day. One hundred days over a 2-year period were randomly selected and the "average" weather for that day and the previous day were looked up. The following data were obtained. What do you conclude?

WEATHER ON THE PARTICULAR DAY	WEATHER ON THE PREVIOUS DAY		
	SUNNY	CLOUDY	PRECIPITATION
Sunny	15	14	5
Cloudy	8	25	8
Precipitation	5	5	15

11.5 The kind of nuts in 3 differently priced containers of mixed nuts were counted. Are price and type of nuts independent?

PRICE	$1	$2	$3
Peanut	250	185	190
Cashews	45	55	70
Filberts	30	40	25
Brazils	3	10	7
Almonds	5	10	20
Pecans	7	20	28

How would the results change if you were to have only 3 categories of nuts—peanut, cashews, others?

11.6 The age breakdown for the 94th Congress was as follows.

AGE	HOUSE	SENATE
25–34	24	1
35–44	90	11
45–54	157	33
55–64	110	31
65–74	43	18
75–84	7	6

Is there any relationship between age and branch of Congress?

11.7 At the end of 1 month the number of good and bad TV sets were listed according to the production shift that manufactured them. Are defectives independent of shift (note also that this problem can be essentially thought of as test of three proportions, and in many cases the χ^2 test is actually testing multiple proportions?

SHIFT	GOOD TVs	DEFECTIVE TVs
1	821	41
2	793	33
3	785	65

11.8 The doctor at a large ski resort declared that the percent of people who broke legs was the same regardless of how good they were. The resort had the following people for a week.
2000 beginner, 1500 novices, 1000 intermediates, 500 advanced
Of these people the following number broke legs:

41 beginners, 24 novices, 18 intermediates, 9 advanced

Do you agree with the doctor? (Test at the 10% level.)

11.9 If you examine a large collection of cars you would expect the number of dented fenders to be approximately the same for all four fenders. Three hundred cars were examined and we found a total of 560 dented fenders. The distribution was:

Passenger front	165
Passenger rear	170
Driver front	120
Driver rear	105

Is there any indication that certain fenders are dented more than others? What is the implication for a manufacturer or distributor of automobile parts?

11.10 The number of suicides might be expected to be constant and random throughout the year. The following numbers of suicides were recorded bimonthly in a large metropolitan area.

J–F	M–A	M–J	J–A	S–O	N–D
16	22	28	20	11	17

Is there reason to believe that time of the year might be related to the number of suicides? Test at the 5% level of significance.

11.11 A box of grass seed is labeled as having:

30% Kentucky Blue (very good)
30% Perenial Ryegrass (cheaper)
30% Fescue (okay)
10% Annual Ryegrass (cheapest—only lasts a year)

A curious gardener counted out 1000 seeds and planted them. When all sprouted, the gardener counted the different seedlings. There were

260 Kentucky Blue
320 Perennial Ryegrass
240 Fescue
180 Annual Ryegrass

Is there reason to doubt the labeling (is the gardener being gypped)?

11.12 The number of fire alarms experienced in the past in a small town were distributed approximately.

30% false alarms
40% 1-alarm
15% 2-alarm
10% 3-alarm
5% 4-alarm

One hundred and sixty fires reported the next year were as follows.

50 false alarms
62 1-alarm
26 2-alarm
18 3-alarm
4 4-alarm

Is this distribution unusual? (Test at the 0.10 level.)

11.13 Reconsider Example 11.3. One possible explanation for the difference in cars owned by students and the sales in 1976 could be that the survey was taken in early 1977 and few students have new cars. If we were to use the 1975 automobiles sales instead, what would be your conclusion? Recalculate the χ^2-value.

SALES OF CARS FOR 1975	
General Motors	3,747,009
Ford	1,983,723
Chrysler	997,116
American Motors	322,272
Imports	1,578,090
Total	8,628,210

11.14 The χ^2 goodness-of-fit method is often used to try to determine if a set of data is approximately normally distributed. To do this, we first find the mean and standard deviation of the sample data. These then serve as the μ and σ for the theoretical population. We then subdivide the sample into a small set of intervals with approximately equal frequencies. In Example 1.1 these might be the 5 categories: less than 7.5, 7.5 – 9.5, 9.5 – 11.5, 11.5 – 13.5, and more than 13.5.

a. Determine the actual relative frequencies for each of these categories.
b. Next determine the theoretical proportions that would fit in each of these categories for a population with mean $\mu = 10.11$ and $\sigma = 3.41$. These are the theoretical frequencies.
c. You then perform a χ^2-statistic using these 5 actual and theoretical frequencies.

SECTION 11.3

11.15 A collection of smoke detectors was tested to determine whether there was a difference in response between photoelectric-type or ionization-type detectors. Six photoelectric and 8 ionization detectors were placed in a room where a fire was started. The order of response was

I, I, P, I, I, P, I, I, P, P, I, I, P, P

Is there a difference? (Use the Mann-Whitney test at the 5% level.)

11.16 The main method of testing hardness of minerals is by scratching. The granite from two mines, V and M, were compared, and the order of hardness was as follows.

Softest M, M, V, M, M, V, V, M, V, V Hardest

Is there a difference in hardness?

11.17 Two brands of calculators T and H were used by a group of students. When they went dead and needed a recharger, the instructor noted the brand that failed. Eighteen failed, as follows.

T, T, H, H, T, T, T, H, T, T, H, T, H, H, T, T, H, T

Is there a difference in the length of time before failure? What other questions and other statistical analyses might be performed?

11.18 The age in years and months that the first permanent tooth appeared was reported for 35 children as follows: (B is boy and G is girl, 6-8 means 6 years, 8 months).

B 6-8	B 6-2	G 5-12	B 6-8	G 5-12
B 5-10	B 5-12	B 6-3	B 5-12	B 6-11
G 7-3	G 6-3	G 5-7	G 5-6	G 5-6
G 6-9	G 5-10	B 7-0	G 7-1	B 7-2
B 6-1	B 5-11	G 6-2	B 6-9	G 5-10
G 5-2	B 5-9	G 6-3	G 8-0	B 5-9
B 5-12	B 6-3	B 5-10	G 6-8	B 6-4

Is there a difference between boys and girls for the time of the first permanent tooth?

11.19 If the first two columns in Problem 11.18 contained the list of children who had molars as their first tooth and the last three columns list those who had incisors first, test to determine whether there is a significant difference in which tooth emerges first.

11.20 The number of home runs hit by the Home Run Champions from 1935 to 1976 are listed by league. Use a Mann-Whitney test to decide if one league tends to have home run leaders that hit more home runs.

American League Champions:

36, 49, 46, 58, 35, 41, 37, 36, 34, 22, 24, 44, 32, 39, 43, 37, 33, 32, 43
32, 37, 52, 42, 42, 42, 40, 61, 48, 45, 49, 32, 49, 44, 44, 49, 44, 33, 37,
32, 32, 36, 32

National League Champions:

34, 33, 31, 36, 28, 43, 34, 30, 29, 33, 28, 23, 51, 40, 54, 47, 42, 37, 47,
49, 51, 43, 44, 47, 46, 41, 46, 49, 44, 47, 52, 44, 39, 36, 45, 45, 48, 40,
44, 36, 38, 38

SECTION 11.4

11.21 Refer to Problem 2.6. Is the number of customers unusual for any particular days?

11.22 Refer to Problem 3.1. Are any of the values in those sets of data that are unusually large or small at the 5% level?

11.23 Refer to Problem 3.3. Are there any unusually large numbers of ants on a particular day?

11.24 Refer to Problem 3.5. Are there any unusually large or small numbers of sales of doughnuts?

11.25 Refer to Problem 3.6. Does any one have an abnormal weight loss (compared to the rest of the group)?

11.26 Refer to Problem 3.7. Are any houses unusually high priced?

11.27 Refer to Problem 3.8. Are the numbers of accidents in any year unusually large or small?

11.28 Refer to Problem 3.9. Are any of the lumber prices unusually high or low?

CHAPTER TWELVE MULTIPLE SAMPLE CASES—ANALYSIS OF VARIANCE

We have already done some testing of pairs of variables to decide if there are significant differences between them. We now want to enlarge this type of problem to include multiple sets of data. In a great many cases we do not want to be limited to only two sets of data or two conditions or two characteristics.

12.1 INVESTIGATING MULTIPLE SETS OF DATA

Let me pose a question that might serve as an area of investigation. A practical inquiry that any automobile driver should be interested in would be, "What factors significantly affect the gasoline cost per mile?"

Immediately, we should be able to write down a collection of criteria that would appear to be of practical and reasonable importance.

A. Several factors affecting cost per mile depend on the gasoline.
1. Cost per gallon.
2. Brand of gasoline.
3. Kind of gasoline (premium, regular, or intermediate).

B. A few factors involve the kind of automobile and its condition.
1. Engine (including such things as number of cylinders, engine cubic inches, horsepower).
2. Make of car.
3. Length of time since tuning the engine.
4. Size of tires or air pressure.

C. A third group of factors involve the driving conditions.
1. The actual driver.
2. Weather.
3. Kind of roads (including city versus rural, city versus superhighway).
4. Speed.

We have now listed about a dozen important considerations and I suppose you might be able to add at least several more. Go back to the previous chapters and try to think about how we might investigate some of these variables. I hope you realize the hopelessness of such a request. At best we can look at a large number of pairs of conditions, each of which would require a great number of tests. So we might hold everything constant by picking 1 individual car and testing 2 different brands of premium gasoline. We would need to take a fairly large number of test measurements before we would be able to make a decision. Then we might pick a third brand and test it against each of the others by running many more tests. And then you would *rerun* the previous sets of values for the gasolines already checked. If we wanted to compare a different make car we would again have to run a large number of trials in order to perform a satisfactory statistical test. This technique is often called the "hold everything constant" method. It is unfortunate that this method is very common. It requires an excessive number of tests and far more time and effort than should be necessary. In addition, the more times you use the same set of data in calculating a statistic of performing a statistical test, the greater is your chance of making an error. (For example, we may have set alpha at 0.05 and calculated a t-value. If our sample t exceeded the limit, we say there is a difference, and we assume there is only a 5% chance of making an error. Now, if you use one of these same sets of data over again, and compare it to a third set of values by calculating another t-value, then the *same limit* would not be giving you an alpha of 0.05, but alpha would be almost 0.10.* Your chance of making an error has doubled.) If you perform enough pairs of tests you will certainly get a few to be significant, even if there are no actual significant differences. This is one of the serious drawbacks of the "hold everything constant" procedure, unless you perform all of the reruns.

A second problem is that certain relationships may be left unidentified. There often exist subtle interdependencies that we cannot identify without a more elaborate kind of statistical manipulation. I am referring to a relation called an *interaction.* An example of interaction is the kind of situation where the brands of gas may work differently with 2 cars. Perhaps Brand A does well in my car while Brand X does well in your car. Is it the brands that are different, or the cars? Or is it just that there is a peculiar interaction or interrelation between the cars and the brands? This interaction confuses our ability to arrive at a straightforward decision. Interdependencies become even more difficult to recognize when we look at numerous sets of data or broaden the problem to include several more of the variables.

Thus, you should be aware of some of the major limitations of the previous approaches to statistical analysis. For practical problems and true research work we must expand our horizons. The general category or heading that covers the cases of looking at more than two samples is called the *design of experiments.* The most common approach to statistical design of experiments is the application of a technique called the *Analysis of Variance* (abbreviated as ANOVA). As the name implies, we will be doing some kind of analysis, determination, or comparison of the variances associated with some sets of data.

* This is found as $1 - (0.95)^2 = 1 - 0.902 \approx 0.098$.

12.2 THE BASIC PRINCIPLE OF ANALYSIS OF VARIANCE (ANOVA)

We originally set the stage for the analysis of variance in Chapter Two when it was demonstrated how you could add together two different distributions to obtain a third, more variable distribution.

Let us go back to the discussion of the traffic tickets. I will now reveal some additional information. The values listed were not listed in random order in each column (recall that in Chapter Seven we observed that each column contained 2 sets of 4 readings taken on different days). The four readings were also listed in an order, such that the first 4 rows were:

Tickets in eastbound lane, morning.
Tickets in westbound lane, morning.
Tickets in westbound lane, afternoon.
Tickets in eastbound lane, afternoon.

The next 4 rows repeated this sequence. Thus, on day 1, there were 14 tickets issued in the morning in the eastbound lane, 11 tickets in the morning in the westbound lane, 13 tickets in the afternoon in the westbound lane, and 9 tickets in the afternoon to cars going east (see the upper left portion of Table 7.2.)

In Figure 12.1*a* we have four histograms, one for each of the four different conditions listed above. Observe that *each of these* small samples has a mean *and* a spread or variance—and we can calculate both of these quantities for *each* sample.

The mean, variance, and standard deviation of the samples are as follows:

	MEAN	VARIANCE	STANDARD DEVIATION	NUMBER OF VALUES
East —morning	12.43	6.11	2.47	28
West—morning	8.68	9.56	3.09	28
West—afternoon	11.89	6.10	2.47	28
East —afternoon	7.43	7.62	2.76	28

Observe that the 4 variances are quite close to each other. Even more important, however, is the fact that each of these variances represent variation that we can consider only as random variation. These values of variance are the best quantities that we have to judge pure or unidentifiable error.

We frequently identify the different categories as treatments. We then talk about the variances of the samples as the *within treatments variance*. Generally, we will assume that the variances of the different populations from which the samples came are effectively the same. If the variances are to be considered essentially identical,* then we could average the sample variances together to obtain an overall *within treatments* or *within samples*

* Remember though, they are not necessarily really identical. We are assuming that they are from sources that all have the same variance.

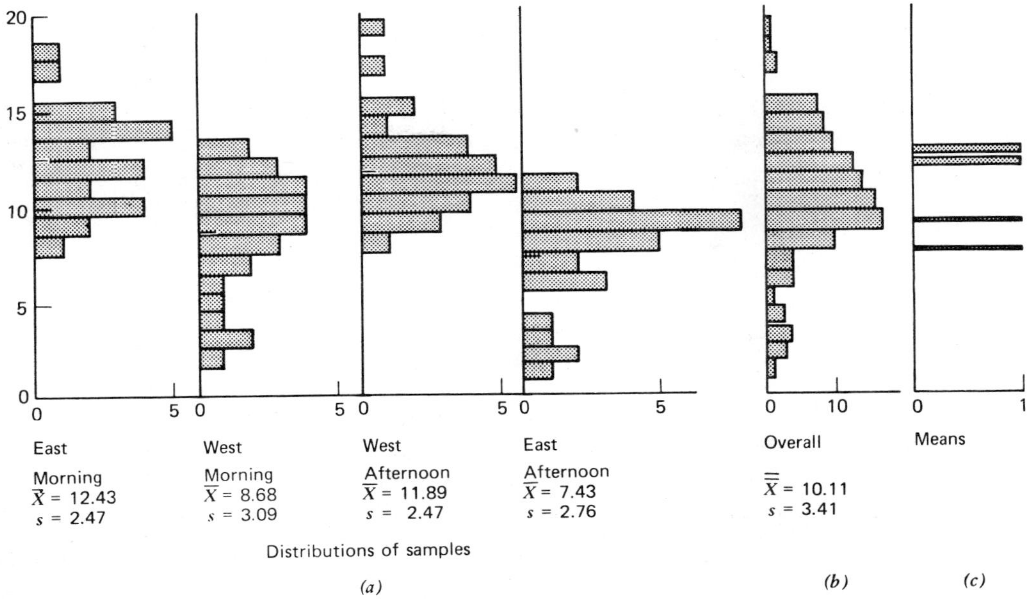

FIGURE 12.1 Comparing the histograms of the small samples within the overall distribution of tickets issued. We are to determine if the means of the smaller samples differ by comparing the spreads within the samples (*a*) to the spread of the sample means (*c*) and the overall distribution (*b*).

variance. This then becomes one important measure of variability that we will use. The averaged within sample variance is 7.35 (and is computed by weighting the variances according to the size of the sample used in computing the respective variances).

Another histogram that we can draw is for the overall distribution. It has a mean of 10.11 and a variance of 11.63. Now comes a key question. If the small distributions all had the *same population mean*—were centered at the same location—how would the variances of the overall distribution compare to the variance of the small samples?

The two variances would essentially be the same. If there was a difference in the locations of each of the small samples, the *variance* of the *overall*, combined distribution would be *larger than* the variance of the small distributions. Then the two estimates of the true population variance (the one based on the small samples and the other based on the overall distribution) would differ significantly (see Figure 12.2).

Now, how do we test the variances? We compute an *F*-ratio.* In this case we are concerned only if the variance of the overall population is significantly larger than our estimation based on the little groups. Thus we are looking at *one-tailed* tests.

* Refer back to Section 9.7 for a discussion of the *F*-test. Instructions on finding values from the tables are provided with Appendix Table A.4.

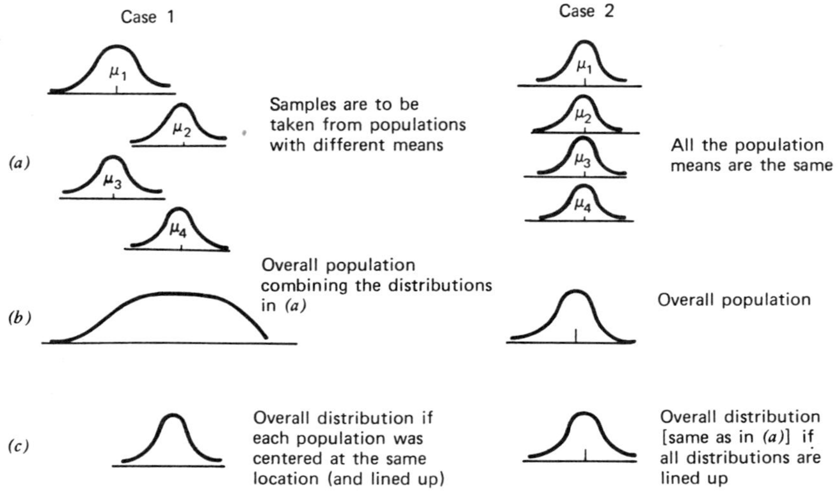

FIGURE 12.2 The F-test compares the variation of (*b*) to the variation of (*c*) for the possible distributions. If the samples all came from populations with the same mean we would have Case 2, and the spreads of the typical distribution would be the same as the overall distribution (not adjusted). If the samples came from populations with different means the two distributions have different spreads—Case 1.

The value we compute is

$$F = \frac{\text{Estimate of population variance based on Overall distribution}}{\text{Estimate of population variance based on within samples variance}} \tag{12.1}$$

or, for this example,

$$F = \frac{11.63}{7.35} = 1.58$$

This quantity is to be compared to the F-values in Table A.4. The degrees of freedom are to be 111 for the numerator (total number of individuals minus 1) and 108 for the denominator (each of the four samples has 28—1 or 27 degrees of freedom; therefore there is a total of 108). For a one-tailed test with $p = 0.95$ (5% chance of error), the table gives the following F-values.

$$F(60, 60) = 1.53 \qquad \text{and} \qquad F(120, 120) = 1.35$$

Thus, $F(111, 108)$ should lie somewhere between these values. But 1.58 exceeds both of these, so we can confidently say (with less than a 5% chance of being wrong) that there *is* a difference in the variances—and we may infer that the differences is a consequence of a difference in locations or values of the means.

We are not finished yet. There is still another way to estimate the true population variance and to look at the data. If you refer to Figure 12.1*c* you will find a third type of distribution. This is a histogram of the *means* of the samples. Observe that they are not all located at the same point.

Should they be?

Of course not!

But how should they be distributed? Well, let us look first at the samples themselves. We can calculate the variance of the *means* as differences from the overall grand mean ($\overline{\overline{X}}$ or X double bar) as follows.

$$\overline{\overline{X}} = \text{overall mean} = 10.11$$

$$s_{\bar{X}}^2 = \frac{\sum (\bar{X} - \overline{\overline{X}})^2}{k - 1}$$

$$= \frac{(12.43 - 10.11)^2 + (8.68 - 10.11)^2 + (11.89 - 10.11)^2 + (7.43 - 10.11)^2}{4 - 1}$$

$$= \frac{17.78}{3} = 5.93$$

where k = categories or treatments.

This value is called the *between* or *among samples variance.* It is a measure of just how much the samples themselves vary. Now for an important question. If you were to rely on the variance *between the samples*, what would you estimate that the *population variance* would be? We must refer back to the central limit theorem and recall that for samples of size n we have

$$\sigma_{\bar{X}}^2 = \frac{\sigma^2}{n} \qquad \text{or} \qquad \sigma^2 = n\sigma_{\bar{X}}^2$$

Since we are dealing with samples, we estimate σ^2 by an s^2 value. The between sample variance is $s_{\bar{X}}^2$ and n is the sample size (observe that things would not be so simple if the samples have different sizes). We then find our last estimation of the population variance, this one based on between samples variance, as

$$s^2 = ns_{\bar{X}}^2 = (27)(5.93) = 160.1$$

(Note that here n is actually taken as the degree of freedom in each sample, or 27.)

Again we will have to test variances and again we have two different estimates of the overall population variance. We will perform an F-test to compare the population variance estimates based on the within samples variance and the between samples variance.

Let's try to reason out what would be unusual. If the four populations were actually located at the same place, then the sample means would be located within a narrow region. We would feel that either of the variances we would calculate should serve as a good estimator of the true population variance.

But what if the samples were actually spread apart and came from different populations? We now have a different situation. The variances of each of the samples might remain the same (thus having the same pooled or averaged variance). However, the means could be spread apart much more than by chance alone. So the variance *between* the samples would be relatively *large*. If there was a significant shift in the means, the between samples variance would be *very much larger* than the within samples variance. So again we would have a one-tailed test, since we will only be concerned if the between samples estimate of variation is too great in

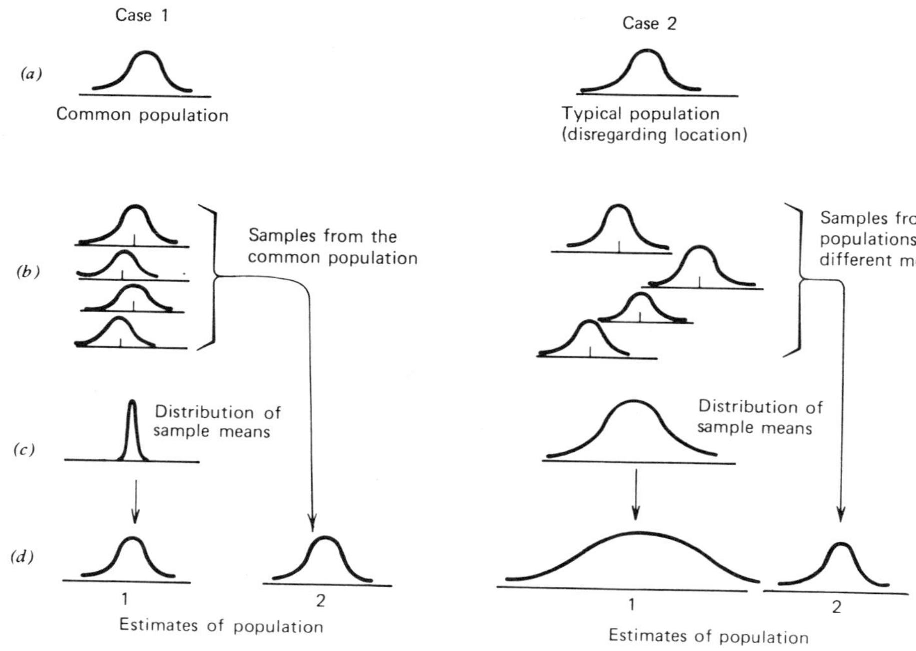

FIGURE 12.3 An alternate way of comparing variances to decide if the samples came from populations with different means. (*a*) The common or the typical populations. (*b*) The samples with means of the samples indicated. (*c*) The distributions of the sample means. These have different spreads depending on whether the populations are the same. (*d*) Two estimates of the population distribution are made for each case. The first estimate is based on the sample means. The second estimate is based on the variation within the typical sample distribution adjusted for differences in mean. If these estimates show significant differences in variance, we conclude that Case 2 is true, that the means are different.

comparison to the within samples variation. We do this with an F-test of

$$F = \frac{\text{Estimate of population variance based on between samples variance}}{\text{Estimate of population variance based on within samples variance}} \tag{12.2}$$

See Figure 12.3 for an illustration of this concept.

From our earlier calculations we found that our estimate of population variance from within samples was 7.35. Thus we find for our example that the F-statistic would be:

$$F = \frac{ns_{\bar{X}}^2}{s^2} = \frac{27(5.93)}{7.35} = \frac{160.1}{7.35} = 21.8$$

This is to be compared to an F-limit. Once more we must determine the degrees of freedom. The denominator we found before to contain 108 degrees of freedom and the numerator contains 3 degrees of freedom (the 4 categories or treatments minus 1). The critical limit with $p = 0.95$ is between

$$F(3, 60) = 2.76 \qquad \text{and} \qquad F(3, 120) = 2.68$$

Our F-value far exceeds these critical values. In fact, not only do we exceed the 0.95 level, but we can go to the 0.99 level where $F(3, 60) = 4.13$ or as far as the 0.999 level* where $F(3, 60)$ is only 6.17. We far exceed even this last value. Thus we can again declare that there are significant differences between the sample means.

The net effect of all of this manipulation is that we have found a very useful and valuable method of comparing the means of several samples. The standard approach, however, will consist of some slight modifications and a formalization of the technique.

12.3 ONE-WAY OR SINGLE-FACTOR ANOVA

We will begin the study of the ANOVA with the simplest type of "many sample" problem. Here we will consider only one factor. The goal at this point may seem to be no more than an extension of the t-test for two samples and, in many respects, this is true.

Refer back to the list of factors that might affect gasoline cost per mile. We can select *one of these factors* and then observe that the reason this factor may be important is that several possible levels or kinds or types of samples can occur within that factor. We are then to determine if there are differences within that factor. Thus, we may look at 4 different brands of gasoline; 3 kinds of gasoline; 5 different drivers; 4 brands or models of cars; and so forth. We will also take several values or repeated measurements to create samples in each of the levels or categories of the factor, but we will only look at one overall classification—brand.

* From another source.

We will approach the general technique by looking at 4 categories or samples, each with a sample size of 3 (3 replications or repeated measurements). In Table 12.1 we have an arrangement showing the values for each of the individuals, and some of the calculations we will need for the $\bar{X}$-values. Note that the table need not be restricted to the number of values listed. For example, the sample sizes could be any amount (and the samples may not necessarily be all the same size). We will determine if there are differences among the 4 brands.

TABLE 12.1 Mile Per Gallon for Four Different Brands of Regular Gasoline (Recorded to Nearest Mile). Some Computations Are Also Included. (Three Separate Readings Taken With Each Brand)

	BRAND OF GASOLINE				
	1 X_1	2 X_2	3 X_3	4 X_2	
	13	11	12	11	
	12	10	11	9	
	14	11	13	10	OVERALL VALUES
Total	$\sum X_1$ or T_1 39	$\sum X_2$ or T_2 32	$\sum X_3$ or T_3 36	$\sum X_4$ or T_4 30	$\sum X$ or T 137
Numbers	n_1 3	n_2 3	n_3 3	n_4 3	n 12
Means	$\bar{X}_1$ 13	$\bar{X}_2$ 10.7	$\bar{X}_3$ 12	$\bar{X}_4$ 10	11.4

We choose the order of purchasing the gasoline randomly by using a random number table. Thus, perhaps, we use gasoline 3 first, then 1, then 4, then 1 again, then 2, and so forth in accordance with random digits chosen from the table. In Table 12.1 we have the miles per gallon for three samples of each gasoline as well as some of the calculations and symbols.

In addition to the individual values, the important quantities that are to be used in further calculations are the totals or the T and the n. We will use these values to calculate variances. Think back to Chapter Three and our discussions of various ways of computing variances, and try to reconsider how we might interpret a variance. As a starter, let us consider a variance as the average of a collection of *sums of squares*. An average requires a division and we will divide by the number of degrees of freedom used in calculating the squares. (the degrees of freedom are equivalent to the $n - 1$ in calculating s^2). Thus, we can state that

$$\text{Variance} = \frac{\text{Sum of squares}}{\text{Degrees of freedom}} = \frac{SS}{df} \tag{12.3}$$

We will now try to work with these two quantities, the sum of squares and the degrees of freedom, separately.

In Chapter Three we found that instead of calculating and squaring many differences, especially from the mean, we could expand the numerator into the form

$$\sum X^2 - \frac{(\sum X)^2}{n}$$

In addition, it does not matter if we subtract a constant from each value. The only consequence of this subtraction is that the numbers become easier to work with. In fact, if we happened to use the mean as the constant then the right term, $(\sum X)^2$, would disappear. So actually that right term is nothing but a *correction factor* that is needed because we are not actually subtracting from the mean. We now need to find three different variances, so we will compute three different sums of squares.

1. The sum of squares for *all* of the values, which is called the *total corrected sum of squares* or just *total SS*. It corresponds to the overall variance and is calculated as

$$\text{Total corrected sum of squares} = \sum X^2 - \frac{(\sum X)^2}{n}$$

$$= \sum X^2 - \frac{(T)^2}{n} \qquad (12.4)$$

For this example,

$$\begin{aligned}\text{Sum of all } X^2 &= \sum X^2 \\ &= (13)^2 + (12)^2 + (14)^2 + (11)^2 + (10)^2 + (11)^2 \\ &\quad + (13)^2 + (11)^2 + (12)^2 + (9)^2 + (11)^2 + (10)^2 \\ &= 1587.0\end{aligned}$$

Since the total $T = T_1 + T_2 + T_3 + T_4 = 137$,

$$\text{Correction factor} = \frac{(T)^2}{n} = \frac{(137)^2}{12} = \frac{18{,}769}{12} = 1564.1$$

$$\begin{aligned}\text{Then Total sum of squares} &= \sum X^2 - \frac{(T)^2}{n} \\ &= 1587 - 1564.1 \\ &= 22.9\end{aligned}$$

2. The sum of squares *between samples* or *between categories*. This corresponds to the variance between or among the means of the categories

and we use the subscripts labeled j to represent different *categories*. We calculate this as

$$\text{Between categories sum of squares} = \sum\left(\frac{T_j^2}{n_j}\right) - \frac{(T)^2}{n}$$

$$= \frac{T_1^2}{n_1} + \frac{T_2^2}{n_2} + \frac{T_3^2}{n_3} + \frac{T_4^2}{n_4} - \frac{(T)^2}{n} \qquad (12.5)$$

$$\text{Between sample or between brands sum of squares} = \sum \frac{T_j^2}{n_j} - \frac{T^2}{n}$$

$$= \left(\frac{(39)^2}{3} + \frac{(32)^2}{3} + \frac{(36)^2}{3} + \frac{(30)^2}{3}\right) - \frac{(137)^2}{12} = 1580.3 - 1564.1$$

$$= 16.2$$

3. The sum of squares for variation *within the samples*. This corresponds to our smallest measure of variance. Here we actually have a separate correction factor for each sample but, since they are all of similar form, we can write the sum of squares as

$$\text{Within samples sums of squares} = \sum X^2 - \sum \frac{T_j^2}{n_j}$$

$$= 1587 - 1580.3$$

$$= 6.7 \qquad (12.6)$$

To complete the variance estimates, we need to determine the denominator values—the degrees of freedom. In each case this is an $n - 1$ term, where n is the number of values that have been used in calculating the squares. Thus we have:

1. The degrees of freedom for all the values, the total *df*. This is the value associated with the total n values. Therefore, total $df = n - 1$ or, in this example, $df = 12 - 1 = 11$.
2. The degrees of freedom between categories. Here, if we call the number of categories by the symbol k, we have k values and $k - 1$ degrees of freedom. Since $k = 4$ (4 brands), we have $k - 1 = 4 - 1 = 3$ *df* for between categories.
3. The degrees of freedom within categories. Here we need to look at each sample and note that there are n_j-values in each sample. Then there would be $n_j - 1$ *df* for each sample. Since each sample in our example has 3 values, we have $3 - 1 = 2$ *df* in each sample. Next we have to add these *df*s for all the samples, and we obtain a total of 8 *df* within samples (4 samples of 2 *df* each). (Note that if we summed up the n_js we would get n *df* but we also lose 1 *df* for each of the k samples, or we lose k *df*. So we would have as a general formula $n - k$ *df*.)

After we calculate the sums of squares and degrees of freedom, we will compute the variance values for the between and the within categories cases. We call these mean squares, since they are essentially averages of the squares. The values are $16.2/3 = 5.4$ for between and $6.7/8 = 0.84$ for within.

To complete our computations, we divide the variance value for *between categories* by the variance value for *within the categories.* We have

$$F = \frac{\text{Variance estimate for between categories}}{\text{Variance estimate for within categories}} = \frac{5.4}{0.84} = 6.44$$

In order to present the computations in a simplified way, a fairly universal approach has been agreed on. This is called an *analysis of variance table.* The general format with the computations is presented in Table 12.2. Although this table may look extremely complicated, if you examine it carefully, you will find it is not as difficult as it might look. The values for our example are summarized in Table 12.3.

You should observe that:

1. The degrees of freedom for *between* and *within* add up to the total. You can quickly find one of them by subtracting. Since n is known, you should immediately be able to write down the total degrees of freedom, and since you know how many categories you are looking at, the degrees of freedom for categories should also be quickly figured.

TABLE 12.2 General Analysis of Variance Table for One-Way ANOVA. There are k Categories and a Total of n values

SOURCE OF VARIATION: YOU IDENTIFY WHAT IS MEANT BY TREATMENT OR CATEGORY	DEGREES OF FREEDOM *df*	SUM OF SQUARES (*SS*) (THESE ARE NEVER NEGATIVE)	MEAN SQUARE (*MS*) (THIS IS *SS* DIVIDED BY *df* AND IS AN ESTIMATE OF VARIANCE TO BE USED IN *F*-RATIO	*F*-RATIO
Between categories or treatments—between means	$k - 1$	$\sum \frac{T_j^2}{n_j} - \frac{(T)^2}{n}$	$\frac{SS \text{ Between}}{k - 1}$	$\frac{MS \text{ Between}}{MS \text{ Within}}$
Within categories or samples, sometimes called *error*	$n - k$	$\sum X^2 - \sum \frac{T_j^2}{n_j}$	$\frac{SS \text{ Within}}{n - k}$	
Totals	$n - 1$	$\sum X^2 - \frac{(T)^2}{n}$		

$n = \sum n_j$

TABLE 12.3 Analysis of Variance Table for Example 12.1

SOURCE OF VARIATION	*df*	SUM OF SQUARES	MEAN SQUARE	*F*-RATIO
Between brands	3	16.2	$\frac{16.2}{3} = 5.4$	$\frac{5.4}{0.84} = 6.44$
Repeated values within brands (error)	8	6.7	$\frac{6.7}{8} = 0.84$	
Total	11	22.9		

5% test limit for F is $F_{0.95}$ (3, 8) = 4.07

2. The *total sum of squares* is equal to the *sum of squares for between* plus *sum of squares for within*. Again, if you calculate two of these, you can find the other by subtracting (or you can use the third as a check). They are *never* negative numbers. If the correction factor is larger than the left term, there is an error in your computation. At most there are only three types or groups of calculations represented in the sum of squares column, and they are all repeated in the different rows.

Finally we perform an F-test to determine if the mean square *between* brands is significantly larger than the mean square *within* brands. We are to use a *one-tailed* test with the following hypothesis.

HYPOTHESIS There is *no* difference between the category means (between brands).

ALTERNATE HYPOTHESIS There is a difference between means of the brands. Choose significance level, α. Let $\alpha = 0.05$.

LIMIT $F_{1-\alpha}(f_1, f_2)$—one-tailed test, $p = 0.95$.

$$f_1 = df \text{ for between brands}$$
$$f_2 = df \text{ within brands}$$

or

$$F(f_1, f_2) = F_{0.95}(3, 8) = 4.07$$

TEST STATISTIC F-value calculated in the F-ratio column; here $F = 6.44$.

DECISION If the test value of F is smaller than the F-limit, we accept the hypothesis of no difference. If the test statistic is larger than the limit, we accept the *alternate* hypothesis that there *is* a difference in means.

Since $F = 6.44$ is greater than the limit of 4.07, we declare that there is a significant difference among these brands of gasoline and that the means are *not* all the same.

12.4 TWO-WAY DESIGN AND INTERACTION

One extension of the previous problem, Example 12.1, is to look at a second factor at the same time. We will look at the same data as before except we will now recognize that the three readings were not all taken under identical conditions. The readings were with three different cars (called A, B, and C). Look in Table 12.4 and observe that the totals and averages for these three cars differ. This tells us that we can go back to the variation that we called "repeated values within brands" and divide it up. We can now assign some of that previously unidentifiable "error" to variations between cars. Thus, we will assume that there is another major classification—between cars. We can calculate the sum of squares for this

TABLE 12.4 Same Data as in Table 12.3 (Example 12.1) Except the Three Values are now Identified as Different Cars. This is a Two-Way Table

		BRANDS OF GASOLINE				TOTALS FOR CARS	MEANS FOR CARS
		1	*2*	*3*	*4*		
	A	13	11	12	11	47	11.75
Cars	B	12	10	11	9	42	10.5
	C	14	11	13	10	48	12
Total for brands		39	32	36	30	137	
Means for brands		13	10.7	12	10		11.4

category in the same manner as we calculated the sum of squares for "between brands." We find

$$\text{Sum of squares between cars} = \sum\left(\frac{T_i^2}{n_i}\right) - \frac{T^2}{n}$$

where the totals to be squared are the totals for cars. Thus we have

$$SS \text{ between cars} = \frac{47^2}{4} + \frac{42^2}{4} + \frac{48^2}{4} - 1564.1$$

$$= 552.5 + 441.0 + 576 - 1564.1$$

$$= 1569.2 - 1564.1 = 5.1$$

The degrees of freedom are 2, since there are 3 categories ($df = k - 1 = 3 - 1 = 2$).

Because we no longer have repeated values, we can no longer directly compute the "within samples sums of squares." We calculate this residual or error variation by subtraction. Table 12.5 is the summary table for this ANOVA problem. The F-values for determining if these classifications are significant are both computed using the mean square for the "error." Both of these F-values are significant at the 0.05 level.

TABLE 12.5 ANOVA Table for Two-Way Analysis. Cars are also Considered

SOURCE OF VARIATION	*df*	*SS*	*MS*	*F*-RATIO	5% *F*-LIMIT
Between brands	3	16.2	5.4	$\frac{5.4}{0.26} = 20.8$	$F(3, 6) = 4.76$
Between cars	2	5.1	2.55	$\frac{2.55}{0.26} = 9.8$	$F(2, 6) = 5.14$
Residual or error	6 [a]	1.6 [a]	0.26		
Total	11	22.9			

[a] By subtraction: For degrees of freedom 3 and 2 are subtracted from 11 to give 6. For sums of squares subtract 16.2 and 5.1 from 22.9.
Both *F*-ratios are significant.

If we were to take repeated measurements for all of the categories we could again obtain a separate independent measure of inherent or smallest variations. We could calculate the sums of squares and degrees of freedom in a way similar to the approach we used in determining the "within samples" variation for one-way analysis. We then find leftover sums of squares and leftover degrees of freedom. These are for another kind of comparison that often is of critical importance. This is a measure of the interrelationship or interdependence among the two different classifications. It is called *interaction.* We will not consider how to perform the calculations for a two-way ANOVA when repeated measurements are taken and an interaction variation would be computed. However, we will look at and attempt to interpret a few ANOVA tables and the corresponding graphs for which these values have been calculated. We would rely on a typical computer printout for the calculations (there are a number of library programs available that can provide the ANOVA table for you).

TABLE 12.6 Two-Way Analysis of Variance with Repeated Measurements of Each Car and Brand. We are able to Identify if there is an Interaction Between Car and Brand.

a. The Data

		BRAND			
		1	2	3	4
Car	A	13	11	12	11
		12	11	11	10
	B	12	10	11	9
		12	9	11	8
	C	14	11	13	10
		13	12	13	12

b. ANOVA Table

SOURCE	*df*	*SS*	*MS*	*F*
Brands	3	25.460	8.486	18.5
Cars	2	16.080	8.042	17.5
Interaction	6	1.917	0.319	0.7
Within	12	5.500	0.458	
Total	23	48.957		

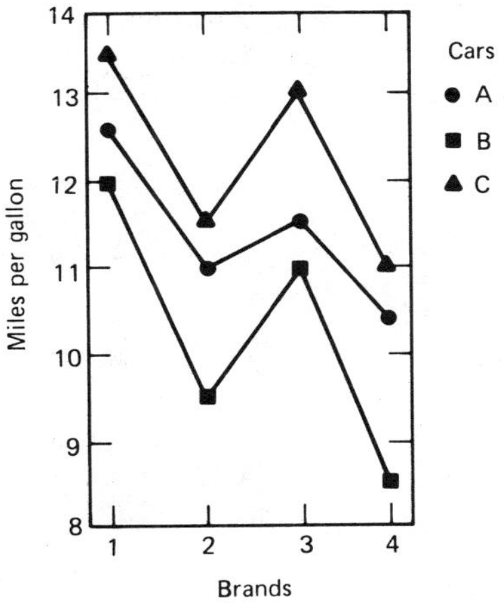

FIGURE 12.4 Graph of the averages of the miles per gallon for different brands of gasoline and several cars. There is no interaction, so that the lines for the different cars are essentially parallel.

Another complete set of gasoline tests were run, and the data are presented in Table 12.6a. The ANOVA table for these values is presented in Table 12.6b. Using an F-test by comparing MS interaction to MS within, we would find that the interaction is *not* considered a significantly large value. We can then proceed to test the main effects, brands and cars, as before.

The important information that the interaction term tells us is that the cars act is a similar fashion from brand to brand (or, correspondingly, the

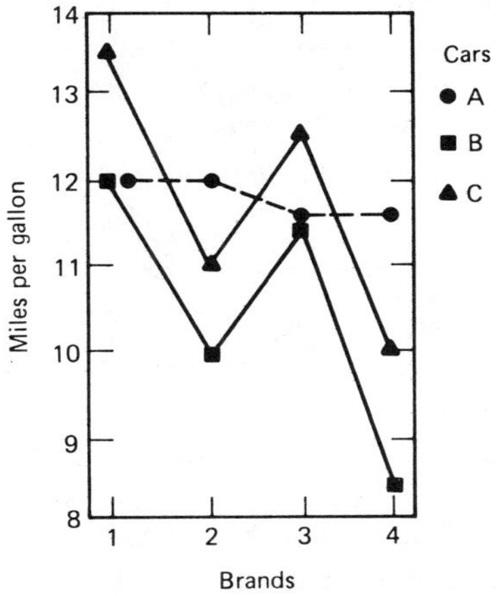

FIGURE 12.5 Graph of the averages of the miles per gallon for different brands and cars listed in Table 12.7. There is a significant interaction which shows up as different lines for cars.

TABLE 12.7 Two-Way Analysis of Variance with Repeated Measurements of each Car and Brand. There is a Different Set of Repeated Measurements this Time and there *is* also a Significant Interaction Term this Time.

a. Data

		BRAND			
		1	*2*	*3*	*4*
	A	13	12	12	11
		11	12	11	12
Car	B	12	10	11	9
		12	10	12	8
	C	14	11	13	10
		13	11	12	10

b. ANOVA Table

SOURCE	*df*	*SS*	*MS*	*F*
Brands	3	21.000	7.000	16.8
Cars	2	8.333	4.166	10.0
Interaction	6	9.000	1.500	3.6
Within	12	5.000	0.4167	
Total	23	43.333		

brands act in the same way from car to car). We will plot the averages for all of the samples. We will select *one* of the factors to be used as the X-axis. In order to distinguish the cars we will use different marks to show the averages for each *car* and then connect the averages for each car. This is done in Figure 12.4. Observe *that all three connecting lines* are essentially identical. When there is *no* interaction you should expect to see a set of similar lines (usually with no lines crossing over each other).

Table 12.7 shows a different collection of sample values, with the interaction term being significant. Now if we examine the graph (see Figure 12.5), we find that the cars act differently. Car A appears not to do any differently with different gasoline, while the other two cars act very similarly. Cars B and C get higher gas mileage with Brand 1 and 3 and get very low mileage with Brand 4. There is a definite interrelation between the cars and brands, and you cannot make any strong statements about gasoline unless you also qualify your conclusions by stating with which car you are dealing. (You can do the same sort of comparison with cars along the X-axis and lines for each brand, and you will also get a set of nonparallel lines.) Thus, performing additional F-tests is somewhat meaningless.

EXAMPLE 12.2

Math and English College Board scores for a group of male and female students are given in Table 12.8. The averages are also provided. Obviously there are no differences between males and females and essentially there are no differences in math and English scores. Correct? Not really. The interrelationship between any kind of test and sex is completely masking any differences. The ANOVA table, Table 12.8b, indicates that almost all of the variation is in interaction. Figure 12.6 shows why.

Here we see that there is a complete crisscrossing of the scores from one sex to the other for math and English (or vice versa). *Because of the interaction, it is absolutely pointless to talk about* differences *between sexes* or

TABLE 12.8 College Board Score for Students of Different Sexes, on the Math and English Portions. A Significant Interaction is Present. The Math and English Averages for Opposite Sexes are Almost the Same, and the Sex Averages are Identical.

a. The Data

SEX	TEST		
	MATH	ENGLISH	AVERAGE
Male	550	480	
	570	460	515
	540	490	
Female	480	540	
	490	540	515
	470	570	
Average	516	514	

b. ANOVA Table

SOURCE	*df*	*SS*	*MS*	*F*
Sex	1	0.0	0.0	
Test	1	33.3	33.3	
Interaction	1	16133.0	16133.0	74.8
Within	8	1733.0	216.7	
	11	17899.3		

differences *between tests.* One must discuss tests for the particular sex, or sexes for the particular test.

We have just scratched the surface of a vast field. The types of problems that the ANOVA can be used with can be, and generally are, much more elaborate. For example, we can analyze designs that can involve many factors at a time, may include estimating the variances (error analysis) or may have ways of overcoming difficulties in randomizing the values.

In some types of advanced problems, regression analysis (which will be discussed in Chapter Thirteen) and ANOVA merge, allowing the categories to become numerical measurements and permitting us to take into account such differences in categories. Much practical work with real data involves statistical analysis, which begin at this point.

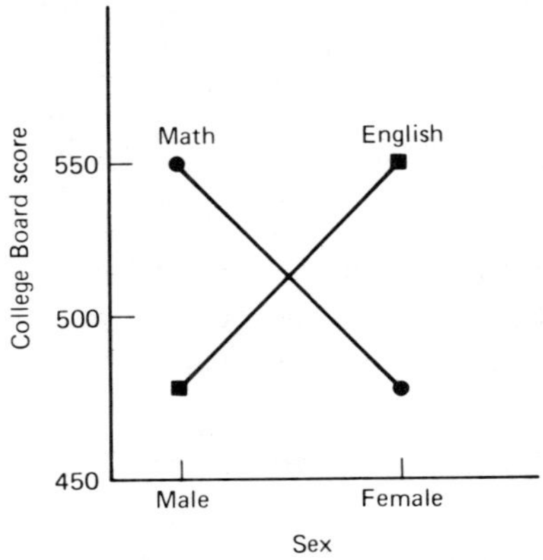

FIGURE 12.6 Graph of sex and College Board scores for the different tests (math and English). There are obvious differences that do not show up as differences in the overall average. The curves for math and English cross over each other, indicating a major interaction.

REVIEW Design of Experiments • ANOVA
Within Sample Variance • Between Samples Variance
Sums of Squares • Mean Squares • *df* • *F*-test
Correction Factor • Analysis of Variance Table
Interaction

PROBLEMS

SECTION 12.3

12.1 Three different groups of preschool children were tested to determine their level of reading preparation. After a year they were tested again, and the increases in the test were recorded as follows.

NURSERY SCHOOL GROUP	SESAME STREET WATCHERS	CONTROL GROUP
6	9	5
7	6	4
6	7	5
8	8	3

Are there significant differences in the amount of improvement that the children have received under the different conditions?

12.2 The number of teenagers per 10,000 who were arrested for possession of drugs were compared for three different types of communities.

SUBURBAN AREA	LARGE CITY	RURAL AREA
12	24	9
16	20	7
15	18	12
22	30	7
10	18	10
6	25	3

Are the means significantly different?

12.3 The level of automotive exhaust emission was measured after using different low-lead or no-lead gasolines. The results in test units were as follows.

BRAND E	BRAND S	BRAND M	BRAND A
6	8	10	12
4	6	8	10
10	8	6	4
5	7	9	11

Do the different brands give different average pollution levels?

12.4 Several drugs are tested on a group of people to determine their effectiveness in reducing blood pressure. The amount of blood pressure reduction in millimeters of mercury was found to be as follows.

DRUG X	DRUG Y	DRUG Z
14	10	11
15	9	11
12	7	10
11	8	11
10	11	8
11	11	7

Do the drugs act differently?

12.5 The average rate of earnings for several companies in different industries were reported in the following percentages.

CHEMICALS	ELECTRICAL	METALS	TRANSPORTATION
8.6	8.1	6.4	7.3
6.2	7.3	6.4	5.7

Are the earnings different for different industries?

12.6 Three different brands of aspirin were to be tested to determine how fast they dissolve in the stomach. Five tablets of each brand were dropped in mild acid (simulating the stomach) and the lengths of time for them to dissolve fully were recorded.

	BRAND		
	BRAG	AND NONE AS FINE	X
	4.3	5.1	5.1
Time to dissolve	4.7	4.7	5.3
(minutes)	5.2	5.1	4.8
	4.2	4.4	5.7

Does there appear to be a significant difference between brands?

12.7 The number of pages of reading assigned by different instructors teaching the same course was recorded for 5 weeks as follows:

MR. A.	MR. P.	MRS. S.	MR. B.	MS. C.
29	30	35	20	35
31	33	30	18	42
34	25	20	14	30
35	27	20	12	29
20	19	17	20	25

Are there significant differences among the teachers?

12.8 Letters were mailed to 3 cities and the number of days before they arrived were determined.

LARGE TOWN NEARBY	LARGE TOWN 1000 MILES AWAY	SMALL TOWN 1000 MILES AWAY
2	2	4
2	3	2
3	2	3
1	2	3
2	3	4

Is there a significant difference in the mail delivery times?

SECTION 12.4

12.9 Examine Table 12.6. Try to verify the value of the sums of squares for the "Within" row. Observe that this is obtained from the variation in repeated values. Also try to verify the total sums of squares and the sums of squares for brands and cars. The interaction term is then found by subtraction.

12.10 Refer to Problem 12.2. Each of the rows was for a different month (for 6 months). Are there significant differences among months? (Extend the one-way analysis into a two-way analysis.) How does the difference in areas change (regarding significance)? You can plot the points to see if an interaction would be somewhat apparent.

12.11 Refer to Problem 12.3. The tests were taken on four different cars. Extend the one-way analysis to a two-way analysis. Plot the values. Is anything unusual about the graph?

12.12 Refer to Problem 12.4. The six readings for each drug were actually pairs of values for three different people. The sum of squares for *within are* 3.500 and for people are 14.78. By subtraction, calculate the sum of squares for interaction. Is the interaction term significant? Plot the averages.

12.13 Refer to Problem 12.5. If the first row of earnings was for 1975 and the second row for 1976, is there a difference for years?

CHAPTER THIRTEEN CORRELATION AND REGRESSION

In Chapter Two we looked at a pictorial way of showing the relationship between a measurement of some quantity and the frequency with which that quantity appeared. This constituted one graphical way of visualizing and describing a distribution of values, one that then told us something about a population that we were interested in. This approach, however, was limited to the distribution of a single set of data. Furthermore, that distribution could only deal with a single characteristic. Now we will explore the idea of looking at two different measurements or *kinds* of measurements simultaneously and observing the ways that they may be related to each other.

13.1 DESCRIPTIONS OF PAIRS OF QUANTITIES

Two sets of data have been provided in Table 13.1. They are measurements of the height and weight of 20 college students. In Figure 13.1 a pair of histograms have been drawn to show the distributions of these two types of measurements.

TABLE 13.1 Heights and Weights of Twenty Students.

HEIGHT				WEIGHT (POUNDS)			
5′ 9″	5′10″	5′10″	5′10″	178	175	150	185
5′11″	5′ 4″	5′ 5″	5′ 3″	171	128	124	118
5′ 6″	5′ 7″	6′ 1″	5′ 9″	138	138	200	154
5′ 8″	5′ 9″	5′ 6″	5′ 6″	142	172	130	144
5′ 7″	5′ 6″	5′ 8″	5′ 8″	163	155	162	133

Observe the general nature of the distributions. The quantities have been recorded in the same order as the measurements were taken; from casually looking at the table or the graphs we have no reason to say that heights and weights are not totally independent of each other.

In order to determine if there is any type of relationship between these two quantities we must bring together the pairs of values for each individual by listing which heights go with which weights. We must say that John is 5 feet 9 inches *and* 178 pounds, Frank is 5 feet 11 inches *and* 171 pounds, Mary is 5 feet 6 inches *and* 138 pounds, and so forth. Table 13.2

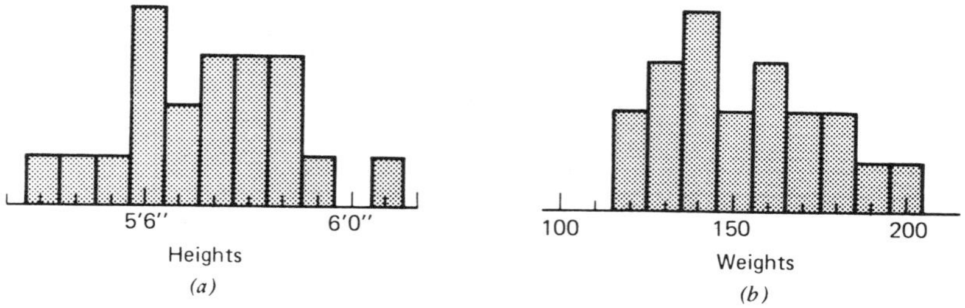

FIGURE 13.1 The distributions of heights and weights. Can you tell if there is any relationship?

lists the heights and weights for the 20 students; now each height is identified with the weight of the same student.

We will now draw a graph for this data. This time the graph will include measurements along *both* the *X*- *and Y*-axes (remember that the histogram only had measurements recorded on the *X*-axis; the *Y*-axis contained frequencies). We will have to locate the two measurements for each student. Heights are to be recorded along the *X*-axis and weights along the *Y*-axis. Every single point on the graph will correspond to some combination of height and weight, and we will place a dot at the point

TABLE 13.2 Heights and Weights of the Twenty Students Listed in Table 13.1. This Time the Heights and Weights are Arranged by Student.

STUDENT	HEIGHT	WEIGHT
1 (John)	5′ 9″	178
2 (Frank)	5′11″	171
3 (Mary)	5′ 6″	138
4 (Jane)	5′ 8″	142
5	5′ 7″	163
6	5′10″	175
7	5′ 4″	128
8	5′ 7″	138
9	5′ 9″	172
10	5′ 6″	155
11	5′10″	150
12	5′ 5″	124
13	6′ 1″	200
14	5′ 6″	130
15	5′ 8″	162
16	5′10″	185
17	5′ 3″	118
18	5′ 9″	154
19	5′ 6″	144
20	5′ 8″	133

where the height and weight of the individuals we observed coincide. Thus the value for John, a height of 5 feet 9 inches and a weight of 178 pounds, is located at the point where the two lines on Figure 13.2*a* intersect (a line going up from a height of 5 feet 9 inches and a line going to the right from 178 pounds). The rest of the values in Table 13.2 have been plotted in Figure 13.2*b* (I have identified the first four points by placing the student's names next to the corresponding points). This type of graph is often called a scatter diagram.

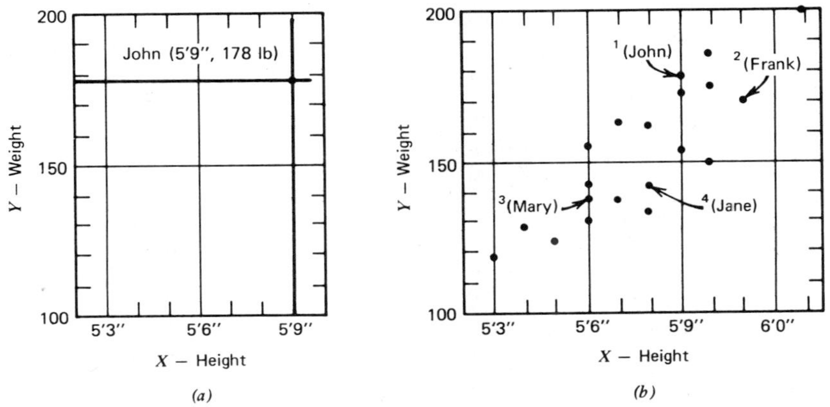

FIGURE 13.2 Scatter diagram or graph of the heights and weights. (*a*) How one point is plotted. (*b*) Contains the points representing all of the 20 students listed in Table 13.2.

13.2 CORRELATION

Now that we have the graph we must define or describe what it is that we will be attempting to observe or learn from the graph. There are two important things that we may look for:

1. We may try to see if there is a *relationship* between these two sets of values; we call this a correlation.
2. We may observe how closely the points seem to fall or *fit* some straight line. This line is then called a regression line.

These two ideas go hand-in-hand and, for the most part, can be treated as one concept. However, there are differences in the underlying theory and perhaps in the way you visualize them.

When you look at the graph, you need to interpret the *number or concentration of dots* in the various locations or areas of the graph as indicating the frequency with which *those pairs of values have simultaneously* occurred. These points show the *distribution of the pairs of quantities*. Depending on the degree of random scatter or the concentration into some pattern, we can state if there is *no* relation or *some* relation between the two quantities. The pattern we are looking for is usually a straight line.

Look carefully at Figure 13.2*b*. There are no short people (under 5 feet 6 inches, who are over 160 pounds) and there are no tall people (over 5 feet 10 inches, who are under 160 pounds). As we look at larger heights, the weights become larger, so there appears to be some relationship between height and weight. If you know a student's height, looking at the graph will provide you with a narrow band of possible values to which the corresponding weights belong (see Figure 13.3*a*). By narrow I really mean narrow with reference to the total possible distribution of weights. Compare the distribution of weights for the 5 feet 9 inch point to the overall distribution of weights plotted along the *Y*-axis of Figure 13.3*a*. The same holds true if you want to "predict" a height from a given weight (see Figure 13.3*b*).

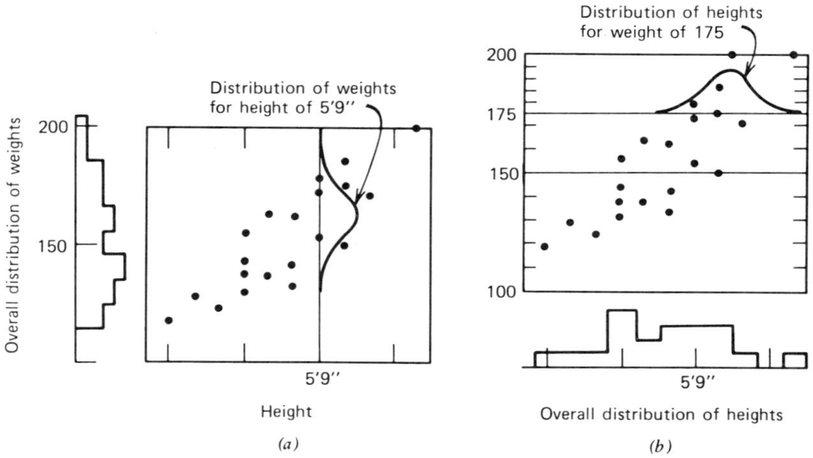

FIGURE 13.3 Two graphs demonstrating the importance of a correlation. (*a*) Given one value, the height, and looking for the weight, you are able to identify the value as being within a distribution that is smaller than the overall distribution of weights. (*b*) Same as in (*a*), except going from weight to height. (Note that the distributions are assumed to be approximately normally distributed.)

At this point we are not discussing actual predictions, but simply trying to gauge if two kinds of data or factors may vary together. We will shortly deal with the problems of predicting weight from height or vice versa. To illustrate further, let us look at two extreme cases of possible "relationships."

The first is a graph of heights and the score that each student received on the first statistics test taken (Figure 13.4). Apparently there is no relation between these two sets of values. If you were to select a height, say 5 feet 8 inches, and look at the distribution of test scores for the students with that particular height, you do not have a distribution that

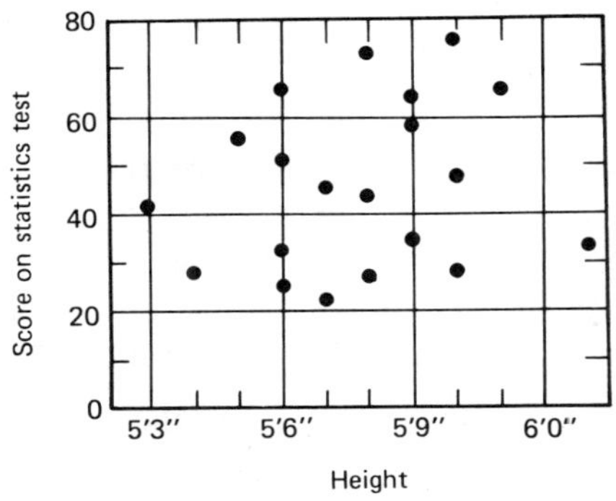

FIGURE 13.4 Graph of student's height and score on the first statistics test. There is no apparent relationship.

is smaller than the overall distribution of tests. So you do not have any predictive possibilities or discrimination. This is the extreme that implies that there is *no relation* or *no correlation.*

The second graph is that of height measured in feet and inches, as before, and height in centimeters. This is merely a simple conversion factor and therefore values should fall on a perfect straight line. Knowing the height as recorded on one scale should quickly and perfectly yield a single value on the second scale. From a practical standpoint, however, if we remeasure the same height we will obtain a different value. In addition, the scales for centimeters will be finer than inches (there are approximately 2 1/2 centimeters for every inch, so the centimeter scale is actually recording to better than the nearest 1/2 inch). Therefore, there is the possibility of slight variation. Nevertheless, the dots will fall about as close to a straight line as we could expect (see Figure 13.5). Pick a value of height along the X-axis, and this corresponds to a distribution that is a mere point, a small location on the broad distribution of possible points on the Y-axis. This is the other extreme, what is essentially a perfect correlation.

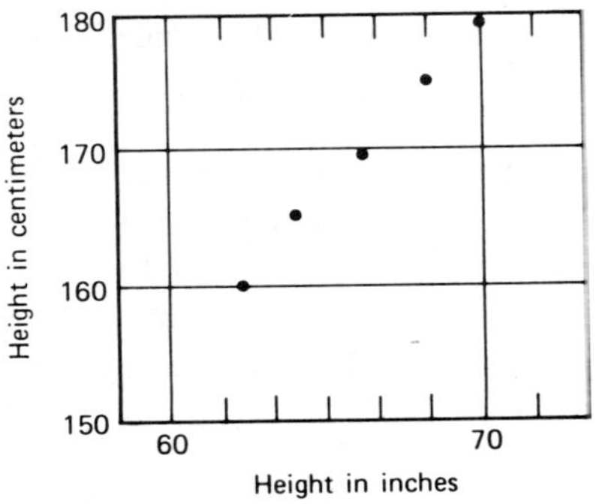

FIGURE 13.5 Graph of height as measured in inches and in centimeters.

13.3 CAUSE AND EFFECT

If we do find that there seems to be a relation between two types of measurements this *does not mean that there is or is not a cause and effect relation.** Any of the following possibilities may exist.

1. X *may* cause Y (large height may cause a large weight).
2. Y *may* cause X (large weight may cause a large height).
3. X and Y may be dependent on a *third factor* (large height and large weight may depend on genetic characteristics and both increase together).
4. The correlation may be an accidental random situation.

The last two possibilities are the critical ones to keep in mind when we discuss correlation. Item 4 pertains to the idea of significance testing and is the general reason why we even deal with the subject of statistics. There always exists the possibility that we have a peculiar situation arising in a set of data, one in which the data just happen to fall close to a straight line.

The third item, that the two sets of measurements may be both dependent on some third factor, is the very reason we have used the word correlation. You might read this word as co-relation, in which case we would emphasize that at best we would envision a mutual relationship. Many absurd correlations are possible (e.g., the way in which an inflationary trend in the economy or population growth may affect totally unrelated subjects). If you were to look up the salaries of teachers or ministers and compare them to the tons of aluminum produced that year, or the federal budget, or the price of books, or the cost of mailing a letter, you would find a high correlation. Obviously they are coincidental and occur because they both depend on a third factor, the economy in this case.

The first two items (X may cause Y, and Y may cause X) involve true cause and effect. However, if the data were obtained merely by the simultaneous recording of both of the measurements, *we cannot know which is the cause and which is the effect*. In order to identify a true causal relationship we must be capable of *precisely controlling one of the variables*. When we have the ability to control one of the variables, this generally means that we are no longer just concerned with *if* a relation exists. We are more interested in the *nature* of this cause and effect relationship. This will mean finding a mathematical or graphical way of *predicting the effect*, and this leads us to the concept of *regression*.

13.4 REGRESSION†

Regression is concerned with replacing the data by a *single line* that also represents a mathematical equation. The line that we are looking for is the same line we previously found for the dots in a correlation problem. The purpose for finding this line generally is to predict. It should be stated here that *there is no necessity for a cause and effect relation to exist*

* In Section 13.6 we will compute a mathematical quantity, called the correlation coefficient, that will be used to assist in deciding if there is a significant relationship between two quantities.

† If you are interested in learning more about correlation at this point, you may turn to Section 13.6.

*in order for you to want to predict one value from the other.** For example, it has generally been found that there is a high correlation or strong linear relation between IQ scores and future success in school. The psychologist recognizes that the score on the IQ test does not cause good grades in school, and it should be obvious that a future success or failure could not influence or cause an earlier test score to be high or low. Thus, cause and effect is to be ruled out as a possibility. However, the importance of knowing the nature of the relationship and being able to accurately *predict* what an individual will do in the future may be considerable.

We will return to our initial height and weight problem and determine the line and formula to be used to predict weight when an individual's height is known. First, look back to Figure 13.2*b* and see how well you can fit *one* line between the dots. One way is to hold a piece of string tightly over the dots and see if you can sort of balance all of the dots above and below the string. Easy? Do you get *one perfect* line? Did you find it difficult to decide which points should go above or below, or whether to go between certain points? In Figure 13.6 I have drawn four possible lines. Which one is the "best"?

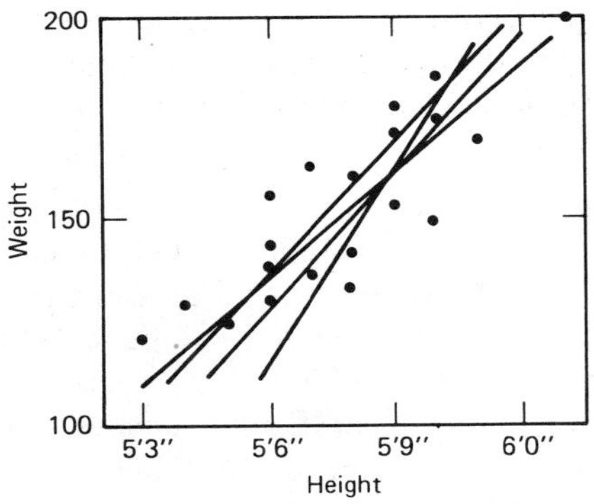

FIGURE 13.6 Four possible regression lines to predict weight from height. The lines are drawn on the scatter diagram of Figure 13.2*b*. Can you pick the best one? Is there a better line?

In order to decide which line to select as the "best" line, we must have some criterion. With any line that we might select there are differences, deviations, or errors, between the points representing the actual values of the heights and weights of the students and that line that we will be selecting. Figure 13.7 shows *one* such line and the differences or distances between this particular line and the points for each student. If we were to move that line in any way, some of these distances would decrease and

* Please note that there are two situations being considered here and above. You will *always* be able to calculate a formula for predicting one value from another. But you cannot say that a particular variable *causes* another to change unless you also can precisely control the predictor during the time you obtain the data to be used in calculating the equation.

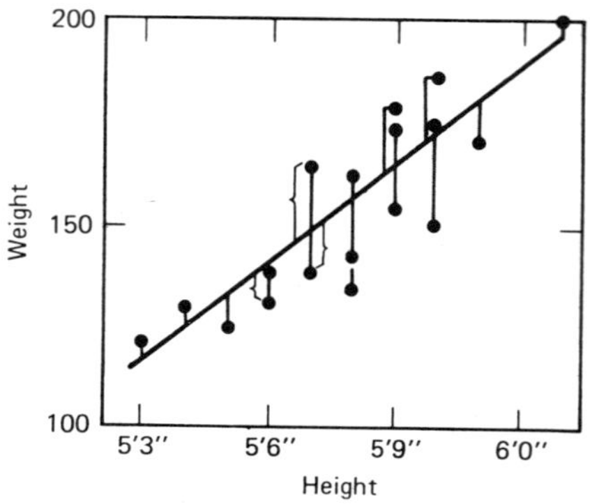

FIGURE 13.7 Differences between one possible regression line and the points of the scatter diagram. The least-squares regression line is the one that has been adjusted so that the sum of the squares of those distances is minimized.

others would increase. The idea is to move the line around until you find the location in which some kind of average distance becomes smallest. The average distance that we try to minimize is the square of the distances (leading to the definition of the method as the "least-squares" method).

Recall from elementary algebra that the equation for a straight line is

$$Y = a + bX \tag{13.1}$$

where a and b are constants for a particular line. The value of a is the same as the Y-value where the line crosses the Y-axis and, hence, is called the Y-intercept. The b-value is the slope of the line.

Obviously, the equation used to estimate or predict values of Y (weights in this case) for any particular X-value (height) must end up in this same form. However, you must also expect that it will involve *all the Xs and Ys* to be utilized in the calculation of the equation. Without going into great detail, the equation is:

$$Y_{\text{predicted}} = \overline{Y} + b(X - \overline{X}) = (\overline{Y} - b\overline{X}) + bX \tag{13.2}$$

or

$$Y_{\text{predicted}} = a + bX \tag{13.3}$$

where

$$b = \frac{\sum (X \cdot Y) - (\sum X)(\sum Y)/n}{\sum X^2 - (\sum X)^2/n} \tag{13.4}$$

$$a = \overline{Y} - b\overline{X} \tag{13.5}$$

$$\overline{X} = \frac{\sum X}{n} \quad \text{and} \quad \overline{Y} = \frac{\sum Y}{n}$$

These formulas may look formidable, especially for calculating b, but they really involve little more than we have already been doing in calculating standard deviations. We will shortly apply these formulas to our data and determine the prediction equation for weight given the height. Before proceeding I want to emphasize one fact. The two values, a and b, are *not* fixed values but, instead are related to the variation in the individual values and are estimates of some population quantities, which we will designate α and β. The a- and b-values therefore have definite distributions associated with them.

Now to our example. In Table 13.3 we have all the necessary sums. Observe that we calculate the squares and the products of the two variables (even though we don't need the Y^2s yet) and take the sums of all these values.

TABLE 13.3 Computations Needed to Calculate a and b for a Regression Line. Data is Height and Weight of 20 Students

	X	Y	X^2	Y^2	XY
STUDENT	HEIGHT[a]	WEIGHT	(HEIGHT)^2	(WEIGHT)^2	(HEIGHT)(WEIGHT)
1	9	178	81	31,684	1602
2	11	171	121	29,241	1881
3	6	138	36	19,044	828
4	8	142	64	20,164	1136
5	7	163	49	26,569	1141
6	10	175	100	30,625	1750
7	4	128	16	16,384	512
8	7	138	49	19,044	966
9	9	172	81	29,584	1551
10	6	155	36	24,025	930
11	10	150	100	22,500	1500
12	5	124	25	15,376	620
13	13	200	169	40,000	2600
14	6	130	36	16,900	780
15	8	162	64	26,244	1296
16	10	185	100	34,225	1850
17	3	118	9	13,924	354
18	9	154	81	23,716	1386
19	6	144	36	20,736	864
20	8	133	64	17,689	1064
Sum	155	3060	1317	477,674	24,608

[a] Values of heights have been converted to inches greater than 5 feet. Thus 5′6″ is shown as 6, 6′1″ is 13 (1 foot 1 inch greater than 5 feet), and so forth.

We will first calculate $\overline{X}$ and $\overline{Y}$.

$$\overline{X} = \frac{\sum X}{n} = \frac{155}{20} = 7.75 \qquad \text{(an average height of 5 feet 7 3/4 inches)}$$

$$\overline{Y} = \frac{\sum Y}{n} = \frac{3060}{20} = 153 \qquad \text{(an average weight of 153 pounds)}$$

Then

$$b = \frac{\sum \text{(heights)(weights)} - \dfrac{(\sum \text{heights})(\sum \text{weights})}{\text{number of students}}}{\sum (\text{heights}^2) - \dfrac{(\sum \text{heights})^2}{\text{number of students}}}$$

$$= \frac{24{,}608 - (155)(3060)/20}{1317 - (155)^2/20} = \frac{24{,}608 - 23{,}715}{1317 - 1201.25}$$

$$= \frac{893}{115.75} = 7.71$$

and

$$a = \overline{Y} - b\overline{X} = 153 - (7.71)(7.75)$$
$$= 93.2$$

Substituting these values into Equation 13.2, we have

$$Y_{\text{predicted}} = a + bX$$
$$= 93.2 + 7.71X \qquad (13.6)$$

This now is the best linear or straight line equation for the data given. There may actually be a nonlinear or curved line that will be even better, but among straight lines using the least squares as the criterion, this is the best one.

If you substitute values for X you will obtain predicted values for Y. For instance, if $X = 10$, meaning a height of 5 feet 10 inches, we estimate or predict the weight to be

$$Y_{\text{predicted}} = 93.2 + 7.71(10) = 93.2 + 77.1 = 170.3$$

This is not too far from the one actual value (individual number 6) which was 5 feet 10 inches and weighed 175 pounds.

You might also observe that if you calculate the predicted Y-value for the point X equal to $\overline{X}$ you will find that the $b(X - \overline{X})$ part of Equation 13.2 becomes 0 and then $Y = \overline{Y}$. Thus, *one point that the line will always go through is the point* $(\overline{X}, \overline{Y})$, in this case the point $(\overline{X} = 7.75, \overline{Y} = 153)$.

Now that we have two points, the $(\overline{X}, \overline{Y})$ point and the previous one $(X = 10, Y = 170.3)$ we can draw a straight line on the graph. This line is called the *regression line*, or more specifically the *regression line of Y on X*,

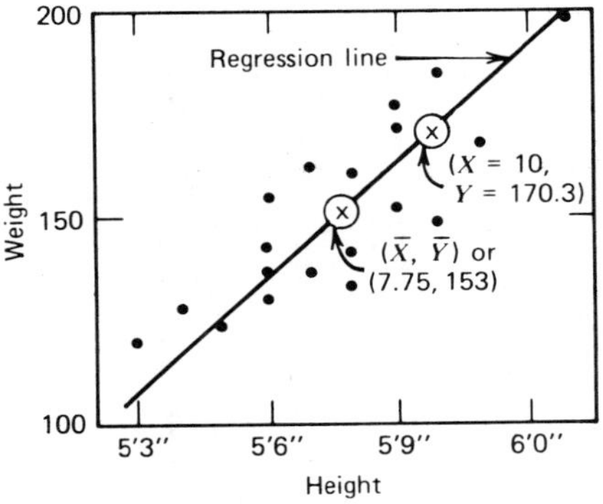

FIGURE 13.8 The least-squares *regression line* (of Y on X). Having found two points, the circled Xs, we can draw a straight line. This line provides the best prediction from height to weight.

or weight on height. In Figure 13.8 this regression line is drawn on the graph. The reason it is stated as the regression of Y on X is to emphasize that *this line is only for predicting Y-values from X-values*. You are not to use the line, nor rework the equation, to use it to go from Y-values to X-values. If you *also* want to predict from Y to X, knowing weight to predict height, you must recalculate b [(with the denominator as $\sum$(weights^2) − ($\sum$ weights)^2/students)] and obtain a new equation. This is necessary because, if you look back to Figure 13.7, you will see that the distances we were working with were vertical distances. In this case we are now working with horizontal distances.

In the most extreme case, where *no* correlation is present, the best line to use to predict Y from X will be a line parallel to the X-axis. This arises because b will essentially be zero and then $Y_{\text{predicted}}$ will be $\bar{Y}$ for *all* of the X-values. Thus no matter what you choose for X you end up with the average value of Y (see Figure 13.9a).

To find the best equation to predict X from Y would merely require us to interchange the Xs and Ys in Equations 13.2 and 13.3. As a result, b would be different. In our example, b would now be 0.094.

$$a = \bar{X} - b\bar{Y} = 7.75 - (0.094)(153) = -6.63$$

and

$$X_{\text{predicted}} = a + bY = -6.63 + 0.094Y \qquad (13.7)$$

This will yield different pairs of X- and Y-values than Equation 13.4. For example, we had found that for $X = 10$, $Y = 170.3$. Now, if we substitute $Y = 170.3$ into Equation 13.5, we will find that $X_{\text{predicted}} = 9.38$.

Thus we have found two possible lines to serve as predictors. In most cases, if true cause and effect relationship exists, we are *only* concerned with the prediction equation *from the cause factor to the effect factor*, not vice-versa. There are some important areas, including error analysis and

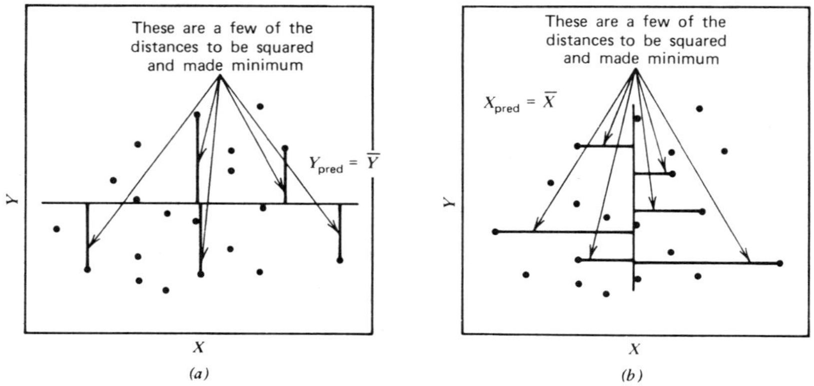

FIGURE 13.9 The best regression lines when the data are not related (or correlated). (*a*) The regression of Y on X, or going from X to Y—the best guess or prediction of the Y-value, regardless of X, is $\overline{Y}$. Observed that the lines to be squared and minimized are those parallel to the Y-axis. (*b*) The regression of X on Y, or going from Y to X. Here the best guess or best prediction is X, regardless of the Y-value. The lines to be squared and minimized this time are those parallel to the X-axis. The two prediction lines, X to Y and Y to X, are perpendicular to each other.

calibration, in which we try to predict in the opposite direction. However, in any given problem you would normally only determine *one* prediction equation.

The next question we should logically ask is how good is the regression line. There will be three ways we will essentially treat this problem. In the next section we will perform one of the possible hypotheses tests (does $\beta = 0.0$) and then we will examine some types of confidence intervals for regression. Later, in Section 13.8, we will look at a quantity called the index or coefficient of determination (r^2), which will help us in assessing how much of the *variation* in one quantity (e.g., weight) may be explained through knowledge of the value of the other quantity (height).

*13.5 EXAMINING THE REGRESSION EQUATION (OPTIONAL)

Up until now we have discussed the regression line in general. Before we conclude that we are willing to use the equation for the purpose of prediction, we should consider whether the equation is actually valid—whether the quantities might actually be independent.

If you look back to Figure 13.9*a* you can see the case when we are trying to predict a Y-value with another variable, X, which actually provides us with no helpful information. Observe that the general equation we would write is $Y_{\text{predicted}} = \overline{Y}$, or *the slope equals zero*. We therefore should ask: Is it possible that the population value of slope, the value called β, is zero? So we will test the hypothesis that $\beta = 0$.

First we will need to compute a few new quantities. We must calculate s_X^2 and s_Y^2, the variance of the Xs and the Ys. These are found as:

$$\begin{aligned} s_Y^2 &= \text{variance of the } Y = \left[\sum Y^2 - \frac{(\sum Y)^2}{n}\right] \Big/ (n-1) \\ s_X^2 &= \text{variance of the } X = \left[\sum X^2 - \frac{(\sum X)^2}{n}\right] \Big/ (n-1) \end{aligned} \tag{13.8}$$

Earlier we observed that knowing an X-value might lead to a smaller distribution of Ys. The standard deviation of the distribution of Ys for a given X is often called the *standard error of the estimate* and may be symbolized by $s_{Y \cdot X}$ (see Figure 13.10). This is computed by first finding $s_{Y \cdot X}^2$ and then taking the square root.

$$s_{Y \cdot X}^2 = \frac{n-1}{n-2}(s_Y^2 - b^2 s_X^2) \tag{13.9}$$

In our example:

$$b = 7.71 \qquad \text{and} \qquad b^2 = 59.44$$

$$n = 20$$

$$\begin{aligned} s_Y^2 &= \left[477{,}674 - \frac{(3060)^2}{20}\right] \Big/ 19 = (477{,}674 - 468{,}180)/19 \\ &= 9494/19 \qquad\qquad = 499.7 \end{aligned}$$

$$\begin{aligned} s_X^2 &= \left(1317 - \frac{(155)^2}{20}\right) \Big/ 19 = (1317 - 1201.25)/19 = 115.79/19 \\ &= 6.092 \end{aligned}$$

$$s_{Y \cdot X}^2 = \frac{19}{18}(499.2 - (59.44)(6.092)) = 1.055(499.7 - 362.1) = 145.2$$

Then the standard error of the estimate $= s_{Y \cdot X} = \sqrt{145.2} = 12.1$.

Now back to the hypothesis test. From statistical theory it can be shown that if β is equal to 0, the statistic

$$t = \frac{b s_X \sqrt{n-1}}{s_{Y \cdot X}} \tag{13.10}$$

follows a t-distribution (with $df = n - 2$), and we essentially perform a type of t-test.

Thus, to test $\beta = 0$ at a 5% two-sided significance level for $n - 2 = 20 - 2 = 18$ df, we find $t = \pm 2.101$.

The test statistic for this problem is:

$$t = \frac{7.71\sqrt{6.092}\sqrt{19}}{12.1} = \frac{(7.71)(2.468)(4.36)}{12.1} = 6.86$$

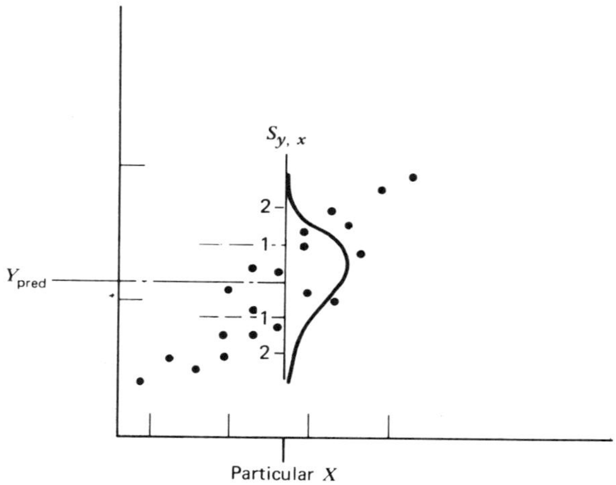

FIGURE 13.10 Illustrating the standard error of the estimate, $s_{Y \cdot X}$ for a predicted Y as the measure of variation of the distribution of possible Y-values for the given X.

This t-value is in the rejection region, and we therefore conclude that β does not equal zero. We find that there is a statistically significant predictive relationship between the two values.

Generally, the next idea is to calculate confidence intervals. And here comes a point of confusion. Two kinds of intervals are possible.

1. There is a confidence interval for the possible values of the *mean* of the Y for a given X.
2. There is a confidence interval* for individual future Y-values for a particular given X-value.

There is an important difference between these two intervals that requires a decision on your part and a statement of your objective. Let us briefly examine and compare these two cases.

In the first case we are recognizing an objective that says we are interested in predicting what the average Y-value will be. The prediction is for a long-range or average estimate of Y, which is basically what the regression line is. However, we also recognize that the line itself is not perfect, but is just an estimate of the possible lines. Essentially, this confidence interval is an interval containing the possible regression lines. Figure 13.11*a* shows a typical pair of confidence intervals for the distribution of means of Y and one typical distribution of means of Y for a particular X-value.

The second type of confidence interval provides an interval for the distribution of the possible actual Y-values. Obviously this should be a wider distribution and a wider confidence interval, since individuals are spread out much more than means. A typical set of confidence limits for

* This is sometimes referred to as a predictive interval.

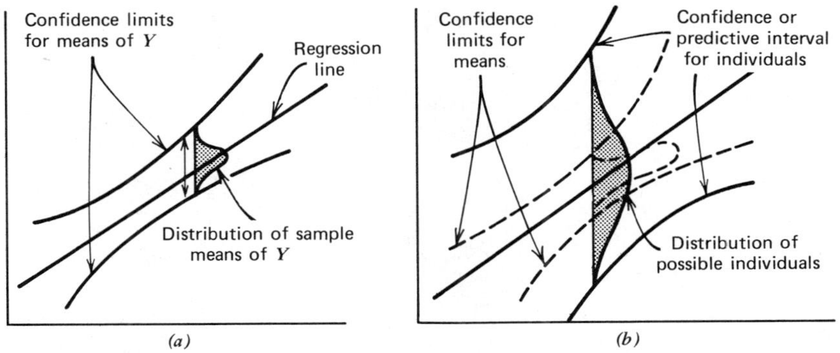

FIGURE 13.11 The two possible confidence intervals for regression. (*a*) Confidence intervals for means. (*b*) Confidence intervals for individuals. (Confidence interval for means is also shown for comparison.)

individuals is shown in Figure 13.11*b*. The confidence interval for means of Y are repeated, this time with dotted lines, for you to see the relative differences in interval sizes.

In both cases you should observe that the confidence intervals get broader as you move from the center of the X-values. Previously we noted that the regression line always goes through the point ($\overline{X}$, $\overline{Y}$). For any of the possible ($\overline{X}$, $\overline{Y}$) points you could have a number of different regression lines going through that point. Pivot the line to obtain a number of different regression lines. As you go further from the center, there becomes a wider spread of possible points that can be on the lines. This is illustrated in Figure 13.12. Of course, this pivoting is true for all of the possible ($\overline{X}$, $\overline{Y}$) points that might occur and therefore we have a wider spread of limits than would be implied from just the one pivot point.

The actual confidence intervals for the two cases are as follows (the significance level or α-value is for one estimate only).

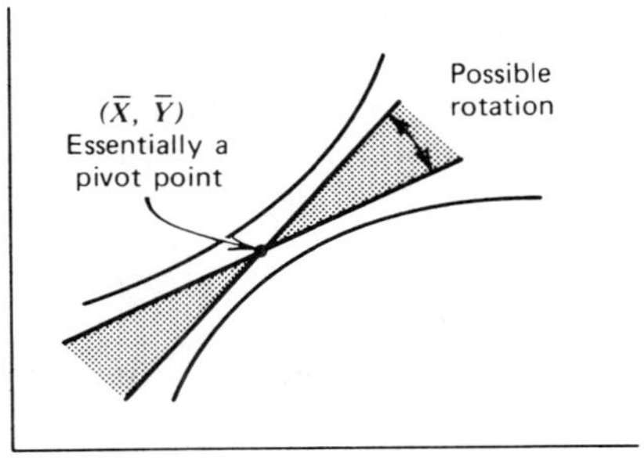

FIGURE 13.12 A graphical illustration for showing wider estimates the further you get from the center of the X and Y. The regression line may pivot about the overall mean.

Confidence interval for mean of Y:

$$Y_{\text{predicted}} + t_{\alpha/2} s_{Y \cdot X} \sqrt{\frac{1}{n} + \frac{(X - \bar{X})^2}{(n-1)s_X^2}} < \text{True mean of } Y$$

$$< Y_{\text{predicted}} + t_{1-\alpha/2} s_{Y \cdot X} \sqrt{\frac{1}{n} + \frac{(X - \bar{X})^2}{(n-1)s_X^2}}. \quad (13.11)$$

Confidence or predictive interval for individual Y:

$$Y_{\text{predicted}} + t_{\alpha/2} s_{Y \cdot X} \sqrt{1 + \frac{1}{n} + \frac{(X - \bar{X})^2}{(n-1)s_X^2}} < \text{Possible values of } Y$$

$$< Y_{\text{predicted}} + t_{1-\alpha/2} s_{Y \cdot X} \sqrt{1 + \frac{1}{n} + \frac{(X - \bar{X})^2}{(n-1)s_X^2}} \quad (13.12)$$

where $Y_{\text{predicted}}$ is the predicted value of Y, and the t-values have $n - 2$ degrees of freedom. (It is $n - 2$ because we now have to estimate 2 variances, the Xs and the Ys.)

For our example of heights and weights we had calculated one prediction of a $\dot{Y}$ for a particular X. For X = height = 10 (actually 5 feet 10 inches) we found Y = weight = 170.3 pounds. We will now find 95% confidence intervals for our predicted value ($\alpha = 0.05$).

Since $n = 20$ the degrees of freedom are $n - 2 = 18$, and the t-values are

$$t_{0.975} = 2.101$$

$$t_{0.025} = -2.101$$

$$\bar{X} = 7.75$$

$$s_{Y \cdot X} = \sqrt{145.2} = 12.1$$

$$s_X^2 = \frac{\sum X^2 - (\sum X)/n}{n - 1} = \frac{115.75}{19} = 6.09$$

Confidence interval of the mean of the predicted Y—from Equation (13.11):

$$170.3 + (-2.101)(12.1)\sqrt{\frac{1}{20} + \frac{(10 - 7.75)^2}{19(6.09)}} < \text{Mean of } Y$$

$$< 170.3 + (2.101)(12.1)\sqrt{\frac{1}{20} + \frac{(10 - 7.75)^2}{19(6.09)}}$$

$$170.3 - 25.4\sqrt{0.05 + 0.04375} < \text{Mean of } Y$$

$$< 170.3 + 25.4\sqrt{0.05 + 0.04375}$$

$$170.3 - 25.4(0.306) < \text{Mean of } Y < 170.3 + 25.4(0.306)$$

$$170.3 - 7.78 < \text{Mean of } Y < 170.3 + 7.78$$

$$162.5 < \text{Mean of } Y < 178.1$$

Confidence interval for possible Y-values—from Equation 13.12:

$$170.3 - (2.101)(12.1)\sqrt{1 + \frac{1}{20} + \frac{(10 - 7.75)^2}{19(6.09)}}$$

$$< Y < 170.3 + (2.101)(12.1)\sqrt{1 + \frac{1}{20} + \frac{(10 - 7.75)^2}{19(6.09)}}$$

$$170.3 - 25.4\sqrt{1 + 0.05 + 0.04375}$$

$$< Y < 170.3 + 25.4\sqrt{1 + 0.05 + 0.04375}$$

$$170.3 - 25.4(1.046) < Y < 170.3 + 25.4(1.046)$$

$$170.3 - 26.6 < Y < 170.3 + 26.6$$

$$143.7 < Y < 196.9$$

You can see the large difference between these two sets of limit.

13.6 CORRELATION COEFFICIENT

Several times in this chapter, and especially in Section 13.3, we observed that in many situations the relationship between two quantities is merely that they tend to vary together. For example, with the problem we have been examining we have already discussed the idea that there is no cause and effect, or no reason to say that one quantity mathematically depends on the other (that one is independently able to be varied or changed in a way that we can observe, measure, or predict a change in the other *dependent* quantity). Instead, we assume that we have two characteristics that probably are both essentially normally distributed and are really related through some other factor. They do, however, tend to vary together, and thus we do observe a correlation between these values. So far what we have done in terms of correlation has been very imprecise, at least as far as the mathematics is concerned. If I gave you a set of different graphs with points scattered all over them, which we called scatter diagrams, and asked you which were more or less correlated, you might have difficulty. The extremes are usually obvious, although the question of no correlation may not always be obvious. However, we will now explore a way of arriving at a numerical value to describe the degree of relation, the degree of closeness to a straight line, or the amount of scatter that exists. This quantity is called a correlation coefficient and is generally given the symbol r.

There are several different types of correlation coefficients. The one we will primarily deal with is called the *Pearson product moment correlation coefficient*.

The value of r can range from -1 to $+1$ with the following general interpretations.

1. If $r = 0$ then we have *no* correlation, or more precisely, we have *no linear* correlation. Usually we just have scatter, but there *may* be other ways

in which the values are related. The distribution in Figure 13.4 is close to an r of 0.

2. If $r = +1$ or $r = -1$ then we have a *perfect* linear correlation; all of the values fall on a straight line. Figure 13.5 is an excellent case.

3. Values of r between 0 and 1 or between 0 and -1 indicate varying degrees of correlation. Thus, the correlation coefficient for the points in Figure 13.2*b* would be greater than 0 but less than 1.

4. *Positive* values of r indicate a line going *up* to the right (as X increases so does Y). Figure 13.2*b* and 13.5 fall into this category.

5. *Negative* values of r indicate a line going *down* to the right (as one of the values increases the other decreases).

Figure 13.13 contains a variety of graphs illustrating the range of correlation and giving the associated correlation coefficients. There are two types of scatter diagrams illustrated. The first set, Figure 13.13*a* to 13.13*c*, contains dots at the points of intersection of the two values. The second set, Figure 13.13*d* to 13.13*f*, resembles frequency tally sheets, except that these frequency charts cover two possibilities at one time. The latter group of diagrams also demonstrates a method that lends itself more readily to graphical presentation of repeated occurrences of the same pairs of values. Thus, if two students have the same height and weight this can easily be recorded (instead of superimposing dots on top of each other, making interpretation difficult).

The calculation of r is the next item. It is a fairly involved procedure, particularly when computation equipment is not available. We will require the standard deviation of the Xs and the Ys and a measure of *variation of the products of the Xs and Ys* (which is called *covariance*). We then define r as:

$$r = \frac{\text{Covariance of the } X\text{s and } Y\text{s}}{(\text{Standard deviation of the } X\text{s})(\text{Standard deviation of the } Y\text{s})}. \tag{13.13}$$

We will not consider the use of frequencies but work with all n sets of X- and Y-values separately. We then write r as follows.

$$r = \frac{\left[\sum (X \cdot Y) - (\sum X)(\sum Y)/n\right]/(n-1)}{\sqrt{\dfrac{\sum X^2 - (\sum X)^2/n}{n-1}} \cdot \sqrt{\dfrac{\sum Y^2 - (\sum Y)^2/n}{n-1}}} \tag{13.14}$$

The $(n - 1)$ all divide out, and this then reduces to

$$r = \frac{\sum (X \cdot Y) - (\sum X)(\sum Y)/n}{\sqrt{\sum X^2 - (\sum X)^2/n} \cdot \sqrt{\sum Y^2 - (\sum Y)^2/n}} \tag{13.15}^*$$

* Observe the similarity to our previous calculations with b. This can also be calculated as

$$r = b(s_X/s_Y). \tag{13.16}$$

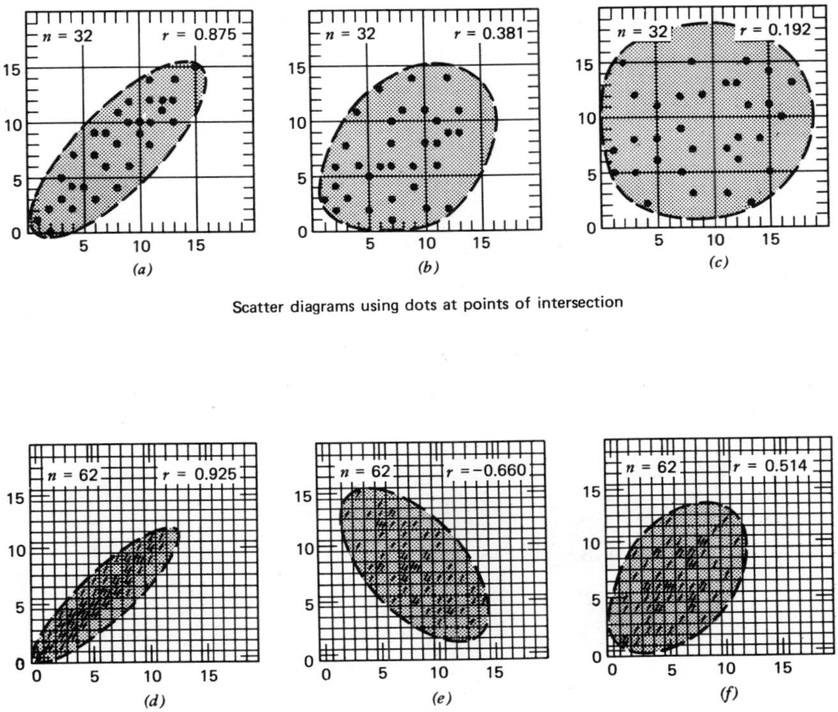

Scatter diagrams using dots at points of intersection

Scatter diagrams containing tally boxes

FIGURE 13.13 Correlation scatter diagrams illustrating different degrees of correlation and two techniques of construction. For both sets of diagrams dashed lines have been drawn around the points to emphasize the general nature of the distributions. Correlation coefficients and the size of the sample are given for all the distributions.
(*a*), (*b*) and (*c*) Intersection of two lines are used to represent the values. The measurements are located along the lines. (*d*), (*e*), and (*f*) Boxes represent the values with the centers of the boxes (or intervals) corresponding to the measurement values.

We will use this equation (Equation 13.15).

For our example of heights and weights we again have heights as *X*s and weights as *Y*s

$$r = \frac{\sum(\text{heights})(\text{weights}) - \dfrac{(\sum \text{heights})(\sum \text{weights})}{\text{number of students}}}{\sqrt{\sum(\text{heights}^2) - \dfrac{(\sum \text{heights})^2}{\text{Number of students}}} \cdot \sqrt{\sum(\text{weights}^2) - \dfrac{(\sum \text{weights})^2}{\text{Number of students}}}} \tag{13.17}$$

Look back to Table 13.3 (page 285) where we have the computations of all of the sums. Substituting the appropriate sums into Equation 13.15, we have

$$r = \frac{24{,}608 - (155)(3060)/20}{\sqrt{1317 - (155)^2/20}\sqrt{477{,}674 - (3060)^2/20}}$$

$$= \frac{24{,}608 - 23{,}715}{\sqrt{1317 - 1201.25}\sqrt{477{,}674 - 468{,}180}} = \frac{893}{\sqrt{115.75}\sqrt{9494}}$$

$$= \frac{893}{(10.8)(97.4)} = \frac{893}{1051.9} = 0.849$$

Our value of r confirms the high degree of correlation between height and weight that was apparent from the graph.

13.7 SIGNIFICANCE TEST FOR CORRELATION COEFFICIENT

The first thing you should question now that we have a value for correlation coefficient is: How large must the correlation coefficient become in order to be considered "large?"

There are two ways of interpreting whether a correlation coefficient is large. The first that will be considered in this section deals with a statistical hypothesis test. Here we need to know the distribution of correlation coefficients for different sample sizes for a true population with *no* correlation. We then compare our value to the distribution of correlation coefficients to see if our value is in the unusual or rejection region. The problem, though, is that as the samples become larger and larger (as n becomes large), what constitutes a very significant correlation coefficient becomes very small.

If we look at only three pairs of values, or have only three points to plot, there is a pretty good chance that they will come close to being on a straight line *regardless of whether there truly is a correlation.* Let's look at a case that would be absurd. Pick three books, a medium-size book, a fairly long book, and a very short book. Now count the number of words in the first sentence of each of these books. There is a pretty good chance that you will just randomly end up with the second book having quite a few more words in its first sentence than the first book had, and the third book having more than either of the other two, or vice-versa. In either case (as well as some other possible cases) you will end up with varying degrees of correlation, sometimes as high as 0.7 or 0.8. Yet this "high" value perhaps should not be considered important.

Now if we were to increase the sample by picking up another two dozen books, thus giving us a sample of 27, we should start to be suspicious that a relation may actually exist if we were suddenly to find that the correlation coefficient was as large as 0.7 or 0.8.

So the significance of a large value of correlation coefficient depends primarily on the *size of the sample* or the number of pairs of values that we have observed. But you should inquire further; you should want to know just *how large* the coefficient should be for *whatever sample size* you are dealing with. To make this determination, we will consult Appendix Table A.10. For convenience in discussing how to use this table I have reproduced some parts of it in Figure 13.14. We have a number of columns in the table. The first is a value related to the size of the sample. The value is labeled *df*, which stands for degrees of freedom. You need only to know that it is $n - 2$. So, since the values of n that have just been discussed were 3 and 27, the degrees of freedom would be 1 for the first case and 25 for the second case. Both are found in the table and the figure.

$df = n - 2$	Level of significance for two-tailed test				
		0.05 (5%)		0.01 (1%)	
1		0.9969		0.9999	
18		0.4438		0.5614	
25		0.3809		0.4869	

FIGURE 13.14 Part of Appendix Table A.10 showing what are considered to be significantly high correlation coefficients for several different sample sizes.

In Table A.10 there are five more columns. For this discussion we will only concern ourselves with two of them: those labeled as 0.05 and 0.01 below the heading of "significance level for two-tailed tests." These are reproduced in part as the second and third columns of Figure 13.14.

These columns list values of correlation coefficient. For an n of 27 we have listed in the second column, the column labeled 0.05 level, the number 0.3809. The 0.3809 is a value of correlation coefficient that is considered unusually large. If the correlation coefficient I obtain is as large as the 0.3809 or larger than 0.3809, then it is sufficiently large enough to say that there is less than a 5% chance that *no* relation exists, or that the odds are about 19 to 1 in favor of some sort of a relation. Note that two tailed here refers essentially to the either positive or negative cases.

The third column also lists correlation coefficients. These are larger values than those in the previous column and, for $n = 27$, it is 0.4869. When I obtain a correlation coefficient larger than 0.4869, I know that there is *less than a* 1% chance that *no* relation exists. The odds are increased to almost 100 to 1 that some relation is present, or I say there is 1 chance in 100 of obtaining a correlation of this sort (larger than 0.4869) when the source of the data is *totally uncorrelated*.

Going back now to our example of the correlation between height and weight, we can determine what the critical or significantly large values of correlation coefficient are for this sample. The sample size was 20, since we measured 20 students. Referring to Table A.10 (or Figure 13.14), we find that the critical values of correlation coefficient corresponding to $n = 20$ are 0.4438 (at the 0.05 or 5% level of significance) and 0.5614 (at the 0.01 or 1% level of significance). Our computed coefficient was 0.849, which far exceeds even the upper value of 0.5614 (the 1 in a 100 chance that there is actually no relation). So we can now state that the *correlation is* actually *very significant* and that there is a *significant linear relationship between height and weight*.

13.8 COEFFICIENT OF DETERMINATION: r^2

The second and perhaps more useful criterion for evaluating the correlation involves the square of r, or r^2. This is called the coefficient or index of determination and is also very useful in deciding how good a regression line is.* For example, if $r = 0.5$, the coefficient of determination, $r^2 = 0.25$. If $r = 0.9$, $r^2 = 0.81$, and if $r = 0.2$, $r^2 = 0.04$.

To see how we will use this number, look back at Figure 13.3 (page 280). At that time we observe that basically the existance of a correlation allowed us to eliminate some of the variation in one distribution by having knowledge of the corresponding value in the other distribution. For example, in Figure 13.3*a*, by knowing height we were able to reduce the variation among weight values, and thus we do not deal with as large a distribution of weights given the information about height. At the one extreme, when we looked at height in inches (see Figure 13.5), we explained *all* of the variation among height value in centimeters. Here we would find $r^2 = 1.00$ and 100% of the variation in centimeters is explained by knowing the value in inches.

At the other extreme we can look at Figure 13.4 (page 281). In this case knowing the height does not contribute to my knowledge of the result on the statistics test. We have an r^2 of about 0, and about 0% of the corresponding variation is explained through the relation.

The coefficient of determination provides a measure of how much of the observed variation can be essentially explained or accounted for through the correlation or the regression line. If we multiply r^2 by 100 we obtain

* Note that if you skipped Section 13.6 on calculating r and only dealt with regression, you can find r by: $r = b(s_X/s_Y)$ (or refer to Section 13.6).

the percent of variation explained. Thus, for an r^2 of 0.25, we explain about 25% of the variation in one value by knowing the other.

In the case of the heights and weights, $r = 0.849$ and $r^2 = 0.721$. Then 72% of the variation in weights of these students can be explained by knowing the height of the student. However, other factors, including just plain random variation, makes up the remaining 28% of the variation.

Look more carefully at Figure 13.13. For example, the r^2 for Figure 13.13*b* is $(0.381)^2 = 0.145$, or 14.5% of the variation in the *Y*-values is explained by the correlation with the *X*-values. Although there is some semblance of a correlation and it is actually statistically significant at the 0.05 level, knowing an *X*-value is 10 really doesn't help very much in narrowing down a corresponding *Y*-value. The small value of r^2 supports this visual impression.

However, if you look at Figure 13.13*a*, you find $r^2 = (0.875)^2 = 0.766$, or about 77% of the variation in *Y* is accounted for by knowledge of *X*. In this case, if you know $X = 10$, you can see that the spread of possible *Y*s is substantially reduced. The value of r^2 can therefore provide a useful and often meaningful gauge of how good a relationship is.

13.8 SOME ADDITIONAL CONSIDERATIONS

Before we complete our discussion of correlation and regression, several observations about potential misinterpretations should be made.

Range of prediction

A particularly important comment should be made regarding predictions. The regression equation and its graph are *only valid within the range for which the data have been accumulated*. We cannot validly predict above or below the limits for which we have plotted our graph. We have no basis for stating or believing that the rest of the curve will be a straight line. We don't know if a sudden increase or decrease will take place. We should not make predictions about the future. In Figure 13.15 I have

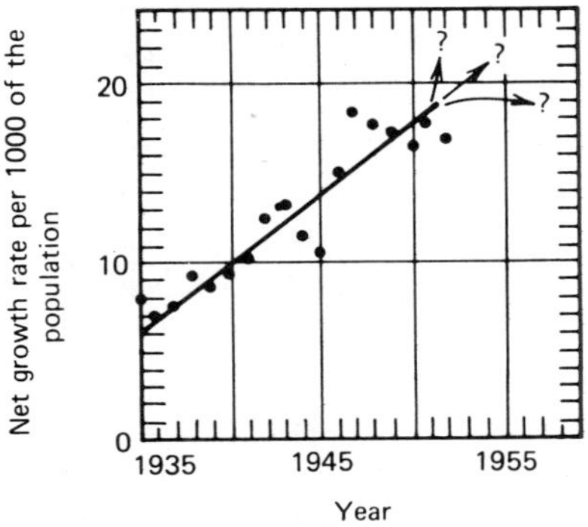

FIGURE 13.15 Net rate of growth of the U.S. population from 1935 to 1952. A rate of 10 means a growth rate of 0.1% per year. A fairly good straight line implies a continuation in the future years.

shown the net growth rate of the U.S. population from 1935 to 1952. There appears to be a high correlation, and it looks like the population should keep increasing. If you extended the curve to 1976 you would have expected about a 30 per 1000 or a 3% a year growth rate to occur.

In Figure 13.16 the graph has been extended to 1976. The straight line suddenly became a curve and our prediction would have been extremely wrong. Now, by the looks of this graph, we soon should hit a zero growth rate and we should not have to worry about a population explosion. Don't trust it.

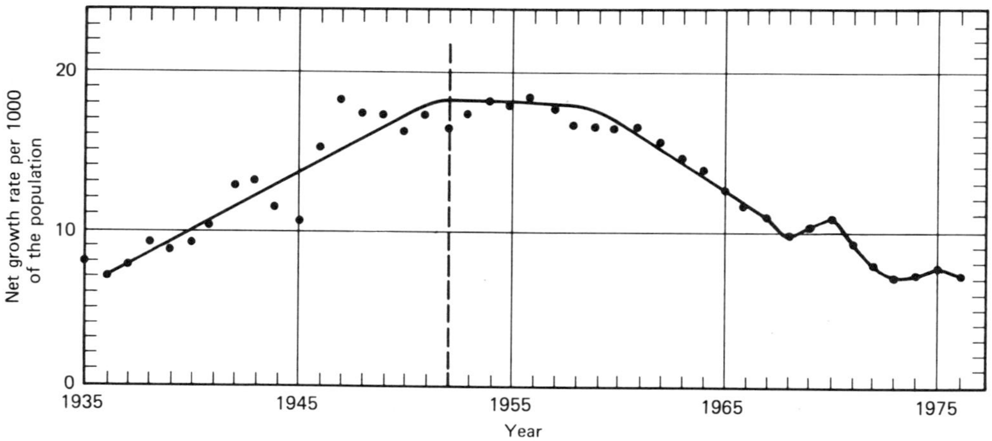

FIGURE 13.16 Net growth rate of the U.S. population now extended to 1976. Observe the complete change in direction of the curve.

Warning: plot the data!

A *zero* correlation may *not* mean *no* relationship.
A *high* correlation may *not* mean *a* relationship.

Let's look at these two odd statements.

A *near zero correlation*, or a poorly fitted straight line, does not necessarily mean that no relation exists. *First*, the relation *may not be linear* to start off with, it may be a curve. Figure 13.16 was one example of a line that is not straight. But the points fall very close to a single definite line, and we can interpret this to mean that there is a relationship here, even though it is not a straight line relation.

Sometimes the low correlation is a result of a collection of many groups of values. A common example of this type of problem is when the data is accumulated over a period of years and changes occur from year to year. Figure 13.17*a* shows a large sample of house costs versus sizes of houses. The correlation seems poor. This time, however, we have costs that are spread over 4 different years (and a span of 6 years). If we identify the house according to the year sold, we find a set of parallel lines. If we discount the increase in cost each year, we will obtain an extremely high correlation (Figure 13.17*b*).

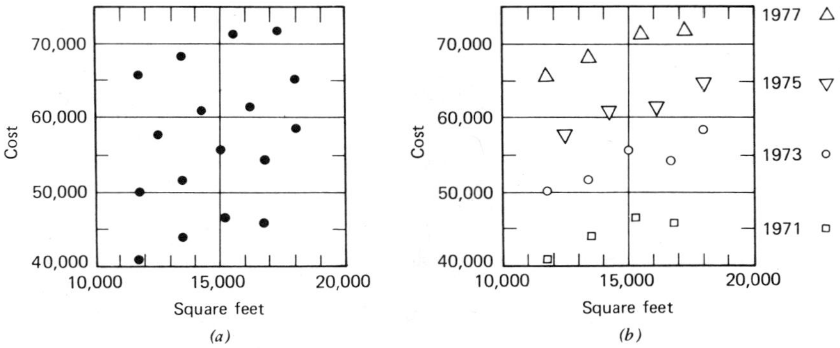

FIGURE 13.17 Scatter diagram of the cost and size of homes in a single development. (*a*) The collection of points are not distinguished in any way. A poor correlation. (*b*) Houses are classified according to year of sale. Observe how good the correlation is if we neglect the increase in cost from year to year.

A *third* circumstance that can lead to low correlation occurs when the range of values is *truncated* or cut short. The argument that says that College Board scores or IQ scores are not good predictors of performance in college might be based on the low correlation between College Board scores and grade point average, as shown in Figure 13.18. The problem with this contention is that it misses the fact that all those who were accepted were already *selected* at least partly on the basis of College Board scores. If, in fact, all the students were admitted randomly, took *all of the same courses*, and also represented a *complete range* of College Board scores, the diagram in Figure 13.19 would be closer to the final result. Here we have a much higher correlation. By cutting off this distribution essentially at the 500 mark, you will be able to approximate the distribution of Figure 13.18.

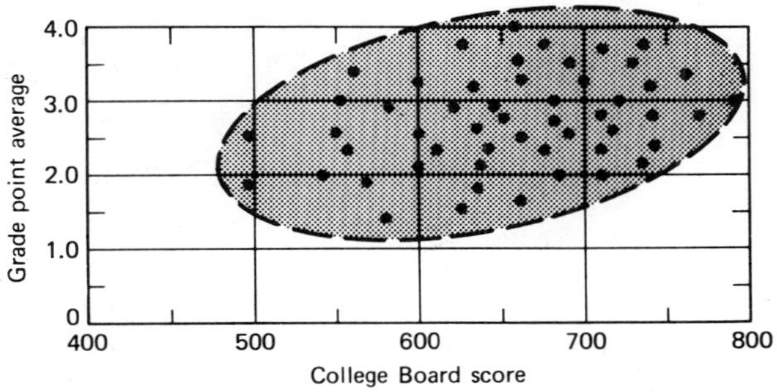

FIGURE 13.18 Grade point average versus College Board score for a group of students at a selective university. 4.0 is highest, 0 is lowest. There appears to be no correlation.

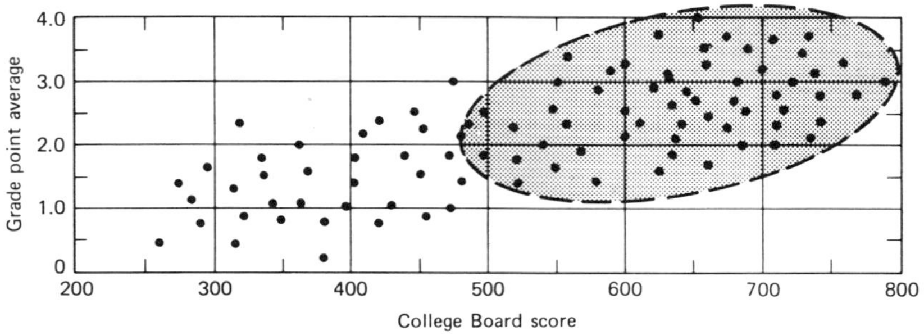

FIGURE 13.19 Grade point average versus College Board score if the selectivity factor were not considered. If all the points to the left of the dotted line were eliminated you would essentially have Figure 13.18. You now have a fairly high correlation.

Now let us examine one case where a *high* correlation is actually a deception. We have already looked at cases when two or more sources or conditions that separately had high correlations were lumped together, only to lose any semblance of correlation. The reverse is also possible. We may have two or more clusters of points, each essentially uncorrelated, but located in such a way that they will cause a high overall correlation.

If you consider the starting salaries for a first job you may find that it is correlated to the age of the person taking the job. Why? It is because the older you get the wiser you become? No, it is because the older job applicants are older generally because they have completed either a 2-year or a 4-year college program. The first beginning salaries to high school graduates as a group are not correlated to age. Neither are first salaries to college graduates correlated to age. But lump the several groups together and you end up with a high correlation as in Figure 13.20.

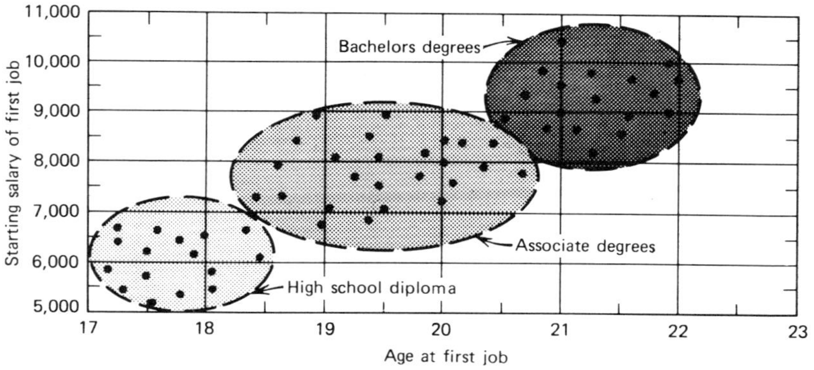

FIGURE 13.20 Starting salary and age at the time of first job. There appears to be a high correlation. This is actually because of the clustering of graduates of different educational levels who enter their first job at similar ages and the fact that more education generally leads to higher starting salary.

REVIEW Scatter Diagram • Correlation • Regression Line
Cause and Effect • Prediction • Least Squares
$Y_{\text{predicted}} = a + bX$ • α and β

$a = \bar{Y} - b\bar{X}$ • $b = \dfrac{\sum XY - (\sum X)(\sum Y)/n}{\sum X^2 - (\sum X)^2/n}$

* Standard Error or Estimate • $\beta = 0$ • Confidence Intervals
Correlation Coefficient •

$$r = \frac{\sum XY - (\sum X)(\sum Y)/n}{\sqrt{\sum X^2 - (\sum X)^2/n}\sqrt{\sum Y^2 - (\sum Y)^2/n}}$$

Coefficient of Determination • Percent Variation Explained = $r^2(100)$

PROBLEMS

SECTION 13.4

13.1 For the following sets of X- and Y-values, calculate a and b and the least squares regression line. In each case determine the predicted value of Y for an X of 2 and an X of 7.

a.	X	1	2	3	4	4	5	6
	Y	2	1	3	4	3	6	8
b.	X	0	−2	3	4	6	8	5
	Y	0	3	2	5	2	6	8
c.	X	1	2	3	4	5	6	7
	Y	−1	3	8	3	2	6	4
d.	X	2	2	1	3	1	5	4
	Y	2	4	6	5	6	3	7
e.	X	3	4	2	7	9	10	14
	Y	7	9	5	3	0	−4	3

13.2 For the data in Problem 13.1, calculate the regression equation to predict X from Y (or $X = a + bY$). Compare the results to the previous equation. What values of X are predicted for $Y = 2$ and for $Y = 7$?

13.3 The quality of coffee is to be gauged by two coffee tasters. They apply a rating from 0 to 10 (with good coffee rated 10, poor coffee rated 0). The coffee has been perked from 1 to 8 minutes and the ratings were as follows.

TIME (MINUTES)	RATING
1	1,1
2	3,0
3	2,2
4	4,2

TIME (MINUTES)	RATING
5	4,4
6	7,5
7	8,7
8	8,6

Plot these values. Calculate the regression equation. What taste value do you expect for coffee perked 7 1/2 minutes? What rating do you predict for coffee perked 5 minutes? How close is this to the actual values? Note that there are a total of 16 pairs of readings.

13.4 A group of secretaries were given a typing test. The speed and number of errors were recorded as follows.

Speed	65	60	55	75	100	35	45	60	70	50
Errors	3	4	2	5	5	2	3	3	2	3

Plot a scatter diagram. What is the regression line for predicting errors from speed? Speed from errors? What is the prediction of errors for a 90 word per minute typist? What speed would a typist with five errors be typing at?

13.5 A study was done for the number of staples (out of 25) that stapled properly through different numbers of pages of paper. The results were as follows.

Number of staples	23	23	20	10	8	24	21	18	10	15
Number of pages	4	3	8	12	15	2	5	7	13	10

Which value belongs on the X-axis? Calculate the regression equation. How many staples do you predict would properly staple 20 sheets? How many staples do you predict would properly staple 10 sheets? How does this value compare to the actual result?

13.6 At the time of the 1976 Super Bowl the Oakland Raiders had 43 players. The number and weight of each player were announced as follows.

NUMBER	WEIGHT	NUMBER	WEIGHT	NUMBER	WEIGHT
72	270	74	271	60	273
41	225	58	240	39	225
83	220	26	205	24	210
43	185	32	206	63	255
21	170	78	265	50	250
64	270	75	260	87	228
25	190	12	215	28	195
30	225	8	195	11	184
14	230	15	190	20	205
29	185	31	205	36	200
40	210	46	235	47	215
49	195	51	215	52	225
54	220	61	245	66	262
70	273	77	270	79	252
81	195				

For this data:

$$\sum(\text{numbers}) = 2022 \qquad \sum(\text{numbers}^2) = 116{,}936$$
$$\sum(\text{weights}) = 9659 \qquad \sum(\text{weights}^2) = 2{,}206{,}623$$
$$\sum(\text{numbers} \cdot \text{weights}) = 473{,}861$$

Plot these data and calculate a regression line to predict weight given number. What do you predict the weight of a new player, number 55, would be? How close is player 40 to his actual weight?

13.7 The income tax rate schedule for a single person after deductions (for the income year of 1976) was as follows.

INCOME	TAX	INCOME	TAX
1000	147	10,000	2097
1500	227	14,000	3218
2000	312	20,000	5230
4000	695	26,000	7590
6000	1116	50,000	20,190
8000	1596	100,000	53,090

Plot these values. What is the linear regression equation relating the amount of tax for different incomes? What do you predict the tax on a \$10,000 income (after deductions) would be? On a \$50,000 income? On a \$2000 income? How much different are the actual and predicted values? (You might simplify some of the calculations by working in thousands of dollars.)

13.8 The median years of school completed for people in different income levels was reported as follows.

INCOME	YEARS OF SCHOOL	INCOME	YEARS OF SCHOOL
\$1500–1999	8.4	\$6000– 6999	12.1
2000–2499	8.6	7000– 7999	12.3
2500–2999	8.7	8000– 9999	12.5
3000–3999	9.0	10000–14999	12.8
4000–4999	10.6	15000–24999	16.1
5000–5999	11.7	25000–over	16.3

Using the appropriate values for income calculate the regression equation. What does your equation predict for a \$10,000 income? \$20,000? How close is the predicted value to actual (in the latter case)?

SECTION 13.5 For each of the following problems test the hypothesis that $\beta = 0$.

13.9 Refer to the sets of data in Problem 13.1.
13.10 Refer to Problem 13.2.
13.11 Refer to Problem 13.3.
13.12 Refer to Problem 13.4.
13.13 Refer to Problem 13.5.

13.14 Refer to Problem 13.6.
13.15 Refer to Problem 13.7.
13.16 Refer to Problem 13.8.

For Problems 13.17 to 13.20 determine the appropriate 95% confidence intervals.

13.17 Refer to Problem 13.1. For each case determine the confidence interval for the mean of Y when $X = 2$. Determine the interval for individual Ys for $X = 7$.
13.18 Refer to Problem 13.4. Determine the interval for mean number of errors at a speed of 75 words per minute. Determine the confidence interval for individuals at 50 words per minute.
13.19 Refer to Problem 13.3. Determine the interval for individuals when the coffee was perked for 6 minutes.
13.20 Refer to Problem 13.8. What interval do you predict an individual's years at school for a \$10,000 earning?

SECTION 13.6 For Problems 13.21 to 13.27 calculate the correlation coefficient.

13.21 Refer to the data in Problem 13.1.
13.22 Refer to Problem 13.2. Should there be any difference between this problem and Problem 13.21?
13.23 Refer to the data in Problem 13.3.
13.24 Refer to the data in Problem 13.4.
13.25 Refer to the data in Problem 13.5.
13.26 Refer to the data in Problem 13.6.
13.27 Refer to the data in Problem 13.7.
13.28 Verify the correlation coefficients of Figure 13.13. For the last three diagrams (Figure 13.13*d*, 13.13*e*, and 13.13*f*) you will find it easier to use the following columns.

X	Y	f	$f \cdot X$	$f \cdot Y$	$f \cdot X^2$	$f \cdot Y^2$	$f \cdot X \cdot Y$
		sum	sum	sum	sum	sum	sum

The frequencies, f, are the number of tally marks in the respective boxes. The correlation coefficient would then be calculated using the following equation.

$$r = \frac{\sum f \cdot X \cdot Y - \dfrac{(\sum f \cdot X)(\sum f \cdot Y)}{(\sum f)}}{\sqrt{\sum f \cdot X^2 - \dfrac{(\sum f \cdot X)^2}{(\sum f)}}\sqrt{\sum f \cdot Y^2 - \dfrac{(\sum f \cdot Y)^2}{(\sum f)}}} \tag{13.18}$$

Thus, for Figure 13.13 instead of 62 sets of calculations, you only need 37.

SECTIONS 13.7 and 13.8 For Problems 13.29 to 13.35 determine if the correlation is significant and calculate r^2 to determine the percent variation explained by the correlation (or regression)

13.29 Refer to the data in Problem 13.1.

13.30 Refer to Problem 13.2. Is there a difference between results in Problem 13.29?

13.31 Refer to Problem 13.3 or Problem 13.23.

13.32 Refer to Problem 13.4 or Problem 13.24.

13.33 Refer to Problem 13.5 or Problem 13.25.

13.34 Refer to Problem 13.6 of Problem 13.26.

13.35 Refer to Problem 13.7 or Problem 13.27.

13.36 Figure 13.6 has 6 scatter diagrams with the respective correlation coefficients (as well as n). Which have a significant correlation (0.05)? Which are *very* highly significant? Which are essentially not significant?

13.37 To understand the need for a large enough sample size, compute a correlation coefficient for just the first 5 students listed in Table 13.2 (use Table 13.3 to ease your computations). Is this correlation significant? Why is there such a difference between this value and the one for all of the 20 students?

SECTION 13.9

13.38 A basic college algebra test was administered to a group of 20 students. Along with the test grade, the heights of the students were also recorded. The 20 grades (out of a possible 100 points) and heights were as follows.

GRADE	HEIGHT	GRADE	HEIGHT	GRADE	HEIGHT
85	5′ 9″	90	5′11″	75	5′ 6″
15	5′ 2″	95	5′ 8″	45	5′ 1″
60	5′ 4″	55	5′ 4″	75	5′ 3″
5	4′ 8″	35	5′ 1″	80	5′ 8″
30	4′ 9″	20	5′ 0″	40	5′ 1″
30	5′ 1″	60	5′ 7″	35	5′ 3″
20	4′11″	25	4′11″		

Plot a graph. Calculate a correlation coefficient and explain this unusual correlation.

13.39 Go back to Problem 13.3. What would you predict that the average taste value would be for coffee perked for 11 minutes? How does this compare to a maximum possible reading of 10?

Several more tests were run and they were as follows.

TIME (MINUTES)	RATING	TIME (MINUTES)	RATING
9	6,7	12	1,4
10	5,7	13	1,3
11	3,5	14	0,3

Add these values to your previous graph. Explain what has happened.

13.40 Refer to the data in Problem 13.8. Why is it improper to use this data to predict income if you know the years of schooling completed?

13.41 Examine the graph and the equation of Problem 13.7. Explain why the apparent large differences between the predicted values and the actual values. What is the difficulty with assuming a linear regression? Try plotting the data on other graph paper such as semilogarithmic or log-log paper.

APPENDIX A
APPENDIX TABLES

List of tables

A.1 NORMAL DISTRIBUTION

These tables are used for finding areas of a normal distribution with mean, $\mu = 0$, and standard deviation, $\sigma = 1.00$. Values along the X axis are designated as z values. The entries in Column A of Table A.1a are z values. Column B contains the corresponding areas between 0 and z, and Column C contains the area above z (or from z to infinity). The area between $-z$ and 0 is the same as from 0 to z (because of the symmetry of the distribution).

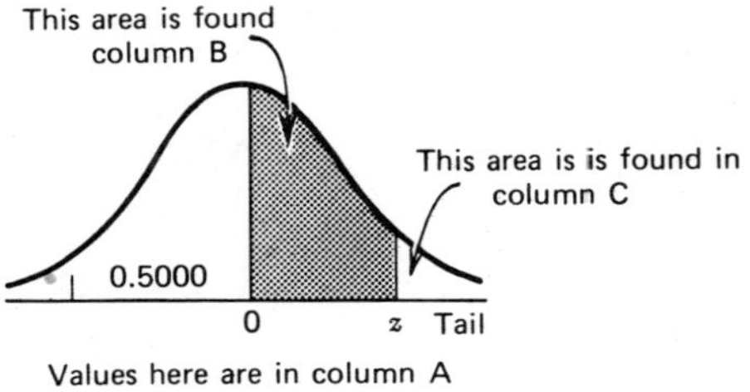

Table A.1b gives the z values for certain commonly encountered areas. Two-tailed and one-tailed values are separated.

Detailed examples and instructions are given in Section 5.4 (pages 114–121).

TABLE A.1a Normal Distribution

(A) z	(B) AREA BETWEEN MEAN AND z	(C) AREA BEYOND z	(A) z	(B) AREA BETWEEN MEAN AND z	(C) AREA BEYOND z	(A) z	(B) AREA BETWEEN MEAN AND z	(C) AREA BEYOND z
0.00	0.0000	0.5000	0.35	0.1368	0.3632	0.70	0.2580	0.2420
0.01	0.0040	0.4960	0.36	0.1406	0.3594	0.71	0.2611	0.2389
0.02	0.0080	0.4920	0.37	0.1443	0.3557	0.72	0.2642	0.2358
0.03	0.0120	0.4880	0.38	0.1480	0.3520	0.73	0.2673	0.2327
0.04	0.0160	0.4840	0.39	0.1517	0.3483	0.74	0.2704	0.2296
0.05	0.0199	0.4801	0.40	0.1554	0.3446	0.75	0.2734	0.2266
0.06	0.0239	0.4761	0.41	0.1591	0.3409	0.76	0.2764	0.2236
0.07	0.0279	0.4721	0.42	0.1628	0.3372	0.77	0.2794	0.2206
0.08	0.0319	0.4681	0.43	0.1664	0.3336	0.78	0.2823	0.2177
0.09	0.0359	0.4641	0.44	0.1700	0.3300	0.79	0.2852	0.2148
0.10	0.0398	0.4602	0.45	0.1736	0.3264	0.80	0.2881	0.2119
0.11	0.0438	0.4562	0.46	0.1772	0.3228	0.81	0.2910	0.2090
0.12	0.0478	0.4522	0.47	0.1808	0.3192	0.82	0.2939	0.2061
0.13	0.0517	0.4483	0.48	0.1844	0.3156	0.83	0.2967	0.2033
0.14	0.0557	0.4443	0.49	0.1879	0.3121	0.84	0.2995	0.2005
0.15	0.0596	0.4404	0.50	0.1915	0.3085	0.85	0.3023	0.1977
0.16	0.0636	0.4364	0.51	0.1950	0.3050	0.86	0.3051	0.1949
0.17	0.0675	0.4325	0.52	0.1985	0.3015	0.87	0.3078	0.1922
0.18	0.0714	0.4286	0.53	0.2019	0.2981	0.88	0.3106	0.1894
0.19	0.0753	0.4247	0.54	0.2054	0.2946	0.89	0.3133	0.1867
0.20	0.0793	0.4207	0.55	0.2088	0.2912	0.90	0.3159	0.1841
0.21	0.0832	0.4168	0.56	0.2123	0.2877	0.91	0.3186	0.1814
0.22	0.0871	0.4129	0.57	0.2157	0.2843	0.92	0.3212	0.1788
0.23	0.0910	0.4090	0.58	0.2190	0.2810	0.93	0.3238	0.1762
0.24	0.0948	0.4052	0.59	0.2224	0.2776	0.94	0.3264	0.1736
0.25	0.0987	0.4013	0.60	0.2257	0.2743	0.95	0.3289	0.1711
0.26	0.1026	0.3974	0.61	0.2291	0.2709	0.96	0.3315	0.1685
0.27	0.1064	0.3936	0.62	0.2324	0.2676	0.97	0.3340	0.1660
0.28	0.1103	0.3897	0.63	0.2357	0.2643	0.98	0.3365	0.1635
0.29	0.1141	0.3859	0.64	0.2389	0.2611	0.99	0.3389	0.1611
0.30	0.1179	0.3821	0.65	0.2422	0.2578	1.00	0.3413	0.1587
0.31	0.1217	0.3783	0.66	0.2454	0.2546	1.01	0.3438	0.1562
0.32	0.1255	0.3745	0.67	0.2486	0.2514	1.02	0.3461	0.1539
0.33	0.1293	0.3707	0.68	0.2517	0.2483	1.03	0.3485	0.1515
0.34	0.1331	0.3669	0.69	0.2549	0.2451	1.04	0.3508	0.1492
0.35	0.1368	0.3632	0.70	0.2580	0.2420	1.05	0.3531	0.1469

TABLE A.1a (*continued*)

(A) z	(B) AREA BETWEEN MEAN AND z	(C) AREA BEYOND z	(A) z	(B) AREA BETWEEN MEAN AND z	(C) AREA BEYOND z	(A) z	(B) AREA BETWEEN MEAN AND z	(C) AREA BEYOND z
1.05	0.3531	0.1469	1.40	0.4192	0.0808	1.75	0.4599	0.0401
1.06	0.3554	0.1446	1.41	0.4207	0.0793	1.76	0.4608	0.0392
1.07	0.3577	0.1423	1.42	0.4222	0.0778	1.77	0.4616	0.0384
1.08	0.3599	0.1401	1.43	0.4236	0.0764	1.78	0.4625	0.0375
1.09	0.3621	0.1379	1.44	0.4251	0.0749	1.79	0.4633	0.0367
1.10	0.3643	0.1357	1.45	0.4265	0.0735	1.80	0.4641	0.0359
1.11	0.3665	0.1335	1.46	0.4279	0.0721	1.81	0.4649	0.0351
1.12	0.3686	0.1314	1.47	0.4292	0.0708	1.82	0.4656	0.0344
1.13	0.3708	0.1292	1.48	0.4306	0.0694	1.83	0.4664	0.0336
1.14	0.3729	0.1271	1.49	0.4319	0.0681	1.84	0.4671	0.0329
1.15	0.3749	0.1251	1.50	0.4332	0.0668	1.85	0.4678	0.0322
1.16	0.3770	0.1230	1.51	0.4345	0.0655	1.86	0.4686	0.0314
1.17	0.3790	0.1210	1.52	0.4357	0.0643	1.87	0.4693	0.0307
1.18	0.3810	0.1190	1.53	0.4370	0.0630	1.88	0.4699	0.0301
1.19	0.3830	0.1170	1.54	0.4382	0.0618	1.89	0.4706	0.0294
1.20	0.3849	0.1151	1.55	0.4394	0.0606	1.90	0.4713	0.0287
1.21	0.3869	0.1131	1.56	0.4406	0.0594	1.91	0.4719	0.0281
1.22	0.3888	0.1112	1.57	0.4418	0.0582	1.92	0.4726	0.0274
1.23	0.3907	0.1093	1.58	0.4429	0.0571	1.93	0.4732	0.0268
1.24	0.3925	0.1075	1.59	0.4441	0.0559	1.94	0.4738	0.0262
1.25	0.3944	0.1056	1.60	0.4452	0.0548	1.95	0.4744	0.0256
1.26	0.3962	0.1038	1.61	0.4463	0.0537	1.96	0.4750	0.0250
1.27	0.3980	0.1020	1.62	0.4474	0.0526	1.97	0.4756	0.0244
1.28	0.3997	0.1003	1.63	0.4484	0.0516	1.98	0.4761	0.0239
1.29	0.4015	0.0985	1.64	0.4495	0.0505	1.99	0.4767	0.0233
1.30	0.4032	0.0968	1.65	0.4505	0.0495	2.00	0.4772	0.0228
1.31	0.4049	0.0951	1.66	0.4515	0.0485	2.01	0.4778	0.0222
1.32	0.4066	0.0934	1.67	0.4525	0.0475	2.02	0.4783	0.0217
1.33	0.4082	0.0918	1.68	0.4535	0.0465	2.03	0.4788	0.0212
1.34	0.4099	0.0901	1.69	0.4545	0.0455	2.04	0.4793	0.0207
1.35	0.4115	0.0885	1.70	0.4554	0.0446	2.05	0.4798	0.0202
1.36	0.4131	0.0869	1.71	0.4564	0.0436	2.06	0.4803	0.0197
1.37	0.4147	0.0853	1.72	0.4573	0.0427	2.07	0.4808	0.0192
1.38	0.4162	0.0838	1.73	0.4582	0.0418	2.08	0.4812	0.0188
1.39	0.4177	0.0823	1.74	0.4591	0.0409	2.09	0.4817	0.0183
1.40	0.4192	0.0808	1.75	0.4599	0.0401	2.10	0.4821	0.0179

TABLE A.1a (*continued*)

(A) z	(B) AREA BETWEEN MEAN AND z	(C) AREA BEYOND z	(A) z	(B) AREA BETWEEN MEAN AND z	(C) AREA BEYOND z	(A) z	(B) AREA BETWEEN MEAN AND z	(C) AREA BEYOND z
2.10	0.4821	0.0179	2.40	0.4918	0.0082	2.90	0.4981	0.0019
2.11	0.4826	0.0174	2.41	0.4920	0.0080	2.92	0.4982	0.0018
2.12	0.4830	0.0170	2.42	0.4922	0.0078	2.94	0.4984	0.0016
2.13	0.4834	0.0166	2.43	0.4925	0.0075	2.96	0.4985	0.0015
2.14	0.4838	0.0162	2.44	0.4927	0.0073	2.98	0.4986	0.0014
2.15	0.4842	0.0158	2.45	0.4929	0.0071	3.00	0.4987	0.0013
2.16	0.4846	0.0154	2.46	0.4931	0.0069	3.02	0.4987	0.0013
2.17	0.4850	0.0150	2.47	0.4932	0.0068	3.04	0.4988	0.0012
2.18	0.4854	0.0146	2.48	0.4934	0.0066	3.06	0.4989	0.0011
2.19	0.4857	0.0143	2.49	0.4936	0.0064	3.08	0.4990	0.0010
2.20	0.4861	0.0139	2.50	0.4938	0.0062	3.10	0.4990	0.0010
2.21	0.4864	0.0136	2.52	0.4941	0.0059	3.12	0.4991	0.0009
2.22	0.4868	0.0132	2.54	0.4945	0.0055	3.14	0.4992	0.0008
2.23	0.4871	0.0129	2.56	0.4948	0.0052	3.16	0.4992	0.0008
2.24	0.4875	0.0125	2.58	0.4951	0.0049	3.20	0.4993	0.0007
2.25	0.4878	0.0122	2.60	0.4953	0.0047	3.25	0.4994	0.0006
2.26	0.4881	0.0119	2.62	0.4956	0.0044	3.30	0.4995	0.0005
2.27	0.4884	0.0116	2.64	0.4959	0.0041	3.35	0.4996	0.0004
2.28	0.4887	0.0113	2.66	0.4961	0.0039	3.40	0.4997	0.0003
2.29	0.4890	0.0110	2.68	0.4963	0.0037	3.45	0.4997	0.0003
2.30	0.4893	0.0107	2.70	0.4965	0.0035	3.50	0.4998	0.0002
2.31	0.4896	0.0104	2.72	0.4967	0.0033	3.55	0.4998	0.0002
2.32	0.4898	0.0102	2.74	0.4969	0.0031	3.60	0.4998	0.0002
2.33	0.4901	0.0099	2.76	0.4971	0.0029	3.65	0.4999	0.0001
2.34	0.4904	0.0096	2.78	0.4973	0.0027	3.70	0.4999	0.0001
2.35	0.4906	0.0094	2.80	0.4974	0.0026	3.75	0.4999	0.0001
2.36	0.4909	0.0091	2.82	0.4976	0.0024	3.80	0.4999	0.0001
2.37	0.4911	0.0089	2.84	0.4977	0.0023	3.85	0.4999	0.0001
2.38	0.4913	0.0087	2.86	0.4979	0.0021	3.90	0.49995	0.00005
2.39	0.4916	0.0084	2.88	0.4980	0.0020	3.95	0.49996	0.00004
2.40	0.4918	0.0082	2.90	0.4981	0.0019	4.00	0.49997	0.00003

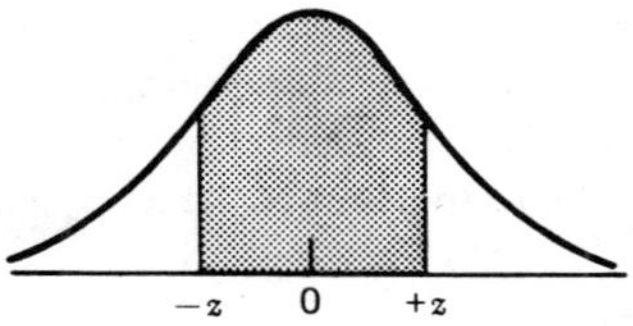

Two–tailed: Areas between the limits of $-z$ and $+z$

TABLE A.1b Normal Distribution (z-values for selected areas)

PERCENT BETWEEN LIMITS	AREA BETWEEN LIMITS	TOTAL AREA OUTSIDE LIMITS	AREA IN EACH TAIL	z
25	0.25	0.75	0.375	0.319
50	0.50	0.50	0.25	0.674
60	0.60	0.40	0.20	0.842
70	0.70	0.30	0.15	1.036
75	0.75	0.25	0.125	1.150
80	0.80	0.20	0.10	1.282
90	0.90	0.10	0.05	1.645
95	0.95	0.05	0.025	1.960
98	0.98	0.02	0.01	2.326
99	0.99	0.01	0.005	2.576
99.5	0.995	0.005	0.0025	2.810
99.9	0.999	0.001	0.0005	3.291

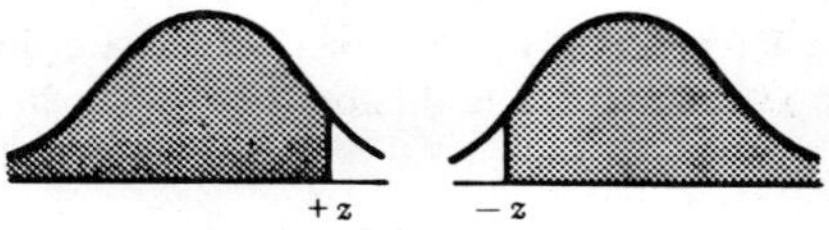

One tailed: Area is below $+z$ or above $-z$

TABLE A.1b (*continued*)

PERCENT BELOW z	AREA BELOW z (SHADED) AREA)	AREA ABOVE z (UNSHADED AREAS)	z (OR $-z$)
50	0.50	0.50	0.000
60	0.60	0.40	0.253
70	0.70	0.30	0.524
75	0.75	0.25	0.674
80	0.80	0.20	0.842
90	0.90	0.10	1.282
95	0.95	0.05	1.645
97.5	0.975	0.025	1.960
98	0.98	0.02	2.054
99	0.99	0.01	2.326
99.5	0.995	0.005	2.576
99.9	0.999	0.001	3.090
99.95	0.9995	0.0005	3.291

A.2 *t*-DISTRIBUTION

This is a symmetric distribution with center equal to 0. *df* is *degrees of freedom* and *generally equals n* − 1. For one-tailed tests or one-sided levels of significance you should consult row labeled "level of significance for one-tailed test." The subscript designates the proportion of the area that is below that value (see Table A.2).

EXAMPLE 1

df = 5. What is the limit below which are 95% *of the values*? This is a one-tailed test. It also corresponds to the value of $t_{0.95}$.

		$t_{0.95}$	
df	*Level . . . one-tailed test*		
		0.05	
5		2.015	

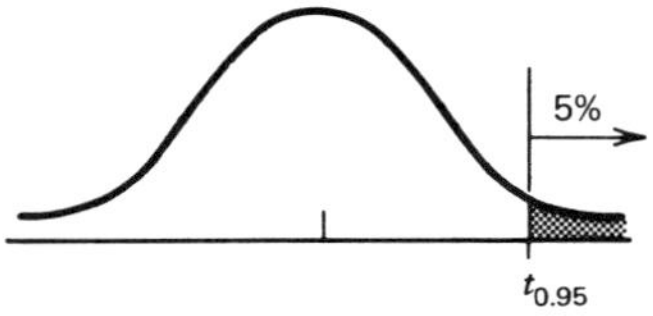

The limit is 2.015.

EXAMPLE 2

df = 12. What are the limits between which lie 95% of the values? This is a two-tailed test for which the 5% in the tails are divided between the two tails. Thus $2\frac{1}{2}$% or 0.025 are in each tail and the *t*-values are $t_{0.025}$ and $t_{0.975}$.

		$t_{0.975}$	
df	*Level . . . two-tailed test*		
		0.05	
12		2.179	

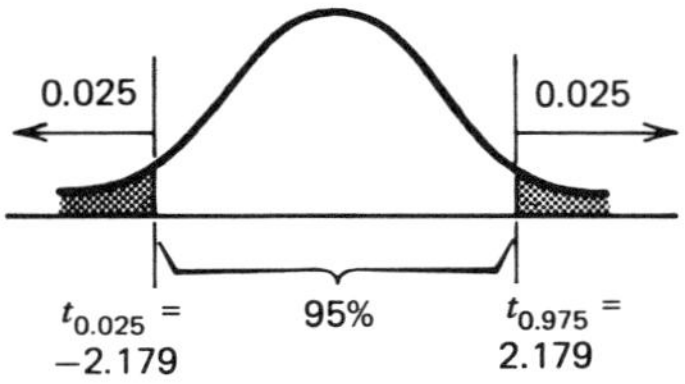

The limits are ±2.179.

TABLE A.2 *t*-Distribution

	$t_{0.90}$	$t_{0.95}$	$t_{0.975}$	$t_{0.99}$	$t_{0.995}$	$t_{0.9995}$
	LEVEL OF SIGNIFICANCE FOR ONE-TAILED TEST					
	0.10	0.05	0.025	0.01	0.005	0.0005
df	LEVEL OF SIGNIFICANCE FOR TWO-TAILED TEST					
	0.20	0.10	0.05	0.02	0.01	0.001
1	3.078	6.314	12.706	31.821	63.657	636.619
2	1.886	2.920	4.303	6.965	9.925	31.598
3	1.638	2.353	3.182	4.541	5.841	12.941
4	1.533	2.132	2.776	3.747	4.604	8.610
5	1.476	2.015	2.571	3.365	4.032	6.859
6	1.440	1.943	2.447	3.143	3.707	5.959
7	1.415	1.895	2.365	2.998	3.499	5.405
8	1.397	1.860	2.306	2.896	3.355	5.041
9	1.383	1.833	2.262	2.821	3.250	4.781
10	1.372	1.812	2.228	2.764	3.169	4.587
11	1.363	1.796	2.201	2.718	3.106	4.437
12	1.356	1.782	2.179	2.681	3.055	4.318
13	1.350	1.771	2.160	2.650	3.012	4.221
14	1.345	1.761	2.145	2.624	2.977	4.140
15	1.341	1.753	2.131	2.602	2.947	4.073
16	1.337	1.746	2.120	2.583	2.921	4.015
17	1.333	1.740	2.110	2.567	2.898	3.965
18	1.330	1.734	2.101	2.552	2.878	3.992
19	1.328	1.729	2.093	2.539	2.861	3.883
20	1.325	1.725	2.086	2.528	2.845	3.850
21	1.323	1.721	2.080	2.518	2.831	3.819
22	1.321	1.717	2.074	2.508	2.819	3.792
23	1.319	1.714	2.069	2.500	2.807	3.767
24	1.318	1.711	2.064	2.492	2.797	3.745
25	1.316	1.708	2.060	2.485	2.787	3.725
26	1.315	1.706	2.056	2.479	2.779	3.707
27	1.314	1.703	2.052	2.473	2.771	3.690
28	1.313	1.701	2.048	2.467	2.763	3.674
29	1.311	1.699	2.045	2.462	2.756	3.659
30	1.310	1.697	2.042	2.457	2.750	3.646
40	1.303	1.684	2.021	2.423	2.704	3.551
60	1.296	1.671	2.000	2.390	2.660	3.460
120	1.289	1.658	1.980	2.358	2.617	3.373
∞	1.282	1.645	1.960	2.326	2.576	3.291

Table A.2 is taken from Table III of Fisher and Yates: *Statistical Tables for Biological, Agricultural and Medical Research*, published by Longman Group Ltd., London (previously published by Oliver & Boyd, Edinburgh), and by permission of the authors and publishers.

TABLE A.3 χ^2 (Chi-Square) Distribution

	PROBABILITY THAT A VALUE IS ABOVE THE CRITICAL VALUE												
								0.250	0.100	0.050	0.025	0.010	0.005
	PROBABILITY THAT A VALUE IS BELOW THE CRITICAL VALUE												
Degrees of freedom	0.005	0.010	0.025	0.050	0.100	0.250	0.500	0.750	0.900	0.950	0.975	0.990	0.995
1	0.0000393	0.000157	0.000982	0.00393	0.0158	0.102	0.455	1.32	2.71	3.84	5.02	6.63	7.88
2	0.0100	0.0201	0.0506	0.103	0.211	0.575	1.39	2.77	4.61	5.99	7.38	9.21	10.6
3	0.0717	0.115	0.216	0.352	0.584	1.21	2.37	4.11	6.25	7.81	9.35	11.3	12.8
4	0.207	0.297	0.484	0.711	1.06	1.92	3.36	5.39	7.78	9.49	11.1	13.3	14.9
5	0.412	0.554	0.831	1.15	1.61	2.67	4.35	6.63	9.24	11.1	12.8	15.1	16.7
6	0.676	0.872	1.24	1.64	2.20	3.45	5.35	7.84	10.6	12.6	14.4	16.8	18.5
7	0.989	1.24	1.69	2.17	2.83	4.25	6.35	9.04	12.0	14.1	16.0	18.5	20.3
8	1.34	1.65	2.18	2.73	3.49	5.07	7.34	10.2	13.4	15.5	17.5	20.1	22.0
9	1.73	2.09	2.70	3.33	4.17	5.90	8.34	11.4	14.7	16.9	19.0	21.7	23.6
10	2.16	2.56	3.25	3.94	4.87	6.74	9.34	12.5	16.0	18.3	20.5	23.2	25.2
11	2.60	3.05	3.82	4.57	5.58	7.58	10.3	13.7	17.3	19.7	21.9	24.7	26.8
12	3.07	3.57	4.40	5.23	6.30	8.44	11.3	14.8	18.5	21.0	23.3	26.2	28.3
13	3.57	4.11	5.01	5.89	7.04	9.30	12.3	16.0	19.8	22.4	24.7	27.7	29.8
14	4.07	4.66	5.63	6.57	7.79	10.2	13.3	17.1	21.1	23.7	26.1	29.1	31.3
15	4.60	5.23	6.26	7.26	8.55	11.0	14.3	18.2	22.3	25.0	27.5	30.6	32.8
16	5.14	5.81	6.91	7.96	9.31	11.9	15.3	19.4	23.5	26.3	28.8	32.0	34.3
17	5.70	6.41	7.56	8.67	10.1	12.8	16.3	20.5	24.8	27.6	30.2	33.4	35.7
18	6.26	7.01	8.23	9.39	10.9	13.7	17.3	21.6	26.0	28.9	31.5	34.8	37.2
19	6.84	7.63	8.91	10.1	11.7	14.6	18.3	22.7	27.2	30.1	32.9	36.2	38.6
20	7.43	8.26	9.59	10.9	12.4	15.5	19.3	23.8	28.4	31.4	34.2	37.6	40.0
21	8.03	8.90	10.3	11.6	13.2	16.3	20.3	24.9	29.6	32.7	35.5	38.9	41.4
22	8.64	9.54	11.0	12.3	14.0	17.2	21.3	26.0	30.8	33.9	36.8	40.3	42.8
23	9.26	10.2	11.7	13.1	14.8	18.1	22.3	27.1	32.0	35.2	38.1	41.6	44.2
24	9.89	10.9	12.4	13.8	15.7	19.0	23.3	28.2	33.2	36.4	39.4	43.0	45.6
25	10.5	11.5	13.1	14.6	16.5	19.9	24.3	29.3	34.4	37.7	40.6	44.3	46.9
26	11.2	12.2	13.8	15.4	17.3	20.8	25.3	30.4	35.6	38.9	41.9	45.6	48.3
27	11.8	12.9	14.6	16.2	18.1	21.7	26.3	31.5	36.7	40.1	43.2	47.0	49.6
28	12.5	13.6	15.3	16.9	18.9	22.7	27.3	32.6	37.9	41.3	44.5	48.3	51.0
29	13.1	14.3	16.0	17.7	19.8	23.6	28.3	33.7	39.1	42.6	45.7	49.6	52.3
30	13.8	15.0	16.8	18.5	20.6	24.5	29.3	34.8	40.3	43.8	47.0	50.9	53.7
40	20.7	22.2	24.4	26.5	29.1	33.7	39.3	45.6	51.8	55.8	59.3	63.7	66.8
50	28.0	29.7	32.4	34.8	37.7	42.9	49.3	56.3	63.2	67.5	71.4	76.2	79.5
60	35.5	37.5	40.5	43.2	46.5	52.3	59.3	67.0	74.4	79.1	83.3	88.4	92.0
70	43.3	45.4	48.8	51.7	55.3	61.7	69.3	77.6	85.5	90.5	95.0	100	104
80	51.2	53.5	57.2	60.4	64.3	71.1	79.3	88.1	96.6	102	107	112	116
90	59.2	61.8	65.6	69.1	73.3	80.6	89.3	98.6	108	113	118	124	128
100	67.3	70.1	74.2	77.9	82.4	90.1	99.3	109	118	124	130	136	140

Abridged with kind permission of Biometrika Trustees from *Biometrika Tables for Statisticians*, Volume 1, 3rd Edition, edited by E. S. Pearson and H. O. Hartley (Cambridge, Cambridge University Press, 1966).

A.3 χ^2 (CHI-SQUARE) DISTRIBUTION

Generally the *df* (degrees of freedom) is equal to $n - 1$. The probability level indicates what proportion of the particular χ^2 distribution lies below the corresponding entry in the table (Table A.3).

EXAMPLE 1

$df = 5$. We desire a χ^2 for which there is a 10% chance of a value being larger. This is the same as asking what is the value for which 90% are smaller. This is $\chi^2_{0.90}$.

	Probability of value above		
		0.100	
	Probability of value below		
df		0.900	
5		9.24	

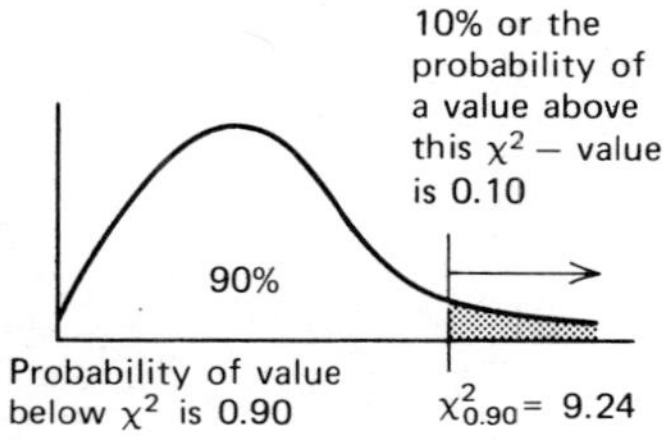

$\chi^2_{0.90} = 9.24$.

EXAMPLE 2

$df = 8$. We desire χ^2 limits (upper and lower) for which 10% are outside. Then we need two limits—a two-tailed case—with 5% in each tail.

	Probability of value below				
df		0.050		0.950	
8		2.73		15.5	

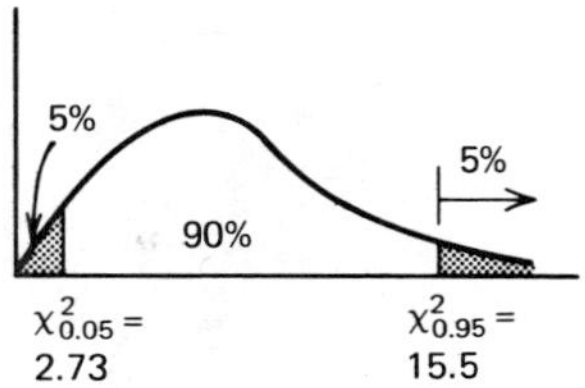

$\chi^2_{0.05} = 2.73$, $\chi^2_{0.95} = 15.5$.

A.4 *F*-DISTRIBUTION

These are all *one-tailed tables* (Table A.4). The way they are used as two-tailed tables is discussed below.

All *F*-values involve a *division*. Symbolically,

$$F(f_1, f_2) = \frac{\text{A numerator with } df = f_1}{\text{A denominator with } df = f_2}$$

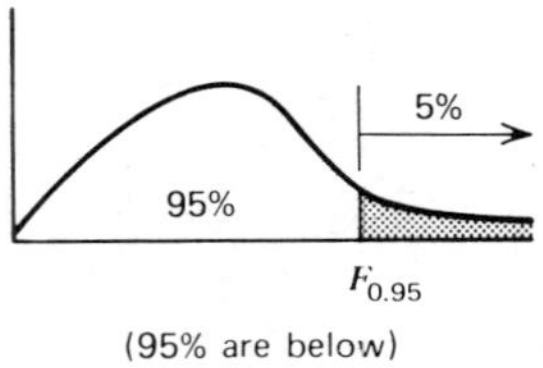

Thus, $F(8, 6)$ means that two numbers were involved in finding the distribution. The number that is (or was) in the numerator has 8 degrees of freedom and the number in the denominator has 6 degrees of freedom. Generally, these are variances where each number of degrees of freedom is referring to a sample size. Typically $df = n - 1$ so that for 2 samples, one with 9 individuals, and the other with 7 individuals, the degrees of freedom would be 8 and 6, respectively.

EXAMPLE 1

Find the F-value below which 95% of the values are found. You are given that the degrees of freedom are $f_1 = 8$ and $f_2 = 6$.

This is to be a *one-tailed* test with probability = 0.95, therefore consult the table, which reads:

One-tailed test $P = 0.95$

f_2 \ f_1		8	
6		4.15	

Two-tailed tests

To determine the limit(s) for a two-tailed test there are two choices.

1. The preferred method (because of its ease). Perform the calculation of the F-value as the following:

$$F = \frac{\text{Larger value with } df = f_1}{\text{Smaller value with } df = f_2}$$

The limit is found by referring to the table labeled "when used as two-tailed test, $P = 0.XX$."

Thus, only *one* limit is found (since all calculated ratios are greater than 1.0).

2. To find the two limits first find the upper limit as a one-tailed test limit. Then find the lower limit by first looking up the F-value with the reversed degrees of freedom. Then calculate the reciprocal of this value (see Example 2).

EXAMPLE 2

Find the limits for a two-sided test between which 95% of the F-values lie. The degrees of freedom are $f_1 = 8$ and $f_2 = 6$.

1. The *preferred method.* We will use the table labeled "when used as two-tailed test, $P = 0.95$." (This is the one-tailed table for $P = 0.975$.)

To find the sample value of F you calculate

$$F = \frac{\text{Larger value}}{\text{Smaller value}}$$

We then look up the appropriate F-value and compare the two. If the calculated quantity is larger than the F-value from the table we would say it has a probability of less than 0.95.

Thus, if the two values were 18.4 with 8 degrees of freedom and 6.4 with 6 degrees of freedom, we would have

$$F = \frac{\text{Larger}}{\text{Smaller}} = \frac{18.4}{6.4} = 2.875$$

The corresponding F would be $F(8, 6)$ and would be 5.60 [see (a) below].

Note that if the two quantities were reversed with the 18.4 having 6 degrees of freedom (instead of 8) and the 6.4 having 8 degrees of freedom, then the ratio of larger/smaller still is 2.875, but the F-value in the table would be found for $f_1 = 6$ and $f_2 = 8$. (These are reversed.) The critical value would be 4.65 [see (b) below].

	f_2 \ f_1	*df* in numerator		
			8	
df in denom.	6		5.60	

(a) $F(8, 6) = 5.60$

	f_2 \ f_1	*df* in numerator		
			6	
df in denom.	8		4.65	

(b) $F(6, 8) = 4.65$

2. To obtain *two limits* (this is necessary if you are finding confidence intervals), refer to the table—one-tailed test, $P = 0.975$ (since 2.5% are in each tail).

UPPER LIMIT $F_{0.95}$ (one-tailed test 0.975 with 8 and 6 degrees of freedom. This is the 5.60 [as in (a) of the figure above].

LOWER LIMIT We find $F_{0.025}$ using the same table as before except *reversing* the degrees of freedom [as in (b) of the figure above]. We obtain the 4.65. Then take the reciprocal of this F-value and obtain:

$$\text{Lower limit} = \frac{1}{F(f_1, f_2)} = \frac{1}{4.65} = 0.215$$

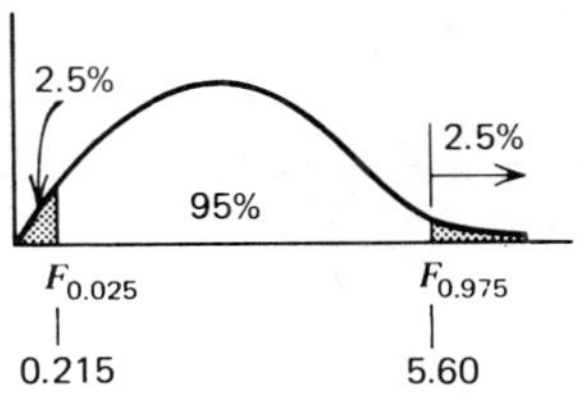

TABLE A.4 *F*-Distribution

a. One-tailed test, $P = 0.95$ (0.05 or 5% in tail). When used as two-tailed test, $P = 0.90$.

Degrees of freedom in numerator

f_2 \ f_1	1	2	3	4	5	6	7	8	9	10	12	15	20	24	30	40	60	120	∞
1	161.4	199.5	215.7	224.6	230.2	234.0	236.8	238.9	240.5	241.9	243.9	245.9	248.0	249.1	250.1	251.1	252.2	253.3	254.3
2	18.51	19.00	19.16	19.25	19.30	19.33	19.35	19.37	19.38	19.40	19.41	19.43	19.45	19.45	19.46	19.47	19.48	19.49	19.50
3	10.13	9.55	9.28	9.12	9.01	8.94	8.89	8.85	8.81	8.79	8.74	8.70	8.66	8.64	8.62	8.59	8.57	8.55	8.53
4	7.71	6.94	6.59	6.39	6.26	6.16	6.09	6.04	6.00	5.96	5.91	5.86	5.80	5.77	5.75	5.72	5.69	5.66	5.63
5	6.61	5.79	5.41	5.19	5.05	4.95	4.88	4.82	4.77	4.74	4.68	4.62	4.56	4.53	4.50	4.46	4.43	4.40	4.36
6	5.99	5.14	4.76	4.53	4.39	4.28	4.21	4.15	4.10	4.06	4.00	3.94	3.87	3.84	3.81	3.77	3.74	3.70	3.67
7	5.59	4.74	4.35	4.12	3.97	3.87	3.79	3.73	3.68	3.64	3.57	3.51	3.44	3.41	3.38	3.34	3.30	3.27	3.23
8	5.32	4.46	4.07	3.84	3.69	3.58	3.50	3.44	3.39	3.35	3.28	3.22	3.15	3.12	3.08	3.04	3.01	2.97	2.93
9	5.12	4.26	3.86	3.63	3.48	3.37	3.29	3.23	3.18	3.14	3.07	3.01	2.94	2.90	2.86	2.83	2.79	2.75	2.71
10	4.96	4.10	3.71	3.48	3.33	3.22	3.14	3.07	3.02	2.98	2.91	2.85	2.77	2.74	2.70	2.66	2.62	2.58	2.54
11	4.84	3.98	3.59	3.36	3.20	3.09	3.01	2.95	2.90	2.85	2.79	2.72	2.65	2.61	2.57	2.53	2.49	2.45	2.40
12	4.75	3.89	3.49	3.26	3.11	3.00	2.91	2.85	2.80	2.75	2.69	2.62	2.54	2.51	2.47	2.43	2.38	2.34	2.30
13	4.67	3.81	3.41	3.18	3.03	2.92	2.83	2.77	2.71	2.67	2.60	2.53	2.46	2.42	2.38	2.34	2.30	2.25	2.21
14	4.60	3.74	3.34	3.11	2.96	2.85	2.76	2.70	2.65	2.60	2.53	2.46	2.39	2.35	2.31	2.27	2.22	2.18	2.13
15	4.54	3.68	3.29	3.06	2.90	2.79	2.71	2.64	2.59	2.54	2.48	2.40	2.33	2.29	2.25	2.20	2.16	2.11	2.07
16	4.49	3.63	3.24	3.01	2.85	2.74	2.66	2.59	2.54	2.49	2.42	2.35	2.28	2.24	2.19	2.15	2.11	2.06	2.01
17	4.45	3.59	3.20	2.96	2.81	2.70	2.61	2.55	2.49	2.45	2.38	2.31	2.23	2.19	2.15	2.10	2.06	2.01	1.96
18	4.41	3.55	3.16	2.93	2.77	2.66	2.58	2.51	2.46	2.41	2.34	2.27	2.19	2.15	2.11	2.06	2.02	1.97	1.92
19	4.38	3.52	3.13	2.90	2.74	2.63	2.54	2.48	2.42	2.38	2.31	2.23	2.16	2.11	2.07	2.03	1.98	1.93	1.88
20	4.35	3.49	3.10	2.87	2.71	2.60	2.51	2.45	2.39	2.35	2.28	2.20	2.12	2.08	2.04	1.99	1.95	1.90	1.84
21	4.32	3.47	3.07	2.84	2.68	2.57	2.49	2.42	2.37	2.32	2.25	2.18	2.10	2.05	2.01	1.96	1.92	1.87	1.81
22	4.30	3.44	3.05	2.82	2.66	2.55	2.46	2.40	2.34	2.30	2.23	2.15	2.07	2.03	1.98	1.94	1.89	1.84	1.78
23	4.28	3.42	3.03	2.80	2.64	2.53	2.44	2.37	2.32	2.27	2.20	2.13	2.05	2.01	1.96	1.91	1.86	1.81	1.76
24	4.26	3.40	3.01	2.78	2.62	2.51	2.42	2.36	2.30	2.25	2.18	2.11	2.03	1.98	1.94	1.89	1.84	1.79	1.73
25	4.24	3.39	2.99	2.76	2.60	2.49	2.40	2.34	2.28	2.24	2.16	2.09	2.01	1.96	1.92	1.87	1.82	1.77	1.71
26	4.23	3.37	2.98	2.74	2.59	2.47	2.39	2.32	2.27	2.22	2.15	2.07	1.99	1.95	1.90	1.85	1.80	1.75	1.69
27	4.21	3.35	2.96	2.73	2.57	2.46	2.37	2.31	2.25	2.20	2.13	2.06	1.97	1.93	1.88	1.84	1.79	1.73	1.67
28	4.20	3.34	2.95	2.71	2.56	2.45	2.36	2.29	2.24	2.19	2.12	2.04	1.96	1.91	1.87	1.82	1.77	1.71	1.65
29	4.18	3.33	2.93	2.70	2.55	2.43	2.35	2.28	2.22	2.18	2.10	2.03	1.94	1.90	1.85	1.81	1.75	1.70	1.64
30	4.17	3.32	2.92	2.69	2.53	2.42	2.33	2.27	2.21	2.16	2.09	2.01	1.93	1.89	1.84	1.79	1.74	1.68	1.62
40	4.08	3.23	2.84	2.61	2.45	2.34	2.25	2.18	2.12	2.08	2.00	1.92	1.84	1.79	1.74	1.69	1.64	1.58	1.51
60	4.00	3.15	2.76	2.53	2.37	2.25	2.17	2.10	2.04	1.99	1.92	1.84	1.75	1.70	1.65	1.59	1.53	1.47	1.39
120	3.92	3.07	2.68	2.45	2.29	2.17	2.09	2.02	1.96	1.91	1.83	1.75	1.66	1.61	1.55	1.50	1.43	1.35	1.25
∞	3.84	3.00	2.60	2.37	2.21	2.10	2.01	1.94	1.88	1.83	1.75	1.67	1.57	1.52	1.46	1.39	1.32	1.22	1.00

Degrees of freedom in denominator

Abridged with the kind permission of the Biometrika Trustees from *Biometrika Tables for Statisticians*, Vo. 1, 3rd Edition, edited by E. S. Pearson and H. O. Hartley (Cambridge, Cambridge University Press, 1966).

TABLE A.4 (*continued*)

b. One-tailed test, $P = 0.975$ (0.025 or 2.5%). When used as two-tailed test, $P = 0.95$.

Degrees of freedom in numerator

Degrees of freedom in denominator

f_2 \ f_1	1	2	3	4	5	6	7	8	9	10	12	15	20	24	30	40	60	120	∞
1	647.8	799.5	864.2	899.6	921.8	937.1	948.2	956.7	963.3	968.6	976.7	984.9	993.1	997.2	1001	1006	1010	1014	1018
2	38.51	39.00	39.17	39.25	39.30	39.33	39.36	39.37	39.39	39.40	39.41	39.43	39.45	39.46	39.46	39.47	39.48	39.49	39.50
3	17.44	16.04	15.44	15.10	14.88	14.73	14.62	14.54	14.47	14.42	14.34	14.25	14.17	14.12	14.08	14.04	13.99	13.95	13.90
4	12.22	10.65	9.98	9.60	9.36	9.20	9.07	8.98	8.90	8.84	8.75	8.66	8.56	8.51	8.46	8.41	8.36	8.31	8.26
5	10.01	8.43	7.76	7.39	7.15	6.98	6.85	6.76	6.68	6.62	6.52	6.43	6.33	6.28	6.23	6.18	6.12	6.07	6.02
6	8.81	7.26	6.60	6.23	5.99	5.82	5.70	5.60	5.52	5.46	5.37	5.27	5.17	5.12	5.07	5.01	4.96	4.90	4.85
7	8.07	6.54	5.89	5.52	5.29	5.12	4.99	4.90	4.82	4.76	4.67	4.57	4.47	4.42	4.36	4.31	4.25	4.20	4.14
8	7.57	6.06	5.42	5.05	4.82	4.65	4.53	4.43	4.36	4.30	4.20	4.10	4.00	3.95	3.89	3.84	3.78	3.73	3.67
9	7.21	5.71	5.08	4.72	4.48	4.32	4.20	4.10	4.03	3.96	3.87	3.77	3.67	3.61	3.56	3.51	3.45	3.39	3.33
10	6.94	5.46	4.83	4.47	4.24	4.07	3.95	3.85	3.78	3.72	3.62	3.52	3.42	3.37	3.31	3.26	3.20	3.14	3.08
11	6.72	5.26	4.63	4.28	4.04	3.88	3.76	3.66	3.59	3.53	3.43	3.33	3.23	3.17	3.12	3.06	3.00	2.94	2.88
12	6.55	5.10	4.47	4.12	3.89	3.73	3.61	3.51	3.44	3.37	3.28	3.18	3.07	3.02	2.96	2.91	2.85	2.79	2.72
13	6.41	4.97	4.35	4.00	3.77	3.60	3.48	3.39	3.31	3.25	3.15	3.05	2.95	2.89	2.84	2.78	2.72	2.66	2.60
14	6.30	4.86	4.24	3.89	3.66	3.50	3.38	3.29	3.21	3.15	3.05	2.95	2.84	2.79	2.73	2.67	2.61	2.55	2.49
15	6.20	4.77	4.15	3.80	3.58	3.41	3.29	3.20	3.12	3.06	2.96	2.86	2.76	2.70	2.64	2.59	2.52	2.46	2.40
16	6.12	4.69	4.08	3.73	3.50	3.34	3.22	3.12	3.05	2.99	2.89	2.79	2.68	2.63	2.57	2.51	2.45	2.38	2.32
17	6.04	4.62	4.01	3.66	3.44	3.28	3.16	3.06	2.98	2.92	2.82	2.72	2.62	2.56	2.50	2.44	2.38	2.32	2.25
18	5.98	4.56	3.95	3.61	3.38	3.22	3.10	3.01	2.93	2.87	2.77	2.67	2.56	2.50	2.44	2.38	2.32	2.26	2.19
19	5.92	4.51	3.90	3.56	3.33	3.17	3.05	2.96	2.88	2.82	2.72	2.62	2.51	2.45	2.39	2.33	2.27	2.20	2.13
20	5.87	4.46	3.86	3.51	3.29	3.13	3.01	2.91	2.84	2.77	2.68	2.57	2.46	2.41	2.35	2.29	2.22	2.16	2.09
21	5.83	4.42	3.82	3.48	3.25	3.09	2.97	2.87	2.80	2.73	2.64	2.53	2.42	2.37	2.31	2.25	2.18	2.11	2.04
22	5.79	4.38	3.78	3.44	3.22	3.05	2.93	2.84	2.76	2.70	2.60	2.50	2.39	2.33	2.27	2.21	2.14	2.08	2.00
23	5.75	4.35	3.75	3.41	3.18	3.02	2.90	2.81	2.73	2.67	2.57	2.47	2.36	2.30	2.24	2.18	2.11	2.04	1.97
24	5.72	4.32	3.72	3.38	3.15	2.99	2.87	2.78	2.70	2.64	2.54	2.44	2.33	2.27	2.21	2.15	2.08	2.01	1.94
25	5.69	4.29	3.69	3.35	3.13	2.97	2.85	2.75	2.68	2.61	2.51	2.41	2.30	2.24	2.18	2.12	2.05	1.98	1.91
26	5.66	4.27	3.67	3.33	3.10	2.94	2.82	2.73	2.65	2.59	2.49	2.39	2.28	2.22	2.16	2.09	2.03	1.95	1.88
27	5.63	4.24	3.65	3.31	3.08	2.92	2.80	2.71	2.63	2.57	2.47	2.36	2.25	2.19	2.13	2.07	2.00	1.93	1.85
28	5.61	4.22	3.63	3.29	3.06	2.90	2.78	2.69	2.61	2.55	2.45	2.34	2.23	2.17	2.11	2.05	1.98	1.91	1.83
29	5.59	4.20	3.61	3.27	3.04	2.88	2.76	2.67	2.59	2.53	2.43	2.32	2.21	2.15	2.09	2.03	1.96	1.89	1.81
30	5.57	4.18	3.59	3.25	3.03	2.87	2.75	2.65	2.57	2.51	2.41	2.31	2.20	2.14	2.07	2.01	1.94	1.87	1.79
40	5.42	4.05	3.46	3.13	2.90	2.74	2.62	2.53	2.45	2.39	2.29	2.18	2.07	2.01	1.94	1.88	1.80	1.72	1.64
60	5.29	3.93	3.34	3.01	2.79	2.63	2.51	2.41	2.33	2.27	2.17	2.06	1.94	1.88	1.82	1.74	1.67	1.58	1.48
120	5.15	3.80	3.23	2.89	2.67	2.52	2.39	2.30	2.22	2.16	2.05	1.94	1.82	1.76	1.69	1.61	1.53	1.43	1.31
∞	5.02	3.69	3.12	2.79	2.57	2.41	2.29	2.19	2.11	2.05	1.94	1.83	1.71	1.64	1.57	1.48	1.39	1.27	1.00

TABLE A.4 (*continued*)

c. One-tailed test, $P = 0.99$ (0.01 or 1 %). When used as two-tailed test, $P = 0.98$.

Degrees of freedom in numerator

f_2 \ f_1 (Degrees of freedom in denominator)	1	2	3	4	5	6	7	8	9	10	12	15	20	24	30	40	60	120	∞
1	4052	4999.5	5403	5625	5764	5859	5928	5982	6022	6056	6106	6157	6209	6235	6261	6287	6313	6339	6366
2	98.50	99.00	99.17	99.25	99.30	99.33	99.36	99.37	99.39	99.40	99.42	99.43	99.45	99.46	99.47	99.47	99.48	99.49	99.50
3	34.12	30.82	29.46	28.71	28.24	27.91	27.67	27.49	27.35	27.23	27.05	26.87	26.69	26.60	26.50	26.41	26.32	26.22	26.13
4	21.20	18.00	16.69	15.98	15.52	15.21	14.98	14.80	14.66	14.55	14.37	14.20	14.02	13.93	13.84	13.75	13.65	13.56	13.46
5	16.26	13.27	12.06	11.39	10.97	10.67	10.46	10.29	10.16	10.05	9.89	9.72	9.55	9.47	9.38	9.29	9.20	9.11	9.02
6	13.75	10.92	9.78	9.15	8.75	8.47	8.26	8.10	7.98	7.87	7.72	7.56	7.40	7.31	7.23	7.14	7.06	6.97	6.88
7	12.25	9.55	8.45	7.85	7.46	7.19	6.99	6.84	6.72	6.62	6.47	6.31	6.16	6.07	5.99	5.91	5.82	5.74	5.65
8	11.26	8.65	7.59	7.01	6.63	6.37	6.18	6.03	5.91	5.81	5.67	5.52	5.36	5.28	5.20	5.12	5.03	4.95	4.86
9	10.56	8.02	6.99	6.42	6.06	5.80	5.61	5.47	5.35	5.26	5.11	4.96	4.81	4.73	4.65	4.57	4.48	4.40	4.31
10	10.04	7.56	6.55	5.99	5.64	5.39	5.20	5.06	4.94	4.85	4.71	4.56	4.41	4.33	4.25	4.17	4.08	4.00	3.91
11	9.65	7.21	6.22	5.67	5.32	5.07	4.89	4.74	4.63	4.54	4.40	4.25	4.10	4.02	3.94	3.86	3.78	3.69	3.60
12	9.33	6.93	5.95	5.41	5.06	4.82	4.64	4.50	4.39	4.30	4.16	4.01	3.86	3.78	3.70	3.62	3.54	3.45	3.36
13	9.07	6.70	5.74	5.21	4.86	4.62	4.44	4.30	4.19	4.10	3.96	3.82	3.66	3.59	3.51	3.43	3.34	3.25	3.17
14	8.86	6.51	5.56	5.04	4.69	4.46	4.28	4.14	4.03	3.94	3.80	3.66	3.51	3.43	3.35	3.27	3.18	3.09	3.00
15	8.68	6.36	5.42	4.89	4.56	4.32	4.14	4.00	3.89	3.89	3.67	3.52	3.37	3.29	3.21	3.13	3.05	2.96	2.87
16	8.53	6.23	5.29	4.77	4.44	4.20	4.03	3.89	3.78	3.69	3.55	3.41	3.26	3.18	3.10	3.02	2.93	2.84	2.75
17	8.40	6.11	5.18	4.67	4.34	4.10	3.93	3.79	3.68	3.59	3.46	3.31	3.16	3.08	3.00	2.92	2.83	2.75	2.65
18	8.29	6.01	5.09	4.58	4.25	4.01	3.84	3.71	3.60	3.51	3.37	3.23	3.08	3.00	2.92	2.84	2.75	2.66	2.57
19	8.18	5.93	5.01	4.50	4.17	3.94	3.77	3.63	3.52	3.43	3.30	3.15	3.00	2.92	2.84	2.76	2.67	2.58	2.49
20	8.10	5.85	4.94	4.43	4.10	3.87	3.70	3.56	3.46	3.37	3.23	3.09	2.94	2.86	2.78	2.69	2.61	2.52	2.42
21	8.02	5.78	4.87	4.37	4.04	3.81	3.64	3.51	3.40	3.31	3.17	3.03	2.88	2.80	2.72	2.64	2.55	2.46	2.36
22	7.95	5.72	4.82	4.31	3.99	3.76	3.59	3.45	3.35	3.26	3.12	2.98	2.83	2.75	2.67	2.58	2.50	2.40	2.31
23	7.88	5.66	4.76	4.26	3.94	3.71	3.54	3.41	3.30	3.21	3.07	2.93	2.78	2.70	2.62	2.54	2.45	2.35	2.26
24	7.82	5.61	4.72	4.22	3.90	3.67	3.50	3.36	3.26	3.17	3.03	2.89	2.74	2.66	2.58	2.49	2.40	2.31	2.21
25	7.77	5.57	4.68	4.18	3.85	3.63	3.46	3.32	3.22	3.13	2.99	2.85	2.70	2.62	2.54	2.45	2.36	2.27	2.17
26	7.72	5.53	4.64	4.14	3.82	3.59	3.42	3.29	3.18	3.09	2.96	2.81	2.66	2.58	2.50	2.42	2.33	2.23	2.13
27	7.68	5.49	4.60	4.11	3.78	3.56	3.39	3.26	3.15	3.06	2.93	2.78	2.63	2.55	2.47	2.38	2.29	2.20	2.10
28	7.64	5.45	4.57	4.07	3.75	3.53	3.36	3.23	3.12	3.03	2.90	2.75	2.60	2.52	2.44	2.35	2.26	2.17	2.06
29	7.60	5.42	4.54	4.04	3.73	3.50	3.33	3.20	3.09	3.00	2.87	2.73	2.57	2.49	2.41	2.33	2.23	2.14	2.03
30	7.56	5.39	4.51	4.02	3.70	3.47	3.30	3.17	3.07	2.98	2.84	2.70	2.55	2.47	2.39	2.30	2.21	2.11	2.01
40	7.31	5.18	4.31	3.83	3.51	3.29	3.12	2.99	2.89	2.80	2.66	2.52	2.37	2.29	2.20	2.11	2.02	1.92	1.80
60	7.08	4.98	4.13	3.65	3.34	3.12	2.95	2.82	2.72	2.63	2.50	2.35	2.20	2.12	2.03	1.94	1.84	1.73	1.60
120	6.85	4.79	3.95	3.48	3.17	2.96	2.79	2.66	2.56	2.47	2.34	2.19	2.03	1.95	1.86	1.76	1.66	1.53	1.38
∞	6.63	4.61	3.78	3.32	3.02	2.80	2.64	2.51	2.41	2.32	2.18	2.04	1.88	1.79	1.70	1.59	1.47	1.32	1.00

A.5 GRAPH FOR FINDING $\sigma/\sqrt{n}$

The accompanying graph (Table A.5) is for finding the standard error, $\sigma/\sqrt{n}$, given σ and n or for finding the necessary sample size n given σ and the desired standard error, $\sigma/\sqrt{n}$.

To find $\sigma/\sqrt{n}$ given σ and n: Locate σ along the right axis (labeled σ). Find and follow the slanted line going to the lower left from that point. Locate the value of n on the bottom axis. Go straight up from that point (n) until you intersect the slanted line from σ. Note the point of intersection (a). From the intersection (point a) proceed horizontally to the $\sigma/\sqrt{n}$ axis (the left axis) and read the standard error ($\sigma/\sqrt{n}$) directly.

EXAMPLE 1

If $\sigma = 10$, $n = 4$, find the standard error.

Figure A.5.1 shows how you follow the line emanating from $\sigma = 10$ down to the left. The second line from $n = 4$ going up intersects at point a. Going to the left from that point, we find $\sigma/\sqrt{n} = 5$.

EXAMPLE 2

If $\sigma = 1.2$, $n = 7$, find the standard error.

No line comes from the point $\sigma = 1.2$, so we must approximate the line. It should go between the lines for $\sigma = 1.0$ and 1.5 (if you have a ruler you can move up or down parallel to the slanted lines until it reaches the 1.2 point on the σ axis.) You follow up the line for $n = 7$ until you are the proper distance above the slanted line from $\sigma = 1.0$ (or until you reach your ruler). This corresponds to point a. Follow this point horizontally until you reach $\sigma/\sqrt{n}$ at approximately 0.45 (see Figure A.5.2).

To find n, the necessary sample size, when σ is known.

EXAMPLE 3

Given that $\sigma = 5$ and we desire a standard error of 1, how large a sample is necessary?

Follow the slanted line from $\sigma = 5$ until it intersects the horizontal line from the point $\sigma/\sqrt{n} = 1$. From this point drop down to the n scale and read $n = 25$ (see Figure A.5.3).

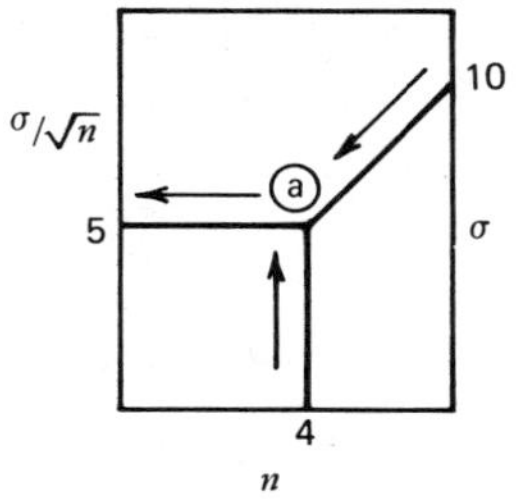

FIGURE A.5.1

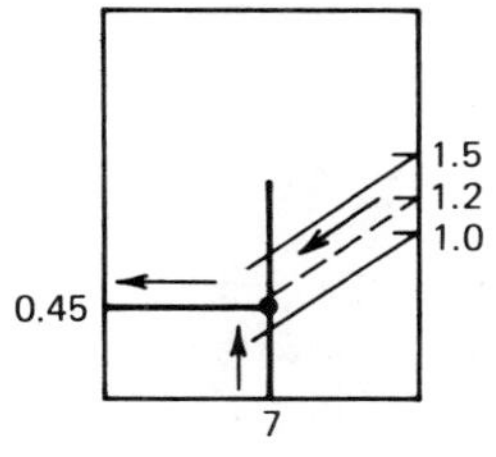

FIGURE A.5.2

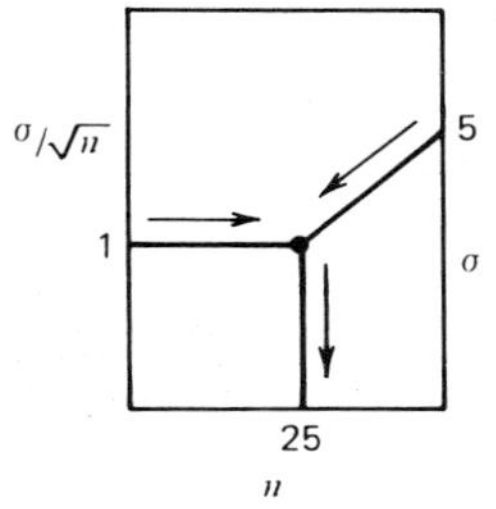

FIGURE A.5.3

TABLE A.5 Graph for Finding $\sigma/\sqrt{n}$

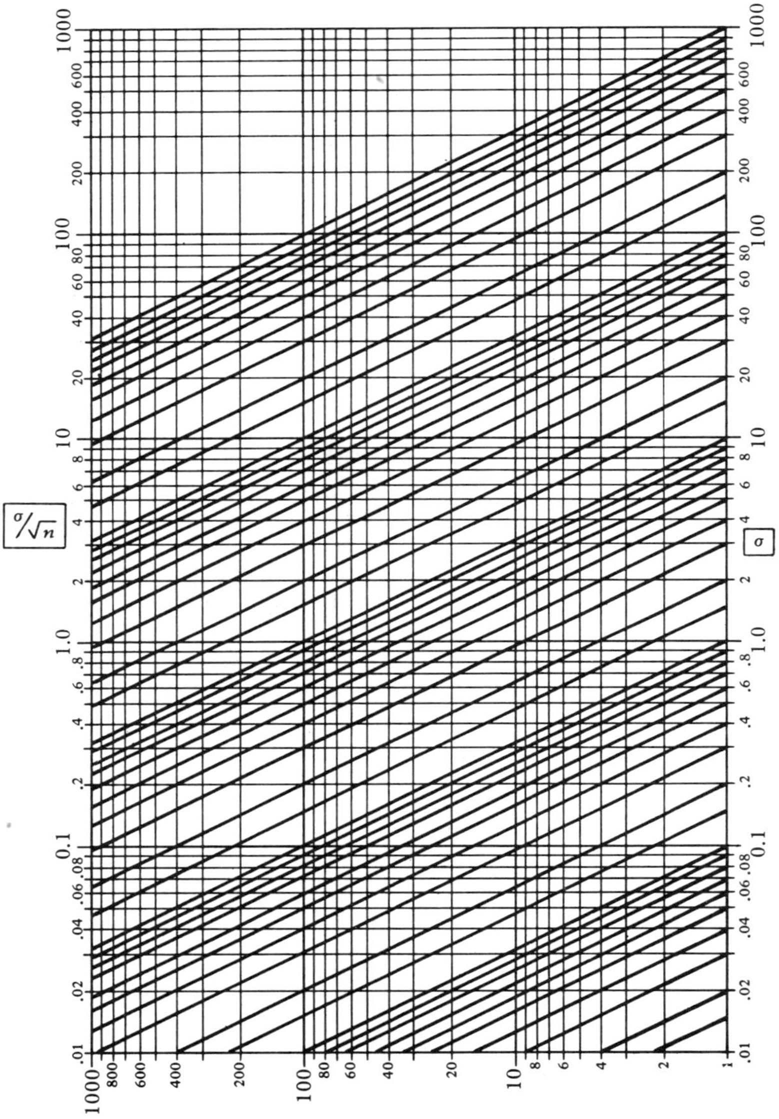

A.6 BINOMIAL DISTRIBUTION

To find the probabilities for selected binomial distributions given n and p (see Table A.6).

For p values *up to 0.50* read the p on the top and use the x values in the left column. For p values *greater than 0.50* read the p values from the bottom of the table and use the x values in the far right column.

EXAMPLE 1

Determine the probabilities for each X-value in the binomial distribution with $n = 6$ and $p = 0.35$.

Note that

$$q = 1 - p = 1 - 0.35 = 0.65$$

Under the column for n locate $n = 6$. All of the values from 0 to 6 in the adjacent column (labeled X column) correspond to possibilities when $n = 6$.

Under the $p = 0.35$ column you will find the probabilities for each X-value. Thus the probability of $X = 0$ is 0.0754, the probability that $X = 1$ is 0.2437, and so forth.

n	X	p 0.35
5		
→ 6	0	0.0754
	1	0.2437
	2	0.3280
	3	0.2355
	4	0.0951
	5	0.0205
	6	0.0018

Note that when the probability of a particular value of X is recorded as 0.0000 this is *not* to be interpreted as impossible but just less than a probability of 0.00005.

EXAMPLE 2

For a binomial distribution with $n = 6$ and $p = 0.35$ determine the probability of a value equal to or greater than 5.

Add up the probabilities for *5 and 6* or $0.0205 + 0.0018 = 0.0223$.

EXAMPLE 3

For a binomial distribution with $n = 15$ and $p = 0.40$ determine the limit above which 5% of the values lie.

Add up the various probabilities until you obtain approximately 0.0500.

$X = 15, p = 0.0000$; $X = 14, p = 0.0000$; $X = 13, p = 0.0003$
$X = 12, p = 0.0016$, and the sum for all p from 12 to 15 = 0.0019
$X = 11, p = 0.0074$, and the sum = 0.0093
$X = 10, p = 0.0245$, and the sum = 0.0338
$X = 9, \; p = 0.0612$, and the sum = 0.0950

The closest value to a sum of 5% or 0.05 is the $X = 10$.

TABLE A.6 Individual Terms, Binomial Distribution

n	X	p = 0.05	0.10	0.15	0.20	0.25	0.30	0.35	0.40	0.50	
1	0	0.9500	0.9000	0.8500	0.8000	0.7500	0.7000	0.6500	0.6000	0.5000	1
	1	0.0500	0.1000	0.1500	0.2000	0.2500	0.3000	0.3500	0.4000	0.5000	0
2	0	0.9025	0.8100	0.7225	0.6400	0.5625	0.4900	0.4225	0.3600	0.2500	2
	1	0.0950	0.1800	0.2550	0.3200	0.3750	0.4200	0.4550	0.4800	0.5000	1
	2	0.0025	0.0100	0.0225	0.0400	0.0625	0.0900	0.1225	0.1600	0.2500	0
3	0	0.8574	0.7290	0.6141	0.5120	0.4219	0.3430	0.2746	0.2160	0.1250	3
	1	0.1354	0.2430	0.3251	0.3840	0.4219	0.4410	0.4436	0.4320	0.3750	2
	2	0.0071	0.0270	0.0574	0.0960	0.1406	0.1890	0.2389	0.2880	0.3750	1
	3	0.0001	0.0010	0.0034	0.0080	0.0156	0.0270	0.0429	0.0640	0.1250	0
4	0	0.8145	0.6561	0.5220	0.4096	0.3164	0.2401	0.1785	0.1296	0.0625	4
	1	0.1715	0.2916	0.3685	0.4096	0.4219	0.4116	0.3845	0.3456	0.2500	3
	2	0.0135	0.0486	0.0975	0.1536	0.2109	0.2646	0.3105	0.3456	0.3750	2
	3	0.0005	0.0036	0.0115	0.0256	0.0469	0.0756	0.1115	0.1536	0.2500	1
	4	0.0000	0.0001	0.0005	0.0016	0.0039	0.0081	0.0150	0.0256	0.0625	0
5	0	0.7738	0.5905	0.4437	0.3277	0.2373	0.1681	0.1160	0.0778	0.0312	5
	1	0.2036	0.3280	0.3915	0.4096	0.3955	0.3602	0.3124	0.2592	0.1562	4
	2	0.0214	0.0729	0.1382	0.2048	0.2637	0.3087	0.3364	0.3456	0.3125	3
	3	0.0011	0.0081	0.0244	0.0512	0.0879	0.1323	0.1811	0.2304	0.3125	2
	4	0.0000	0.0004	0.0022	0.0064	0.0146	0.0284	0.0488	0.0768	0.1562	1
	5	0.0000	0.0000	0.0001	0.0003	0.0010	0.0024	0.0053	0.0102	0.0312	0
6	0	0.7351	0.5314	0.3771	0.2621	0.1780	0.1176	0.0754	0.0467	0.0156	6
	1	0.2321	0.3543	0.3993	0.3932	0.3560	0.3025	0.2437	0.1866	0.0938	5
	2	0.0305	0.0984	0.1762	0.2458	0.2966	0.3241	0.3280	0.3110	0.2344	4
	3	0.0021	0.0146	0.0415	0.0819	0.1318	0.1852	0.2355	0.2765	0.3125	3
	4	0.0001	0.0012	0.0055	0.0154	0.0330	0.0595	0.0951	0.1382	0.2344	2
	5	0.0000	0.0001	0.0004	0.0015	0.0044	0.0102	0.0205	0.0369	0.0938	1
	6	0.0000	0.0000	0.0000	0.0001	0.0002	0.0007	0.0018	0.0041	0.0156	0
7	0	0.6983	0.4783	0.3206	0.2097	0.1335	0.0824	0.0490	0.0280	0.0078	7
	1	0.2573	0.3720	0.3960	0.3670	0.3115	0.2471	0.1848	0.1306	0.0547	6
	2	0.0406	0.1240	0.2097	0.2753	0.3115	0.3177	0.2985	0.2613	0.1641	5
	3	0.0036	0.0230	0.0617	0.1147	0.1730	0.2269	0.2679	0.2903	0.2734	4
	4	0.0002	0.0026	0.0109	0.0287	0.0577	0.0972	0.1442	0.1935	0.2734	3
	5	0.0000	0.0002	0.0012	0.0043	0.0115	0.0250	0.0466	0.0774	0.1641	2
	6	0.0000	0.0000	0.0001	0.0004	0.0013	0.0036	0.0084	0.0172	0.0547	1
	7	0.0000	0.0000	0.0000	0.0000	0.0001	0.0002	0.0006	0.0016	0.0078	0
8	0	0.6634	0.4305	0.2725	0.1678	0.1001	0.0576	0.0319	0.0168	0.0039	8
	1	0.2793	0.3826	0.3847	0.3355	0.2670	0.1977	0.1373	0.0896	0.0312	7
		0.95	0.90	0.85	0.80	0.75 (p)	0.70	0.65	0.60	0.50	X

TABLE A.6 (*continued*)

n	X	0.05	0.10	0.15	0.20	p 0.25	0.30	0.35	0.40	0.50	
8	2	0.0515	0.1488	0.2376	0.2936	0.3115	0.2965	0.2587	0.2090	0.1094	6
	3	0.0054	0.0331	0.0839	0.1468	0.2076	0.2541	0.2786	0.2787	0.2188	5
	4	0.0004	0.0046	0.0185	0.0459	0.0865	0.1361	0.1875	0.2322	0.2734	4
	5	0.0000	0.0004	0.0026	0.0092	0.0231	0.0467	0.0808	0.1239	0.2188	3
	6	0.0000	0.0000	0.0002	0.0011	0.0038	0.0100	0.0217	0.0413	0.1094	2
	7	0.0000	0.0000	0.0000	0.0001	0.0004	0.0012	0.0033	0.0079	0.0312	1
	8	0.0000	0.0000	0.0000	0.0000	0.0000	0.0001	0.0002	0.0007	0.0039	0
9	0	0.6302	0.3874	0.2316	0.1342	0.0751	0.0404	0.0207	0.0101	0.0020	9
	1	0.2985	0.3874	0.3679	0.3020	0.2253	0.1556	0.1004	0.0605	0.0176	8
	2	0.0629	0.1722	0.2597	0.3020	0.3003	0.2668	0.2162	0.1612	0.0703	7
	3	0.0077	0.0446	0.1069	0.1762	0.2336	0.2668	0.2716	0.2508	0.1641	6
	4	0.0006	0.0074	0.0283	0.0661	0.1168	0.1715	0.2194	0.2508	0.2461	5
	5	0.0000	0.0008	0.0050	0.0165	0.0389	0.0735	0.1181	0.1672	0.2461	4
	6	0.0000	0.0001	0.0006	0.0028	0.0087	0.0210	0.0424	0.0743	0.1641	3
	7	0.0000	0.0000	0.0000	0.0003	0.0012	0.0039	0.0098	0.0212	0.0703	2
	8	0.0000	0.0000	0.0000	0.0000	0.0001	0.0004	0.0013	0.0035	0.0176	1
	9	0.0000	0.0000	0.0000	0.0000	0.0000	0.0000	0.0001	0.0003	0.0020	0
10	0	0.5987	0.3487	0.1969	0.1074	0.0563	0.0282	0.0135	0.0060	0.0010	10
	1	0.3151	0.3874	0.3474	0.2684	0.1877	0.1211	0.0725	0.0403	0.0098	9
	2	0.0746	0.1937	0.2759	0.3020	0.2816	0.2335	0.1757	0.1209	0.0439	8
	3	0.0105	0.0574	0.1298	0.2013	0.2503	0.2668	0.2522	0.2150	0.1172	7
	4	0.0010	0.0112	0.0401	0.0881	0.1460	0.2001	0.2377	0.2508	0.2051	6
	5	0.0001	0.0015	0.0085	0.0264	0.0584	0.1029	0.1536	0.2007	0.2461	5
	6	0.0000	0.0001	0.0012	0.0055	0.0162	0.0368	0.0689	0.1115	0.2051	4
	7	0.0000	0.0000	0.0001	0.0008	0.0031	0.0090	0.0212	0.0425	0.1172	3
	8	0.0000	0.0000	0.0000	0.0001	0.0004	0.0014	0.0043	0.0106	0.0439	2
	9	0.0000	0.0000	0.0000	0.0000	0.0000	0.0001	0.0005	0.0016	0.0098	1
	10	0.0000	0.0000	0.0000	0.0000	0.0000	0.0000	0.0000	0.0001	0.0010	0
11	0	0.5688	0.3138	0.1673	0.0859	0.0422	0.0198	0.0088	0.0036	0.0004	11
	1	0.3293	0.3835	0.3248	0.2362	0.1549	0.0932	0.0518	0.0266	0.0055	10
	2	0.0867	0.2131	0.2866	0.2953	0.2581	0.1998	0.1395	0.0887	0.0269	9
	3	0.0137	0.0710	0.1517	0.2215	0.2581	0.2568	0.2254	0.1774	0.0806	8
	4	0.0014	0.0158	0.0536	0.1107	0.1721	0.2201	0.2428	0.2365	0.1611	7
	5	0.0001	0.0025	0.0132	0.0388	0.0803	0.1321	0.1830	0.2207	0.2256	6
	6	0.0000	0.0003	0.0023	0.0097	0.0268	0.0566	0.0985	0.1471	0.2256	5
	7	0.0000	0.0000	0.0003	0.0017	0.0064	0.0173	0.0379	0.0701	0.1611	4
	8	0.0000	0.0000	0.0000	0.0002	0.0011	0.0037	0.0102	0.0234	0.0806	3
	9	0.0000	0.0000	0.0000	0.0000	0.0001	0.0005	0.0018	0.0052	0.0269	2
	10	0.0000	0.0000	0.0000	0.0000	0.0000	0.0000	0.0002	0.0007	0.0054	1
	11	0.0000	0.0000	0.0000	0.0000	0.0000	0.0000	0.0000	0.0000	0.0005	0
		0.95	0.90	0.85	0.80	0.75 p	0.70	0.65	0.60	0.50	X

TABLE A.6 (*continued*)

n	X	0.05	0.10	0.15	0.20	p 0.25	0.30	0.35	0.40	0.50	
12	0	0.5404	0.2824	0.1422	0.0687	0.0317	0.0138	0.0057	0.0022	0.0002	12
	1	0.3413	0.3766	0.3012	0.2062	0.1267	0.0712	0.0368	0.0174	0.0029	11
	2	0.0388	0.2301	0.2924	0.2835	0.2323	0.1678	0.1088	0.0639	0.0161	10
	3	0.0173	0.0852	0.1720	0.2362	0.2581	0.2397	0.1954	0.1419	0.0537	9
	4	0.0021	0.0213	0.0683	0.1329	0.1936	0.2311	0.2367	0.2128	0.1208	8
	5	0.0002	0.0038	0.0193	0.0532	0.1032	0.1585	0.2039	0.2270	0.1934	7
	6	0.0000	0.0005	0.0040	0.0155	0.0401	0.0792	0.1281	0.1766	0.2256	6
	7	0.0000	0.0000	0.0006	0.0033	0.0115	0.0291	0.0591	0.1009	0.1934	5
	8	0.0000	0.0000	0.0001	0.0005	0.0024	0.0078	0.0199	0.0420	0.1208	4
	9	0.0000	0.0000	0.0000	0.0001	0.0004	0.0015	0.0048	0.0125	0.0537	3
	10	0.0000	0.0000	0.0000	0.0000	0.0000	0.0002	0.0008	0.0025	0.0161	2
	11	0.0000	0.0000	0.0000	0.0000	0.0000	0.0000	0.0001	0.0003	0.0029	1
	12	0.0000	0.0000	0.0000	0.0000	0.0000	0.0000	0.0000	0.0000	0.0002	0
13	0	0.5133	0.2542	0.1209	0.0550	0.0238	0.0097	0.0037	0.0013	0.0001	13
	1	0.3512	0.3672	0.2774	0.1787	0.1029	0.0540	0.0259	0.0113	0.0016	12
	2	0.1109	0.2448	0.2937	0.2680	0.2059	0.1388	0.0836	0.0453	0.0095	11
	3	0.0214	0.0997	0.1900	0.2457	0.2517	0.2181	0.1651	0.1107	0.0349	10
	4	0.0028	0.0277	0.0838	0.1535	0.2097	0.2337	0.2222	0.1845	0.0873	9
	5	0.0003	0.0055	0.0266	0.0691	0.1258	0.1803	0.2154	0.2214	0.1571	8
	6	0.0000	0.0008	0.0063	0.0230	0.0559	0.1030	0.1546	0.1968	0.2095	7
	7	0.0000	0.0001	0.0011	0.0058	0.0186	0.0442	0.0833	0.1312	0.2095	6
	8	0.0000	0.0000	0.0001	0.0011	0.0047	0.0142	0.0336	0.0656	0.1571	5
	9	0.0000	0.0000	0.0000	0.0001	0.0009	0.0034	0.0101	0.0243	0.0873	4
	10	0.0000	0.0000	0.0000	0.0000	0.0001	0.0006	0.0022	0.0065	0.0349	3
	11	0.0000	0.0000	0.0000	0.0000	0.0000	0.0001	0.0003	0.0012	0.0095	2
	12	0.0000	0.0000	0.0000	0.0000	0.0000	0.0000	0.0000	0.0001	0.0016	1
	13	0.0000	0.0000	0.0000	0.0000	0.0000	0.0000	0.0000	0.0000	0.0001	0
14	0	0.4877	0.2288	0.1028	0.0440	0.0178	0.0068	0.0024	0.0008	0.0001	14
	1	0.3593	0.3559	0.2539	0.1539	0.0832	0.0407	0.0181	0.0073	0.0009	13
	2	0.1229	0.2570	0.2912	0.2501	0.1802	0.1134	0.0634	0.0317	0.0056	12
	3	0.0259	0.1142	0.2056	0.2501	0.2402	0.1943	0.1366	0.0845	0.0222	11
	4	0.0037	0.0349	0.0998	0.1720	0.2202	0.2290	0.2022	0.1549	0.0611	10
	5	0.0004	0.0078	0.0352	0.0860	0.1468	0.1963	0.2178	0.2066	0.1222	9
	6	0.0000	0.0013	0.0093	0.0322	0.0734	0.1262	0.1759	0.2066	0.1833	8
	7	0.0000	0.0002	0.0019	0.0092	0.0280	0.0618	0.1082	0.1574	0.2095	7
	8	0.0000	0.0000	0.0003	0.0020	0.0082	0.0232	0.0510	0.0918	0.1833	6
	9	0.0000	0.0000	0.0000	0.0003	0.0018	0.0066	0.0183	0.0408	0.1222	5
	10	0.0000	0.0000	0.0000	0.0000	0.0003	0.0014	0.0049	0.0136	0.0611	4
	11	0.0000	0.0000	0.0000	0.0000	0.0000	0.0002	0.0010	0.0033	0.0222	3
		0.95	0.90	0.85	0.80	0.75 p	0.70	0.65	0.60	0.50	X

TABLE A.6 (*continued*)

n	X	0.05	0.10	0.15	0.20	p 0.25	0.30	0.35	0.40	0.50	
14	12	0.0000	0.0000	0.0000	0.0000	0.0000	0.0000	0.0001	0.0005	0.0056	2
	13	0.0000	0.0000	0.0000	0.0000	0.0000	0.0000	0.0000	0.0001	0.0009	1
	14	0.0000	0.0000	0.0000	0.0000	0.0000	0.0000	0.0000	0.0000	0.0001	0
15	0	0.4633	0.2059	0.0874	0.0352	0.0134	0.0047	0.0016	0.0005	0.0000	15
	1	0.3658	0.3432	0.2312	0.1319	0.0668	0.0305	0.0126	0.0047	0.0005	14
	2	0.1348	0.2669	0.2856	0.2309	0.1559	0.0916	0.0476	0.0219	0.0032	13
	3	0.0307	0.1285	0.2184	0.2501	0.2252	0.1700	0.1110	0.0634	0.0139	12
	4	0.0049	0.0428	0.1156	0.1876	0.2252	0.2186	0.1792	0.1268	0.0417	11
	5	0.0006	0.0105	0.0449	0.1032	0.1651	0.2061	0.2123	0.1859	0.0916	10
	6	0.0000	0.0019	0.0132	0.0430	0.0917	0.1472	0.1906	0.2066	0.1527	9
	7	0.0000	0.0003	0.0030	0.0138	0.0393	0.0811	0.1319	0.1771	0.1964	8
	8	0.0000	0.0000	0.0005	0.0035	0.0131	0.0348	0.0710	0.1181	0.1964	7
	9	0.0000	0.0000	0.0001	0.0007	0.0034	0.0116	0.0298	0.0612	0.1527	6
	10	0.0000	0.0000	0.0000	0.0001	0.0007	0.0030	0.0096	0.0245	0.0916	5
	11	0.0000	0.0000	0.0000	0.0000	0.0001	0.0006	0.0024	0.0074	0.0417	4
	12	0.0000	0.0000	0.0000	0.0000	0.0000	0.0001	0.0004	0.0016	0.0139	3
	13	0.0000	0.0000	0.0000	0.0000	0.0000	0.0000	0.0001	0.0003	0.0032	2
	14	0.0000	0.0000	0.0000	0.0000	0.0000	0.0000	0.0000	0.0000	0.0005	1
	15	0.0000	0.0000	0.0000	0.0000	0.0000	0.0000	0.0000	0.0000	0.0000	0
16	0	0.4401	0.1853	0.0743	0.0281	0.0100	0.0033	0.0010	0.0003	0.0000	16
	1	0.3706	0.3294	0.2097	0.1126	0.0535	0.0228	0.0087	0.0030	0.0002	15
	2	0.1463	0.2745	0.2775	0.2111	0.1336	0.0732	0.0353	0.0150	0.0018	14
	3	0.0359	0.1423	0.2285	0.2463	0.2079	0.1465	0.0888	0.0468	0.0085	13
	4	0.0061	0.0514	0.1311	0.2001	0.2252	0.2040	0.1553	0.1014	0.0278	12
	5	0.0008	0.0137	0.0555	0.1201	0.1802	0.2099	0.2008	0.1623	0.0667	11
	6	0.0001	0.0028	0.0180	0.0550	0.1101	0.1649	0.1982	0.1983	0.1222	10
	7	0.0000	0.0004	0.0045	0.0197	0.0524	0.1010	0.1524	0.1889	0.1746	9
	8	0.0000	0.0001	0.0009	0.0055	0.0197	0.0487	0.0923	0.1417	0.1964	8
	9	0.0000	0.0000	0.0001	0.0012	0.0058	0.0185	0.0442	0.0840	0.1746	7
	10	0.0000	0.0000	0.0000	0.0002	0.0014	0.0056	0.0167	0.0392	0.1222	6
	11	0.0000	0.0000	0.0000	0.0000	0.0002	0.0013	0.0049	0.0142	0.0667	5
	12	0.0000	0.0000	0.0000	0.0000	0.0000	0.0002	0.0011	0.0040	0.0278	4
	13	0.0000	0.0000	0.0000	0.0000	0.0000	0.0000	0.0002	0.0008	0.0085	3
	14	0.0000	0.0000	0.0000	0.0000	0.0000	0.0000	0.0000	0.0001	0.0018	2
	15	0.0000	0.0000	0.0000	0.0000	0.0000	0.0000	0.0000	0.0000	0.0002	1
	16	0.0000	0.0000	0.0000	0.0000	0.0000	0.0000	0.0000	0.0000	0.0000	0
17	0	0.4181	0.1668	0.0631	0.0225	0.0075	0.0023	0.0007	0.0002	0.0000	17
	1	0.3741	0.3150	0.1893	0.0957	0.0426	0.0169	0.0060	0.0019	0.0001	16
		0.95	0.90	0.85	0.80	0.75 p	0.70	0.65	0.60	0.50	X

TABLE A.6 (*continued*)

n	X	0.05	0.10	0.15	0.20	p 0.25	0.30	0.35	0.40	0.50	
17	2	0.1575	0.2800	0.2673	0.1914	0.1136	0.0581	0.0260	0.0102	0.0010	15
	3	0.0415	0.1556	0.2359	0.2393	0.1893	0.1245	0.0701	0.0341	0.0052	14
	4	0.0076	0.0605	0.1457	0.2093	0.2209	0.1868	0.1320	0.0796	0.0182	13
	5	0.0010	0.0175	0.0668	0.1361	0.1914	0.2081	0.1849	0.1379	0.0472	12
	6	0.0001	0.0039	0.0236	0.0680	0.1276	0.1784	0.1991	0.1839	0.0944	11
	7	0.0000	0.0007	0.0065	0.0267	0.0668	0.1201	0.1685	0.1927	0.1484	10
	8	0.0000	0.0001	0.0014	0.0084	0.0279	0.0644	0.1134	0.1606	0.1855	9
	9	0.0000	0.0000	0.0003	0.0021	0.0093	0.0276	0.0611	0.1070	0.1855	8
	10	0.0000	0.0000	0.0000	0.0004	0.0025	0.0095	0.0263	0.0571	0.1484	7
	11	0.0000	0.0000	0.0000	0.0001	0.0005	0.0026	0.0090	0.0242	0.0944	6
	12	0.0000	0.0000	0.0000	0.0000	0.0001	0.0006	0.0024	0.0081	0.0472	5
	13	0.0000	0.0000	0.0000	0.0000	0.0000	0.0001	0.0005	0.0021	0.0182	4
	14	0.0000	0.0000	0.0000	0.0000	0.0000	0.0000	0.0001	0.0004	0.0052	3
	15	0.0000	0.0000	0.0000	0.0000	0.0000	0.0000	0.0000	0.0001	0.0010	2
	16	0.0000	0.0000	0.0000	0.0000	0.0000	0.0000	0.0000	0.0000	0.0001	1
	17	0.0000	0.0000	0.0000	0.0000	0.0000	0.0000	0.0000	0.0000	0.0000	0
18	0	0.3972	0.1501	0.0536	0.0180	0.0056	0.0016	0.0004	0.0001	0.0000	18
	1	0.3763	0.3002	0.1704	0.0811	0.0338	0.0126	0.0042	0.0012	0.0001	17
	2	0.1683	0.2835	0.2556	0.1723	0.0958	0.0458	0.0190	0.0069	0.0006	16
	3	0.0473	0.1680	0.2406	0.2297	0.1704	0.1046	0.0547	0.0246	0.0031	15
	4	0.0093	0.0700	0.1592	0.2153	0.2130	0.1681	0.1104	0.0614	0.0117	14
	5	0.0014	0.0218	0.0787	0.1507	0.1988	0.2017	0.1664	0.1146	0.0327	13
	6	0.0002	0.0052	0.0301	0.0816	0.1436	0.1873	0.1941	0.1655	0.0708	12
	7	0.0000	0.0010	0.0091	0.0350	0.0820	0.1376	0.1792	0.1892	0.1214	11
	8	0.0000	0.0002	0.0022	0.0120	0.0376	0.0811	0.1327	0.1734	0.1669	10
	9	0.0000	0.0000	0.0004	0.0033	0.0139	0.0386	0.0794	0.1284	0.1855	9
	10	0.0000	0.0000	0.0001	0.0008	0.0042	0.0149	0.0385	0.0771	0.1669	8
	11	0.0000	0.0000	0.0000	0.0001	0.0010	0.0046	0.0151	0.0374	0.1214	7
	12	0.0000	0.0000	0.0000	0.0000	0.0002	0.0012	0.0047	0.0145	0.0708	6
	13	0.0000	0.0000	0.0000	0.0000	0.0000	0.0002	0.0012	0.0045	0.0327	5
	14	0.0000	0.0000	0.0000	0.0000	0.0000	0.0000	0.0002	0.0011	0.0117	4
	15	0.0000	0.0000	0.0000	0.0000	0.0000	0.0000	0.0000	0.0002	0.0031	3
	16	0.0000	0.0000	0.0000	0.0000	0.0000	0.0000	0.0000	0.0000	0.0006	2
	17	0.0000	0.0000	0.0000	0.0000	0.0000	0.0000	0.0000	0.0000	0.0001	1
	18	0.0000	0.0000	0.0000	0.0000	0.0000	0.0000	0.0000	0.0000	0.0000	0
19	0	0.3774	0.1351	0.0456	0.0144	0.0042	0.0011	0.0003	0.0001	0.0000	19
	1	0.3774	0.2852	0.1529	0.0685	0.0268	0.0093	0.0029	0.0008	0.0000	18
		0.95	0.90	0.85	0.80	0.75 p	0.70	0.65	0.60	0.50	X

TABLE A.6 (*continued*)

n	X	0.05	0.10	0.15	0.20	p 0.25	0.30	0.35	0.40	0.50	
19	2	0.1787	0.2852	0.2428	0.1540	0.0803	0.0358	0.0138	0.0046	0.0003	17
	3	0.0533	0.1796	0.2428	0.2182	0.1517	0.0869	0.0422	0.0175	0.0018	16
	4	0.0112	0.0798	0.1714	0.2182	0.2023	0.1491	0.0909	0.0467	0.0074	15
	5	0.0018	0.0266	0.0907	0.1636	0.2023	0.1916	0.1468	0.0933	0.0222	14
	6	0.0002	0.0069	0.0374	0.0955	0.1574	0.1916	0.1844	0.1451	0.0518	13
	7	0.0000	0.0014	0.0122	0.0443	0.0974	0.1525	0.1844	0.1797	0.0961	12
	8	0.0000	0.0002	0.0032	0.0166	0.0487	0.0981	0.1489	0.1797	0.1442	11
	9	0.0000	0.0000	0.0007	0.0051	0.0198	0.0514	0.0980	0.1464	0.1762	10
	10	0.0000	0.0000	0.0001	0.0013	0.0066	0.0220	0.0528	0.0976	0.1762	9
	11	0.0000	0.0000	0.0000	0.0003	0.0018	0.0077	0.0233	0.0532	0.1442	8
	12	0.0000	0.0000	0.0000	0.0000	0.0004	0.0022	0.0083	0.0237	0.0961	7
	13	0.0000	0.0000	0.0000	0.0000	0.0001	0.0005	0.0024	0.0085	0.0518	6
	14	0.0000	0.0000	0.0000	0.0000	0.0000	0.0001	0.0006	0.0024	0.0222	5
	15	0.0000	0.0000	0.0000	0.0000	0.0000	0.0000	0.0001	0.0005	0.0074	4
	16	0.0000	0.0000	0.0000	0.0000	0.0000	0.0000	0.0000	0.0001	0.0018	3
	17	0.0000	0.0000	0.0000	0.0000	0.0000	0.0000	0.0000	0.0000	0.0003	2
	18	0.0000	0.0000	0.0000	0.0000	0.0000	0.0000	0.0000	0.0000	0.0000	1
	19	0.0000	0.0000	0.0000	0.0000	0.0000	0.0000	0.0000	0.0000	0.0000	0
20	0	0.3585	0.1216	0.0388	0.0115	0.0032	0.0008	0.0002	0.0000	0.0000	20
	1	0.3774	0.2702	0.1368	0.0576	0.0211	0.0068	0.0020	0.0005	0.0000	19
	2	0.1887	0.2852	0.2293	0.1369	0.0669	0.0278	0.0100	0.0031	0.0002	18
	3	0.0596	0.1901	0.2428	0.2054	0.1339	0.0716	0.0323	0.0123	0.0011	17
	4	0.0133	0.0898	0.1821	0.2182	0.1897	0.1304	0.0738	0.0350	0.0046	16
	5	0.0022	0.0319	0.1028	0.1746	0.2023	0.1789	0.1272	0.0746	0.0148	15
	6	0.0003	0.0089	0.0454	0.1091	0.1686	0.1916	0.1712	0.1244	0.0370	14
	7	0.0000	0.0020	0.0160	0.0545	0.1124	0.1643	0.1844	0.1659	0.0739	13
	8	0.0000	0.0004	0.0046	0.0222	0.0609	0.1144	0.1614	0.1797	0.1201	12
	9	0.0000	0.0001	0.0011	0.0074	0.0271	0.0654	0.1158	0.1597	0.1602	11
	10	0.0000	0.0000	0.0002	0.0020	0.0099	0.0308	0.0686	0.1171	0.1762	10
	11	0.0000	0.0000	0.0000	0.0005	0.0030	0.0120	0.0336	0.0710	0.1602	9
	12	0.0000	0.0000	0.0000	0.0001	0.0008	0.0039	0.0136	0.0355	0.1201	8
	13	0.0000	0.0000	0.0000	0.0000	0.0002	0.0010	0.0045	0.0146	0.0739	7
	14	0.0000	0.0000	0.0000	0.0000	0.0000	0.0002	0.0012	0.0049	0.0370	6
	15	0.0000	0.0000	0.0000	0.0000	0.0000	0.0000	0.0003	0.0013	0.0148	5
	16	0.0000	0.0000	0.0000	0.0000	0.0000	0.0000	0.0000	0.0003	0.0046	4
	17	0.0000	0.0000	0.0000	0.0000	0.0000	0.0000	0.0000	0.0000	0.0011	3
	18	0.0000	0.0000	0.0000	0.0000	0.0000	0.0000	0.0000	0.0000	0.0002	2
	19	0.0000	0.0000	0.0000	0.0000	0.0000	0.0000	0.0000	0.0000	0.0000	1
	20	0.0000	0.0000	0.0000	0.0000	0.0000	0.0000	0.0000	0.0000	0.0000	0
		0.95	0.90	0.85	0.80	0.75 p	0.70	0.65	0.60	0.50	X

A.7 RANDOM NUMBERS

The collections of numbers are arranged so that all digits may be considered as random, regardless of the point you begin to select numbers from or the direction in which you continue to select values. Individual numbers or groups of more than one number may be selected at one time and you may go to the right, left, up, down, or diagonally. They are grouped in fives merely for convenience (Table A.7).

TABLE A.7 Random Numbers

11164	36318	75061	37674	26320	75100	10431	20418	19228	91792
21215	91791	76831	58678	87054	31687	93205	43685	19732	08468
10438	44482	66558	37649	08882	90870	12462	41810	01806	02977
36792	26236	33266	66583	60881	97395	20461	36742	02852	50564
73944	04773	12032	51414	82384	38370	00249	80709	72605	67497
49563	12872	14063	93104	78483	72717	68714	18078	25005	04151
64208	48237	41701	73117	33242	42314	83049	21933	92813	04763
51486	72875	38605	29341	80749	80151	33835	52602	79147	08868
99756	26360	64516	17971	48478	09610	04638	17141	09227	10606
71325	55217	13015	79207	00431	45117	33827	92873	02953	85474
65285	97198	12138	53010	94601	15838	16805	61004	43516	17020
17264	57327	38224	29301	31381	38109	34976	65692	98566	29550
95639	99754	31199	92558	68368	04985	51092	37780	40261	14479
61555	76404	86210	11808	12841	45147	97438	60022	12645	62000
78137	98768	04689	87130	79225	08153	84967	64539	79493	74917
62490	99215	84987	28759	19177	14733	24550	28067	68894	38490
24216	63444	21283	07044	92729	37284	13211	37485	10415	36457
16975	95428	33226	55903	31605	43817	22250	03918	46999	98501
59138	39542	71168	57609	91510	77904	74244	50940	31553	62562
29478	59652	50414	31966	87912	87154	12944	49862	96566	48825
96155	95009	27429	72918	08457	78134	48407	26061	58754	05326
29621	66583	62966	12468	20245	14015	04014	35713	03980	03024
12639	75291	71020	17265	41598	64074	64629	63293	53307	48766
14544	37134	54714	02401	63228	26831	19386	15457	17999	18306
83403	88827	09834	11333	68431	31706	26652	04711	34593	22561
10011	75004	86054	41190	10061	19660	03500	68412	57812	57929
92420	65431	16530	05547	10683	88102	30176	84750	10115	69220
35542	55865	07304	47010	43233	57022	52161	82976	47981	46588
86595	26247	18552	29491	33712	32285	64844	69395	41387	87195
72115	34985	58036	99137	47482	06204	24138	24272	16196	04393

TABLE A.7 (*continued*)

40603	16152	83235	37361	98783	24838	39793	80954	76865	32713
40941	53585	69958	60916	71018	90561	84505	53980	64735	85140
73505	83472	55953	17957	11446	22618	34771	25777	27064	13526
39412	16013	11442	89320	11307	49396	39805	12249	57656	88686
57994	76748	54627	48511	78646	33287	35524	54522	08795	56273
07428	58863	96023	88936	51343	70958	96768	74317	27176	29600
35379	27922	28906	55013	26937	48174	04197	36074	65315	12537
10982	22807	10920	26299	23593	64629	57801	10437	43965	15344
90127	33341	77806	12446	15444	49244	47277	11346	15884	28131
63002	12990	23510	68774	48983	20481	59815	67248	17076	78910
40779	86382	48454	65269	91239	45989	45389	54847	77919	41105
43216	12608	18167	84631	94058	82458	15139	76856	86019	47928
96167	64375	74108	93643	09204	98855	59051	56492	11933	64958
70975	62693	35684	72607	23026	37004	32989	24843	01128	74658
85812	61876	23570	75754	29090	40264	80399	47254	40135	69911
61834	59199	15469	82285	84164	91333	90954	87186	31598	25942
91402	77227	79516	21007	58602	81418	87838	18443	76162	51146
58299	83880	20125	10794	37780	61705	18276	99041	78135	99661
40684	99948	33880	76413	63839	71371	32392	51812	48248	96419
75978	64298	08074	62055	73864	01926	78374	15741	74452	49954
34556	39861	88267	76068	62445	64361	78685	24246	27027	48239
65990	57048	25067	77571	77974	37634	81564	98608	37224	49848
16381	15069	25416	87875	90374	86203	29677	82543	37554	89179
52458	88880	78352	67913	09245	47773	51272	06976	99571	33365
33007	85607	92008	44897	24964	50559	79549	85658	96865	24186
38712	31512	08588	61490	72294	42862	87334	05866	66269	43158
58722	03678	19186	69602	34625	75958	56869	17907	81867	11535
26188	69497	51351	47799	20477	71786	52560	66827	79419	70886
12893	54048	07255	86149	99090	70958	50775	31768	52903	27645
33186	81346	85095	37282	85536	72661	32180	40229	19209	74939
79893	29448	88392	54211	61708	83452	61227	81690	42265	20310
48449	15102	44126	19438	23382	14985	37538	30120	82443	11152
94205	04259	68983	50561	06902	10269	22216	70210	60736	58772
38648	09278	81313	77400	41126	52614	93613	27263	99381	49500
04292	46028	75666	26954	34979	68381	45154	09314	81009	05114
17026	49737	85875	12139	59391	81830	30185	83095	78752	40899
48070	76848	02531	97737	10151	18169	31709	74842	85522	74092
30159	95450	83778	46115	99178	97718	98440	15076	21199	20492
12148	92231	31361	60650	54695	30035	22765	91386	70399	79270
73838	77067	24863	97576	01139	54219	02959	45696	98103	78867

From the RAND Corporation, *One Million Random Digits with* 100,000 *Normal Deviates*, published 1955 by the Free Press, reprinted with the kind permission of the RAND Corporation.

A.8 SQUARES, SQUARE ROOTS, AND RECIPROCALS

Table A.8 may be used to find the squares, square roots, and reciprocals of any two-digit number. For any value find the first two digits in the column labeled n.

Squares: The square of the number n is found in the column n^2.

EXAMPLE 1

Find the square of 8.3.

From Table A.8,

n	n^2
8.3	68.89

$$n^2 = 68.89$$

If the number is not between 1.0 and 9.9, then you must adjust the decimal point. Table A.8a gives you factors by which to multiply the value in column n^2.

EXAMPLE 2

Find the square of 760.

The first two digits are 76, thus look up for $n = 7.6$ to find $n^2 = 57.76$. From Table A.8a for values from 100 to 900 we multiply n^2 by 10.000. Then $(760)^2 = 10{,}000(57.76) = 577{,}600$.

n	n^2
7.6	57.76

Square Roots: Two columns are provided. Consult Table A.8a to decide which column to use and by what to multiply the value in the table.

EXAMPLE 1 (cont.)

Find the square root of 8.3.

8.3 is between 1.0 and 9.9, therefore from Table A.8a we find $\sqrt{n}$ and multiply by 1.0. From Table A.8b we obtain 2.8810.

n	$\sqrt{n}$
8.3	2.8810

EXAMPLE 2 (cont.)

Find the square root of 760. This is between 100 and 990. Table A.8a refers you to Column $\sqrt{n}$; multiply by 10. From Table A.8b we find

n	$\sqrt{n}$
7.6	2.7568

Therefore, $\sqrt{760} = 10(2.7568) = 27.568$.

EXAMPLE 3

Find the square root of 0.37. This is between 0.10 and 0.99. From Table A.8a we find that we should use the column labeled $\sqrt{10n}$ and multiply by 0.1. From Table A.8b we find

n	$\sqrt{10n}$
3.7	6.083

Therefore,

$$\sqrt{0.37} = 0.1(6.083) = 0.6083.$$

Reciprocals: For any value of n we read the reciprocal or $1/n$ in the $1/n$ column. Table A.8a lists multipliers for values not between 1 and 9.9.

TABLE A.8a

FOR NUMBERS BETWEEN	TO FIND SQUARES MULTIPLY n^2 BY	TO FIND SQUARE ROOTS		TO FIND RECIPROCALS MULTIPLY BY THIS
		USE COLUMN LABELED	MULTIPLY BY THIS VALUE	
1 and 9.9	1	$\sqrt{n}$	1	1.0
10 and 99	100	$\sqrt{10n}$	1	0.1
100 and 990	10,000	$\sqrt{n}$	10	0.01
1,000 and 9,900	1,000,000	$\sqrt{10n}$	10	0.001
10,000 and 99,000	100,000,000	$\sqrt{n}$	100	0.0001
0.10 and 0.99	0.01	$\sqrt{10n}$	0.1	10
0.010 and 0.099	0.0001	$\sqrt{n}$	0.01	100
0.0010 and 0.0099	0.000001	$\sqrt{10n}$	0.001	1000

TABLE A.8b Squares, Square Roots, and Reciprocals

n	n^2	$\sqrt{n}$	$\sqrt{10n}$	$1/n$	n	n^2	$\sqrt{n}$	$\sqrt{10n}$	$1/n$
1.0	1.00	1.0000	3.162	1.000					
1.1	1.21	1.0488	3.317	0.909	1.6	2.56	1.2649	4.000	0.625
1.2	1.44	1.0954	3.464	0.833	1.7	2.89	1.3038	4.123	0.588
1.3	1.69	1.1402	3.606	0.769	1.8	3.24	1.3416	4.243	0.556
1.4	1.96	1.1832	3.742	0.714	1.9	3.61	1.3784	4.359	0.526
1.5	2.25	1.2247	3.873	0.667	2.0	4.00	1.4142	4.472	0.500

TABLE A.8b (*continued*)

n	n^2	$\sqrt{n}$	$\sqrt{10n}$	$1/n$	n	n^2	$\sqrt{n}$	$\sqrt{10n}$	$1/n$
2.1	4.41	1.4491	4.583	0.476	6.1	37.21	2.4698	7.810	0.164
2.2	4.84	1.4832	4.690	0.455	6.2	38.44	2.4900	7.874	0.161
2.3	5.29	1.5166	4.796	0.435	6.3	39.69	2.5100	7.937	0.159
2.4	5.76	1.5492	4.899	0.417	6.4	40.96	2.5298	8.000	0.156
2.5	6.25	1.5811	5.000	0.400	6.5	42.25	2.5495	8.062	0.154
2.6	6.76	1.6125	5.099	0.385	6.6	43.56	2.5690	8.124	0.152
2.7	7.29	1.6432	5.196	0.370	6.7	44.89	2.5884	8.185	0.149
2.8	7.84	1.6733	5.292	0.357	6.8	46.24	2.6077	8.246	0.147
2.9	8.41	1.7029	5.385	0.345	6.9	47.61	2.6268	8.307	0.145
3.0	9.00	1.7321	5.477	0.333	7.0	49.00	2.6458	8.367	0.143
3.1	9.61	1.7607	5.568	0.323	7.1	50.41	2.6646	8.426	0.141
3.2	10.24	1.7889	5.657	0.312	7.2	51.84	2.6833	8.485	0.139
3.3	10.89	1.8166	5.745	0.303	7.3	53.29	2.7019	8.544	0.137
3.4	11.56	1.8439	5.831	0.294	7.4	54.76	2.7203	8.602	0.135
3.5	12.25	1.8708	5.916	0.286	7.5	56.25	2.7386	8.660	0.133
3.6	12.96	1.8974	6.000	0.278	7.6	57.76	2.7568	8.718	0.132
3.7	13.69	1.9235	6.083	0.270	7.7	59.29	2.7749	8.775	0.130
3.8	14.44	1.9494	6.164	0.263	7.8	60.84	2.7928	8.832	0.128
3.9	15.21	1.9748	6.245	0.256	7.9	62.41	2.8107	8.888	0.127
4.0	16.00	2.0000	6.325	0.250	8.0	64.00	2.8284	8.944	0.125
4.1	16.81	2.0248	6.403	0.244	8.1	65.61	2.8460	9.000	0.123
4.2	17.64	2.0494	6.481	0.238	8.2	67.24	2.8636	9.055	0.122
4.3	18.49	2.0736	6.557	0.233	8.3	68.89	2.8810	9.110	0.120
4.4	19.36	2.0976	6.633	0.227	8.4	70.56	2.8983	9.165	0.119
4.5	20.25	2.1213	6.708	0.222	8.5	72.25	2.9155	9.220	0.118
4.6	21.16	2.1448	6.782	0.217	8.6	73.96	2.9326	9.274	0.116
4.7	22.09	2.1679	6.856	0.213	8.7	75.69	2.9496	9.327	0.115
4.8	23.04	2.1909	6.928	0.208	8.8	77.44	2.9665	9.381	0.114
4.9	24.01	2.2136	7.000	0.204	8.9	79.21	2.9833	9.434	0.112
5.0	25.00	2.2361	7.071	0.200	9.0	81.00	3.0000	9.487	0.111
5.1	26.01	2.2583	7.141	0.196	9.1	82.81	3.0166	9.539	0.110
5.2	27.04	2.2804	7.211	0.192	9.2	84.64	3.0332	9.592	0.109
5.3	28.09	2.3022	7.280	0.189	9.3	86.49	3.0496	9.644	0.108
5.4	29.16	2.3238	7.348	0.185	9.4	88.36	3.0659	9.695	0.106
5.5	30.25	2.3452	7.416	0.182	9.5	90.25	3.0822	9.747	0.105
5.6	31.36	2.3664	7.483	0.179	9.6	92.16	3.0984	9.798	0.104
5.7	32.49	2.3875	7.550	0.175	9.7	94.09	3.1145	9.849	0.103
5.8	33.64	2.4083	7.616	0.172	9.8	96.04	3.1305	9.899	0.102
5.9	34.81	2.4290	7.681	0.169	9.9	98.01	3.1464	9.950	0.101
6.0	36.00	2.4495	7.746	0.167	10.0	100.00	3.1623	10.000	0.100

A.9 SAMPLING PIECES

Table A.9 is a table of the 200 chips referred to in the beginning of Chapter 7. Each block contains a pair of numbers. One number is to go on each side of the chip that the block represents. For example, the block in the upper right corner is $\boxed{\begin{matrix}6 & \\ & 2\end{matrix}}$. The 6 goes on one side of the chip (cardboard, paper, plastic, or whatever you have available). The 2 goes on the other side. It is also necessary to use two different colors to identify the different sets of numbers. Thus, all of the upper numbers might be drawn in red and the lower numbers in black. The two sets of numbers represent two different distributions.

The chips or blocks have been arranged in the order that they were originally sampled and have an order that has been referred to in Chapter 7 (see page 145). This order is from right to left and starts in the upper right corner.

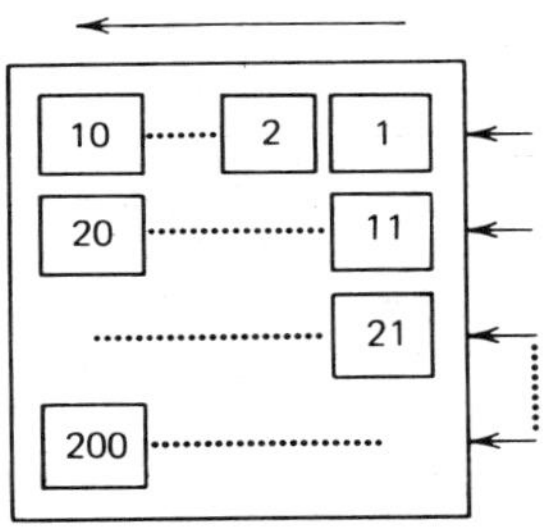

Thus the first chip is $\boxed{\begin{matrix}6 & \\ & 2\end{matrix}}$, the second is $\boxed{\begin{matrix}8 & \\ & 5\end{matrix}}$, the tenth is $\boxed{\begin{matrix}5 & \\ & 2\end{matrix}}$, the eleventh $\boxed{\begin{matrix}4 & \\ & 1\end{matrix}}$, etc., with the very last one $\boxed{\begin{matrix}5 & \\ & 3\end{matrix}}$.

The two distributions are normal and:

Upper distribution—mean = 5, σ = 1.72
Lower distribution—mean = 2, σ = 1.72

TABLE A.9 Sampling Pieces

5 / 2	6 / 2	7 / 4	5 / 2	3 / 0	5 / 1	5 / 3	3 / 1	8 / 5	6 / 2
8 / 5	3 / 1	7 / 3	4 / 0	3 / 1	3 / 2	0 / −3	6 / 3	6 / 4	4 / 1
7 / 3	9 / 5	5 / 0	6 / 1	5 / 3	6 / 3	5 / 2	8 / 4	9 / 6	3 / 0
6 / 4	2 / 0	7 / 2	5 / 2	5 / 4	4 / 1	5 / 1	1 / −2	4 / 2	5 / 3

TABLE A.9 (*continued*)

7 5	5 2	6 2	3 0	7 4	7 2	3 2	7 4	3 1	5 1
4 1	3 1	8 4	4 0	7 3	5 2	7 4	6 3	4 2	6 3
6 3	7 5	6 1	6 2	4 1	5 2	4 1	5 1	5 2	2 −1
4 3	3 −1	4 2	7 4	8 5	8 3	5 2	3 −1	1 −2	5 3
5 1	6 3	5 4	4 0	7 3	7 4	4 2	5 2	6 3	6 4
6 5	4 2	5 2	2 0	5 3	6 3	4 −1	6 4	6 2	9 6
3 1	5 1	6 4	6 3	4 0	7 5	6 3	8 5	4 1	4 1
5 2	2 −2	4 3	3 0	7 3	3 −1	4 2	4 1	6 3	6 2
5 3	4 1	6 2	3 0	5 2	5 1	4 1	4 2	8 4	5 1
5 3	4 1	8 4	4 0	6 1	5 2	4 2	6 3	4 1	5 1
5 2	3 2	7 3	7 2	5 2	4 0	3 0	7 3	4 1	8 6
3 2	5 2	3 0	6 2	2 −1	3 −1	4 1	6 2	2 −1	7 4
6 4	4 2	5 3	5 2	6 3	5 0	5 2	7 3	5 0	5 2
6 4	4 3	4 0	3 0	6 2	5 3	6 3	2 −1	6 4	10 7
6 3	4 0	7 4	4 2	2 0	5 1	4 3	4 1	7 5	2 0
5 3	5 1	3 1	1 −1	5 4	6 1	2 1	4 2	6 2	5 3

A.10 CRITICAL VALUES FOR CORRELATION COEFFICIENT, *r*. PEARSON PRODUCT MOMENT CORRELATION COEFFICIENT

Table A.10 gives critically large values of the correlation coefficient for various sample sizes (degrees of freedom). Several levels of significance are given.

$$n = \text{Number of } pairs \text{ of values}$$

Generally: *Degrees of freedom* $= df = n - 2$.

Critical values are given for both one- and two-tailed tests. A *one-tailed* test would mean that we are *only* interested in a positive correlation or we are *only* interested in a negative correlation. A two-tailed test means we are simultaneously considering *either* a positive or a negative correlation.

EXAMPLE 1

Given a sample of 22, what are the significantly large values or limits for the correlation coefficient? Consider a two-tailed test and 5% and 1% limits.

$$n = 22$$

Therefore

$$df = n - 2 = 20$$

From the table for $df = 20$ we read:

1. The 5% limit or 0.05 limit is 0.4227 (there is less than a 5% chance that r should randomly be as large as 0.4227).
2. The 1% limit or 0.01 limit is 0.5368 (there is less than a 1% chance that r should randomly be as large as 0.5368).

df	LEVEL OF SIGNIFICANCE FOR TWO-TAILED TEST	
	0.05	0.01
20	0.4227	0.5368

TABLE A.10 Critical Values of the Pearson Product Moment Correlation Coefficient

	LEVEL OF SIGNIFICANCE FOR ONE-TAILED TEST				
	0.05	0.025	0.01	0.005	0.0005
	LEVEL OF SIGNIFICANCE FOR TWO-TAILED TEST				
$df = n - 2$	0.10	0.05	0.02	0.01	0.001
1	0.9877	0.9969	0.9995	0.9999	1.0000
2	0.9000	0.9500	0.9800	0.9900	0.9990
3	0.8054	0.8783	0.9343	0.9587	0.9912
4	0.7293	0.8114	0.8822	0.9172	0.9741
5	0.6694	0.7545	0.8329	0.8745	0.9507
6	0.6215	0.7067	0.7887	0.8343	0.9249
7	0.5822	0.6664	0.7498	0.7977	0.8982
8	0.5494	0.6319	0.7155	0.7646	0.8721
9	0.5214	0.6021	0.6851	0.7348	0.8471
10	0.4973	0.5760	0.6581	0.7079	0.8233
11	0.4762	0.5529	0.6339	0.6835	0.8010
12	0.4575	0.5324	0.6120	0.6614	0.7800
13	0.4409	0.5139	0.5923	0.6411	0.7603
14	0.4259	0.4973	0.5742	0.6226	0.7420
15	0.4124	0.4821	0.5577	0.6055	0.7246
16	0.4000	0.4683	0.5425	0.5897	0.7084
17	0.3887	0.4555	0.5285	0.5751	0.6932
18	0.3783	0.4438	0.5155	0.5614	0.6787
19	0.3687	0.4329	0.5034	0.5487	0.6652
20	0.3598	0.4227	0.4921	0.5368	0.6524
25	0.3233	0.3809	0.4451	0.4869	0.5974
30	0.2960	0.3494	0.4093	0.4487	0.5541
35	0.2746	0.3246	0.3810	0.4182	0.5189
40	0.2573	0.3044	0.3578	0.3932	0.4896
45	0.2428	0.2875	0.3384	0.3721	0.4648
50	0.2306	0.2732	0.3218	0.3541	0.4433
60	0.2108	0.2500	0.2948	0.3248	0.4078
70	0.1954	0.2319	0.2737	0.3017	0.3799
80	0.1829	0.2172	0.2565	0.2830	0.3568
90	0.1726	0.2050	0.2422	0.2673	0.3375
100	0.1638	0.1946	0.2301	0.2540	0.3211

Table A.10 is taken from Table VII of Fisher and Yates: *Statistical Tables for Biological, Agricultural and Medical Research*, published by Longman Group Ltd., London (previously published by Oliver & Boyd, Edinburgh), and by permission of the authors and publishers.

A.11 TEST FOR EXTREME VALUES: CRITICAL VALUES

Table A.11 provides test statistics and critical limits for determining if one of the values of a small set of numbers appears excessively large or unusual. *The data must be ordered.*

The particular test statistic depends on the size of the sample, n. Look down the column for n, the test statistic is given in the left column. Two sets of statistics are given, depending on whether the extreme is the largest or the smallest value. Select the appropriate statistic. Critical values are found in the columns to the right.

EXAMPLE 1

The following 8 values of times were recorded.

26, 31, 29, 49, 30, 34, 27, 29

Is the 49 unusual? Placing them in order, we have

X_1	X_2	X_3	X_4	X_5	X_6	X_7	X_8
26	27	29	29	30	31	34	49

From Table A.11, for $n = 8$ the statistic for a *largest* extreme value is

$$r_{11} = \frac{X_n - X_{n-1}}{X_n - X_2}$$

Since $n = 8$,

$$r_{11} = \frac{X - X_{8-1}}{X_8 - X_2} = \frac{X_8 - X_7}{X_8 - X_2}$$

$$= \frac{49 - 34}{49 - 26} = \frac{15}{23} = 0.653.$$

Consulting the table for $n = 8$ we find this value exceeds the 0.02 critical value of 0.631 but is less than the 0.01 critical value of 0.683. It is significant but not very highly significant.

TEST STATISTIC	n	PROBABILITY LEVEL 0.02	0.01
$r_{11} = \frac{X_n - X_{n-1}}{X_n - X_2}$	8	0.631	0.683

TABLE A.11 Test for Extreme Values: Critical Values

TEST STATISTIC		*n*	PROBABILITY OF A VALUE AS LARGE AS THE TABLE						
EXTREME SMALL	EXTREME LARGE		0.30	0.20	0.10	0.05	0.02	0.01	0.005
		3	0.684	0.781	0.886	0.941	0.976	0.988	0.994
		4	0.471	0.560	0.679	0.765	0.846	0.889	0.926
$r_{10} = \dfrac{X_2 - X_1}{X_n - X_1}$	$r_{10} = \dfrac{X_n - X_{n-1}}{X_n - X_1}$	5	0.373	0.451	0.557	0.642	0.729	0.780	0.821
		6	0.318	0.386	0.482	0.560	0.644	0.698	0.740
		7	0.281	0.344	0.434	0.507	0.586	0.637	0.680
		8	0.318	0.385	0.479	0.554	0.631	0.683	0.725
$r_{11} = \dfrac{X_2 - X_1}{X_{n-1} - X_1}$	$r_{11} = \dfrac{X_n - X_{n-1}}{X_n - X_2}$	9	0.288	0.352	0.441	0.512	0.587	0.635	0.677
		10	0.265	0.325	0.409	0.477	0.551	0.597	0.639
		11	0.391	0.442	0.517	0.576	0.638	0.679	0.713
$r_{21} = \dfrac{X_3 - X_1}{X_{n-1} - X_1}$	$r_{21} = \dfrac{X_n - X_{n-2}}{X_n - X_2}$	12	0.370	0.419	0.490	0.546	0.605	0.642	0.675
		13	0.351	0.399	0.467	0.521	0.578	0.615	0.649
		14	0.370	0.421	0.492	0.546	0.602	0.641	0.674
		15	0.353	0.402	0.472	0.525	0.579	0.616	0.647
		16	0.338	0.386	0.454	0.507	0.559	0.595	0.624
		17	0.325	0.373	0.438	0.490	0.542	0.577	0.605
		18	0.314	0.361	0.424	0.475	0.527	0.561	0.589
		19	0.304	0.350	0.412	0.462	0.514	0.547	0.575
$r_{22} = \dfrac{X_3 - X_1}{X_{n-2} - X_1}$	$r_{22} = \dfrac{X_n - X_{n-2}}{X_n - X_3}$	20	0.295	0.340	0.401	0.450	0.502	0.535	0.562
		21	0.287	0.331	0.391	0.440	0.491	0.524	0.551
		22	0.280	0.323	0.382	0.430	0.481	0.514	0.541
		23	0.274	0.316	0.374	0.421	0.472	0.505	0.532
		24	0.268	0.310	0.367	0.413	0.464	0.497	0.524
		25	0.262	0.304	0.360	0.406	0.457	0.489	0.516

Entries are extracted from W. J. Dixon, "Processing Data for Outliers," *Biometrics*, Vol. 9 (1952), p. 74, with the kind permission of the author and *Biometrics* trustees.

A.12 CRITICAL VALUES OF *U* AND *U′* FOR THE MANN-WHITNEY TEST

This test is used to determine if there is a difference between two populations. Two independent samples, Sample A of size n_1, and Sample B of size n_2, are ranked together. For each B the number of A that have a smaller rank than the B are counted. The sum for all of the B is called U. U' would be the other sum, the sum of all the B that come before each A.

This set of tables (Table A.12) is good only if n_1 and n_2 are less than or equal to 20. See Chapter 11, Section 11.3, for n greater than 20.

Locate n_1 along the top and n_2 along the side. At the intersection find the critical values of U and U'.

EXAMPLE 1

For a two-tailed test at $\alpha = 0.05$, with $n_1 = 7$ and $n_2 = 8$ what are the critical limits on U and U'?

At the intersection we find

10
46

The top number, 10, stands for the lower limit and the bottom number, *46*, stands for the upper limit. Thus, a sum of less than or equal to 10 and a sum of greater than or equal to 46 would be significant.

		n_1	
		7	
n_2 8		10 *46*	

To use the table for a *one-tailed* test you must decide which of the samples is expected to be the larger. If you classify the expected larger distribution as Sample A and compute the number of A less than each B, you would use only the top values in the table, the smaller numbers. (Thus, for a one tail with a $\alpha = 0.05$ limit and 2 samples of 6, you would consult the first Table A.12 and the critical value of U would be 7.)

TABLE A.12 a. One-Tailed Test at $\alpha = 0.025$ or a Two-Tailed Test at $\alpha = 0.05$[a]

n_2 \ n_1	1	2	3	4	5	6	7	8	9	10	11	12	13	14	15	16	17	18	19	20
1	—	—	—	—	—	—	—	—	—	—	—	—	—	—	—	—	—	—	—	—
2	—	—	—	—	—	—	—	0 *16*	0 *18*	0 *20*	0 *22*	1 *23*	1 *25*	1 *27*	1 *29*	1 *31*	2 *32*	2 *34*	2 *36*	2 *38*
3	—	—	—	—	0 *15*	1 *17*	1 *20*	2 *22*	2 *25*	3 *27*	3 *30*	4 *32*	4 *35*	5 *37*	5 *40*	6 *42*	6 *45*	7 *47*	7 *50*	8 *52*
4	—	—	—	0 *16*	1 *19*	2 *22*	3 *25*	4 *28*	4 *32*	5 *35*	6 *38*	7 *41*	8 *44*	9 *47*	10 *50*	11 *53*	11 *57*	12 *60*	13 *63*	13 *67*
5	—	—	0 *15*	1 *19*	2 *23*	3 *27*	5 *30*	6 *34*	7 *38*	8 *42*	9 *46*	11 *49*	12 *53*	13 *57*	14 *61*	15 *65*	17 *68*	18 *72*	19 *76*	20 *80*
6	—	—	1 *17*	2 *22*	3 *27*	5 *31*	6 *36*	8 *40*	10 *44*	11 *49*	13 *53*	14 *58*	16 *62*	17 *67*	19 *71*	21 *75*	22 *80*	24 *84*	25 *89*	27 *93*
7	—	—	1 *20*	3 *25*	5 *30*	6 *36*	8 *41*	10 *46*	12 *51*	14 *56*	16 *61*	18 *66*	20 *71*	22 *76*	24 *81*	26 *86*	28 *91*	30 *96*	32 *101*	34 *106*
8	—	0 *16*	2 *22*	4 *28*	6 *34*	8 *40*	10 *46*	13 *51*	15 *57*	17 *63*	19 *69*	22 *74*	24 *80*	26 *86*	29 *91*	31 *97*	34 *102*	36 *108*	38 *111*	41 *119*
9	—	0 *18*	2 *25*	4 *32*	7 *38*	10 *44*	12 *51*	15 *57*	17 *64*	20 *70*	23 *76*	26 *82*	28 *89*	31 *95*	34 *101*	37 *107*	39 *114*	42 *120*	45 *126*	48 *132*
10	—	0 *20*	3 *27*	5 *35*	8 *42*	11 *49*	14 *56*	17 *63*	20 *70*	23 *77*	26 *84*	29 *91*	33 *97*	36 *104*	39 *111*	42 *118*	45 *125*	48 *132*	52 *138*	55 *145*
11	—	0 *22*	3 *30*	6 *38*	9 *46*	13 *53*	16 *61*	19 *69*	23 *76*	26 *84*	30 *91*	33 *99*	37 *106*	40 *114*	44 *121*	47 *129*	51 *136*	55 *143*	58 *151*	62 *158*
12	—	1 *23*	4 *32*	7 *41*	11 *49*	14 *58*	18 *66*	22 *74*	26 *82*	29 *91*	33 *99*	37 *107*	41 *115*	45 *123*	49 *131*	53 *139*	57 *147*	61 *155*	65 *163*	69 *171*
13	—	1 *25*	4 *35*	8 *44*	12 *53*	16 *62*	20 *71*	24 *80*	28 *89*	33 *97*	37 *106*	41 *115*	45 *124*	50 *132*	54 *141*	59 *149*	63 *158*	67 *167*	72 *175*	76 *184*
14	—	1 *27*	5 *37*	9 *47*	13 *57*	17 *67*	22 *76*	26 *86*	31 *95*	36 *104*	40 *114*	45 *123*	50 *132*	55 *141*	59 *151*	64 *160*	67 *171*	74 *178*	78 *188*	83 *197*
15	—	1 *29*	5 *40*	10 *50*	14 *61*	19 *71*	24 *81*	29 *91*	34 *101*	39 *111*	44 *121*	49 *131*	54 *141*	59 *151*	64 *161*	70 *170*	75 *180*	80 *190*	85 *200*	90 *210*
16	—	1 *31*	6 *42*	11 *53*	15 *65*	21 *75*	26 *86*	31 *97*	37 *107*	42 *118*	47 *129*	53 *139*	59 *149*	64 *160*	70 *170*	75 *181*	81 *191*	86 *202*	92 *212*	98 *222*
17	—	2 *32*	6 *45*	11 *57*	17 *68*	22 *80*	28 *91*	34 *102*	39 *114*	45 *125*	51 *136*	57 *147*	63 *158*	67 *171*	75 *180*	81 *191*	87 *202*	93 *213*	99 *224*	105 *235*
18	—	2 *34*	7 *47*	12 *60*	18 *72*	24 *84*	30 *96*	36 *108*	42 *120*	48 *132*	55 *143*	61 *155*	67 *167*	74 *178*	80 *190*	86 *202*	93 *213*	99 *225*	106 *236*	112 *248*
19	—	2 *36*	7 *50*	13 *63*	19 *76*	25 *89*	32 *101*	38 *114*	45 *126*	52 *138*	58 *151*	65 *163*	72 *175*	78 *188*	85 *200*	92 *212*	99 *224*	106 *236*	113 *248*	119 *261*
20	—	2 *38*	8 *52*	13 *67*	20 *80*	27 *93*	34 *106*	41 *119*	48 *132*	55 *145*	62 *158*	69 *171*	76 *184*	83 *197*	90 *210*	98 *222*	105 *235*	112 *248*	119 *261*	127 *273*

[a] To be significant, the value of U (or U') must be *less than or equal* to the top value; or the value of U (or U') must be *greater than or equal* to the lower value in the table.

TABLE A.12 (*continued*) b. One-Tailed Test at $\alpha = 0.005$ or a Two-Tailed Test at $\alpha = 0.01$[b]

n_2 \ n_1	1	2	3	4	5	6	7	8	9	10	11	12	13	14	15	16	17	18	19	20
1	—	—	—	—	—	—	—	—	—	—	—	—	—	—	—	—	—	—	—	—
2	—	—	—	—	—	—	—	—	—	—	—	—	—	—	—	—	—	—	0 *38*	0 *40*
3	—	—	—	—	—	—	—	—	0 *27*	0 *30*	0 *33*	1 *35*	1 *38*	1 *41*	2 *43*	2 *46*	2 *49*	2 *52*	3 *54*	3 *57*
4	—	—	—	—	—	0 *24*	0 *28*	1 *31*	1 *35*	2 *38*	2 *42*	3 *45*	3 *49*	4 *52*	5 *55*	5 *59*	6 *62*	6 *66*	7 *69*	8 *72*
5	—	—	—	—	0 *25*	1 *29*	1 *34*	2 *38*	3 *42*	4 *46*	5 *50*	6 *54*	7 *58*	7 *63*	8 *67*	9 *71*	10 *75*	11 *79*	12 *83*	13 *87*
6	—	—	—	0 *24*	1 *29*	2 *34*	3 *39*	4 *44*	5 *49*	6 *54*	7 *59*	9 *63*	10 *68*	11 *73*	12 *78*	13 *83*	15 *87*	16 *92*	17 *97*	18 *102*
7	—	—	—	0 *28*	1 *34*	3 *39*	4 *45*	6 *50*	7 *56*	9 *61*	10 *67*	12 *72*	13 *78*	15 *83*	16 *89*	18 *94*	19 *100*	21 *105*	22 *111*	24 *116*
8	—	—	—	1 *31*	2 *38*	4 *44*	6 *50*	7 *57*	9 *63*	11 *69*	13 *75*	15 *81*	17 *87*	18 *94*	20 *100*	22 *106*	24 *112*	26 *118*	28 *124*	30 *130*
9	—	—	0 *27*	1 *35*	3 *42*	5 *49*	7 *56*	9 *63*	11 *70*	13 *77*	16 *83*	18 *90*	20 *97*	22 *104*	24 *111*	27 *117*	29 *124*	31 *131*	33 *138*	36 *144*
10	—	—	0 *30*	2 *38*	4 *46*	6 *54*	9 *61*	11 *69*	13 *77*	16 *84*	18 *92*	21 *99*	24 *106*	26 *114*	29 *121*	31 *129*	34 *136*	37 *143*	39 *151*	42 *158*
11	—	—	0 *33*	2 *42*	5 *50*	7 *59*	10 *67*	13 *75*	16 *83*	18 *92*	21 *100*	24 *108*	27 *116*	30 *124*	33 *132*	36 *140*	39 *148*	42 *156*	45 *164*	48 *172*
12	—	—	1 *35*	3 *45*	6 *54*	9 *63*	12 *72*	15 *81*	18 *90*	21 *99*	24 *108*	27 *117*	31 *125*	34 *134*	37 *143*	41 *151*	44 *160*	47 *169*	51 *177*	54 *186*
13	—	—	1 *38*	3 *49*	7 *58*	10 *68*	13 *78*	17 *87*	20 *97*	24 *106*	27 *116*	31 *125*	34 *135*	38 *144*	42 *153*	45 *163*	49 *172*	53 *181*	56 *191*	60 *200*
14	—	—	1 *41*	4 *52*	7 *63*	11 *73*	15 *83*	18 *94*	22 *104*	26 *114*	30 *124*	34 *134*	38 *144*	42 *154*	46 *164*	50 *174*	54 *184*	58 *194*	63 *203*	67 *213*
15	—	—	2 *43*	5 *55*	8 *67*	12 *78*	16 *89*	20 *100*	24 *111*	29 *121*	33 *132*	37 *143*	42 *153*	46 *164*	51 *174*	55 *185*	60 *195*	64 *206*	69 *216*	73 *227*
16	—	—	2 *46*	5 *59*	9 *71*	13 *83*	18 *94*	22 *106*	27 *117*	31 *129*	36 *140*	41 *151*	45 *163*	50 *174*	55 *185*	60 *196*	65 *207*	70 *218*	74 *230*	79 *241*
17	—	—	2 *49*	6 *62*	10 *75*	15 *87*	19 *100*	24 *112*	29 *124*	34 *136*	39 *148*	44 *160*	49 *172*	54 *184*	60 *195*	65 *207*	70 *219*	75 *231*	81 *242*	86 *254*
18	—	—	2 *52*	6 *66*	11 *79*	16 *92*	21 *105*	26 *118*	31 *131*	37 *143*	42 *156*	47 *169*	53 *181*	58 *194*	64 *206*	70 *218*	75 *231*	81 *243*	87 *255*	92 *268*
19	—	0 *38*	3 *54*	7 *69*	12 *83*	17 *97*	22 *111*	28 *124*	33 *138*	39 *151*	45 *164*	51 *177*	56 *191*	63 *203*	69 *216*	74 *230*	81 *242*	87 *255*	93 *268*	99 *281*
20	—	0 *40*	3 *57*	8 *72*	13 *87*	18 *102*	24 *116*	30 *130*	36 *144*	42 *158*	48 *172*	54 *186*	60 *200*	67 *213*	73 *227*	79 *241*	86 *254*	92 *268*	99 *281*	105 *295*

Adapted from D. Auble, "Extended Tables for the Mann-Whitney Statistic," Bulletin of the Institute of Educational Research at Indiana University, Vol. 1, No. 2, 1953, with the kind permission of the author and publisher.

[b] To be significant, the value of U (or U') must be *less than or equal* to the top value; or the value of U (or U') must be *greater than or equal* to the lower value in the table.

APPENDIX B SUMMATION AND SUBSCRIPT NOTATION

B.1 ALGEBRAIC SYMBOLS AND NOTATION

In all cases statistics deals with numbers. The numbers themselves are specific or particular measurements of a certain characteristic. However, often we desire to designate or discuss the characteristic without being restricted to an individual measurement of the characteristic. To alleviate this problem, we use a letter to represent whatever the numerical value of the measurement might be.

As an example of data, we will be considering the number of broken dishes in a restaurant. A possible situation of interest might be the keeping of a record of the total number of dishes broken by various employees during one day's work. For instance, Ann might have broken 2 dishes yesterday, Jill 3 dishes, Frank 1, and so forth. It is quite possible to work with such a type of designation (Ann broke 2 dishes yesterday, etc.). However, it is cumbersome and does not lend itself easy to generalizations such as discussing "the number of dishes broken by Ann—whatever they might be."

We will now substitute a symbol to represent the number of dishes broken. In this way we will remove any restriction on the particular or exact number. We might decide that we will designate "the number of dishes broken" with the symbol X. Of course, we could use any symbol, as long as we agree on the meaning of the symbol. Now observe several things.

1. The symbol stands for the measurement of the characteristic (broken dishes).
2. The symbol represents many different possible values. It may be standing for 0, 1, 2, 3, and so forth, depending on, in this case, the actual number of dishes broken.
3. By stating the letter X by itself we do not restrict ourselves to a particular value and thus we have generalized on the idea (dishes broken).

One difficulty now arises—how to designate or distinguish the different individuals. We might be interested in discussing or separately designating dishes broken by Ann, Jill, Frank, and so forth. We then are to be concerned with the X-values, the measurements, for each of these people, and you should realize that the *X-value for Ann is* 2 (2 dishes broken), the X for Jill is 3 (3 dishes broken), and the X for Frank is 1. But this too is awkward.

Although several possibilities for simplifying those statements exist, probably the easiest and most convenient method will be to rename each individual with a number. We will call Ann 1, Jill 2, and Frank 3. We will then distinguish the X by calling one X_1, the next X_2, the third X_3, and so forth. We read these as X sub 1, X sub 2, and X sub 3, with the numbers being called *subscripts*. Thus 1 represents Ann and X_1 represents the number of dishes she broke. Jill is 2 and X_2 is the number of dishes she broke. In general we observe that each subscript corresponds to a particular individual being measured and each X corresponds to the measurement.

An additional advantage of this method of designation is that we do not have to specify *how many* X there are. We can state that the values being considered are:

$$X_1, X_2, X_3, \ldots$$

where the ... means "and so on."

We can end the list of X's by calling the last value X_n. So we obtain

$$X_1, X_2, X_3, \ldots, X_n$$

At any time by just specifying what n is, we also designate the number of X. Thus, if there are 5 employees in our restaurant (David and Sara being the additional two), $n = 5$ and we have 5 values of X.

$$X_1, X_2, X_3, X_4, X_5$$

One possible concern should be mentioned at this point. The order of these subscripts or individuals is not rigid but must be agreed on. We might have scrambled the 5 people's names and ended up with a different order, perhaps Frank as 1, David as 2, and so forth. Now X_1 stands for the number of dishes Frank has broken. In many cases it won't really matter what order we place the individuals in and we can even juggle them around. However, you should always remember that the subscript implies that the value corresponds to a particular individual and if we desire to learn which individual that is, we need some method of locating or finding it.

Often we want to talk about or refer to the measured quantity and recognize that it belongs to a *particular individual, but* we are not concerned about *which* individual. We do not care if the discussion centers about the dishes broken by Frank or Jill but, instead, the dishes *someone* broke. Here we are still considering an X-value and, in a sense, a particular value. The particular one however, will be generalized and called

$$X_i \text{ (read } X \text{ sub } i\text{)}$$

with the assumption that i can mean *any* number from 1 to n.

Thus, I can talk about "the number of dishes broken by *an* employee" with the single symbol X_i, the X meaning some number of dishes, and the i declaring that the number of dishes is to correspond to a particular employee, the ith employee. But we do not have to state *who* or *how many*.

B.2 ADDITION

Often we want to perform the operation of addition. Here we want to know the total number of dishes broken yesterday. So we add up the 2 dishes broken by Ann, the 3 by Jill, the 1 by Frank and the 1 and 2 by David and Sara (the last two were not mentioned previously). The total is 9.

If we wanted to make the statement "add the 5 values" without referring to the particular values themselves we could do this with the algebraic symbols as follows.

$$X_1 + X_2 + X_3 + X_4 + X_5 \tag{B.1}$$

and in our example this would be the same as

$$2 + 3 + 1 + 1 + 2 = 9$$

The advantages of using the algebraic method (using X) should be evident. We do not need to write as much. Juggling the X around does not affect our understanding of the operation. And, most of all, we will be able to generalize one step further. If we do not know how many individuals are to be included in the addition we can end the addition with X_n, yielding

$$X_1 + X_2 + X_3 + X_4 + \cdots + X_n \qquad \text{(B.2)}$$

In this way by merely stating what n is on a particular day we can complete the expression for addition, but we do not *need* to do so in order to discuss the operation. Observe that if yesterday 5 people were working and today 8 people were working, this general equation works. We just state that for yesterday n was 5 and today n is 8.

B.3 SUMMATION NOTATION

In most problems the use of expressions like equation B.2 are still too cumbersome so I will introduce another symbol to simplify the writing. The symbol is a Greek letter $\sum$ (sigma), which is to be read as "the sum of" and tells you to take the sum of the quantities to the right of it. Using this symbol, we will replace Equation B.1 by

$$\sum_{i=1}^{5} X_i \qquad \text{(B.3)}$$

The symbol has something on top of it and something on the bottom of it. These things tell you how many and which individuals you are going to be adding. The things we are adding will be X or terms involving X, but in order to condense Equation B.1 we only want to look at *one of the terms* that sit between the plus signs. The general term in this case is X_i. But which i are to be included? That is where the $i = 1$ and the 5 come in. The number on the bottom of the $\sum$ tells where to start. The number on the top of the $\sum$ tells where to end.

It is also understood that the i increases in increments of 1. Thus,

tells us where to stop

$$\sum_{i=1}^{5} X_i = X_1 + X_2 + X_3 + X_4 + X_5 \qquad \text{(B.4)}$$

first

increased by 1

As another example, where T may mean the size of tips received by waiters,

$$\sum_{i=1}^{3} T_i = T_1 + T_2 + T_3 \qquad \text{(B.5)}$$

The lower limit does *not* have to be 1.

$$\sum_{i=3}^{7} X_i = X_3 + X_4 + X_5 + X_6 + X_7 \qquad \text{(B.6)}$$

The upper limit may be n, so that Equation B.2 may be simplified as:

$$\sum_{i=1}^{n} X_i = X_1 + X_2 + X_3 + \cdots + X_n \qquad \text{(B.7)}$$

Sometimes the numbers on the top and bottom of the $\sum$ are omitted. In these cases it is *understood* that we mean

$$\sum X_i = X_1 + X_2 + X_3 + \cdots + X_n \qquad \text{(B.8)}$$

As I have stated previously, the summation sign refers to *terms* that are to be added. When the terms are more complicated than a single symbol, we often need to use parentheses to identify the complete terms. Perhaps the owner disregards one broken dish, so he or she is really concerned with adding the quantities $X - 1$. Each of the terms would be of the form $X_i - 1$, so the first would be $X_1 - 1$, the second $X_2 - 1$, and so forth, and the sum would be

$$(X_1 - 1) + (X_2 - 1) + (X_3 - 1) + (X_4 - 1) + (X_5 - 1) \tag{B.9}$$

Using the summation notation we could replace this by

$$\sum_{i=1}^{5} (X_i - 1) \tag{B.10}$$

For the particular values this would be

$$\begin{aligned}\sum_{i=1}^{5} (X_i - 1) &= (X_1 - 1) + (X_2 - 1) + (X_3 - 1) + (X_4 - 1) + (X_5 - 1) \\ &= (2 - 1) + (3 - 1) + (1 - 1) + (1 - 1) + (2 - 1) \\ &= 1 \quad + 2 \quad + 0 \quad + 0 \quad + 1 = 4\end{aligned} \tag{B.11}$$

Let me compare this to something that looks very similar.

$$\sum_{i=1}^{5} X_i - 1 \tag{B.12}$$

All I have done is *remove the parentheses*, but the consequence is that the number 1 is subtracted *only once* and plays no part in the summation. Thus,

$$\begin{aligned}\sum_{i=1}^{5} X_i - 1 &= \underbrace{(X_1 + X_2 + X_3 + X_4 + X_5)}_{\text{this is the } \sum X_i} \underbrace{- 1}_{\text{this is the } -1 \text{ part}} \\ &= 2 + 3 + 1 + 1 + 1 + 2 - 1 \\ &= 8\end{aligned} \tag{B.13}$$

The value here is 8 and was only 4 using Equation B.11.

Let us look at a few more examples (using the same values of X).

$$\begin{aligned}\sum_{i=1}^{5} (X_i + 1)^2 &= (X_1 + 1)^2 + (X_2 + 1)^2 + (X_3 + 1)^2 + (X_4 + 1)^2 + (X_5 + 1)^2 \\ &= (2 + 1)^2 + (3 + 1)^2 + (1 + 1)^2 + (1 + 1)^2 + (2 + 1)^2 \\ &= (3)^2 + (4)^2 + (2)^2 + (2)^2 + (3)^2 \\ &= 9 + 16 + 4 + 4 + 9 = 42\end{aligned} \tag{B.14}$$

Observe that each term we added was of the form $(X + 1)^2$ and the squaring was done *before* the terms were added.

$$\sum_{i=1}^{5}(X_i^2 + 1) = (X_1^2 + 1) + (X_2^2 + 1) + (X_3^2 + 1) + (X_4^2 + 1) + (X_5^2 + 1)$$
$$= (2^2 + 1) + (3^2 + 1) + (1^2 + 1) + (1^2 + 1) + (2^2 + 1)$$
$$= (4 + 1) + (9 + 1) + (1 + 1) + (1 + 1) + (4 + 1)$$
$$= 5 + 10 + 2 + 2 + 2 + 5 = 24 \qquad \text{(B.15)}$$

The main case when parenthesis may be omitted is with multiplication.

$$\sum_{i=1}^{5} 2X_i = 2X_1 + 2X_2 + 2X_3 + 2X_4 + 2X_5$$
$$= 2 \cdot 2 + 2 \cdot 3 + 2 \cdot 1 + 2 \cdot 1 + 2 \cdot 2$$
$$= 4 + 6 + 2 + 2 + 4 = 18 \qquad \text{(B.16)}$$

Sometimes different quantities may be involved with the same subscripts. Let X still be dishes broken and let Y be glasses broken. If Ann broke 1 glass, Jill 3, Frank 2, David 2, and Sara 3, then $Y_1 = 1$, $Y_2 = 3$, $Y_3 = 2$, $Y_4 = 2$, and $Y_5 = 3$.

$$\sum_{i=1}^{5}(X_i + Y_i) = (X_1 + Y_1) + (X_2 + Y_2) + (X_3 + Y_3) + (X_4 + Y_4) + (X_5 + Y_5)$$
$$= (2 + 1) + (3 + 3) + (1 + 2) + (1 + 2) + (2 + 3)$$
$$= 3 + 6 + 3 + 3 + 5 = 20 \qquad \text{(B.17)}$$

Several important differences between some types of summations frequently encountered in statistics should be pointed out. The following pairs of summations look similar but are *not* similar. I will drop the $i = 1$ and the n (or 5 for the example) from these.

$$(\sum X)^2$$

and

$$\sum X^2$$

For the example,

$$(\sum X)^2 = (X_1 + X_2 + X_3 + X_4 + X_5)^2 = (9)^2 = 81$$

The addition is done first, then the squaring. On the other hand,

$$\sum X^2 = X_1^2 + X_2^2 + X_3^2 + X_4^2 + X_5^2$$
$$= 2^2 + 3^2 + 1^2 + 1^2 + 2^2$$
$$= 4 + 9 + 1 + 1 + 4 = 19$$

Similarly,

$$(\sum X_i + \sum Y_i)^2$$

is not the same as

$$\sum(X_i + Y_i)^2$$

In our example the first expression is 400 and the second is only 88.

B.4 THREE CONVENIENT THEOREMS OF SUMMATION

I will not prove the following, but I will demonstrate their validity using some examples

1. If C is a constant (say the number 6.4) then

$$\sum_{i=1}^{n} C = n \cdot C \tag{B.18}$$

EXAMPLE 1

$$\sum_{i=1}^{4} 6.4 = C_1 + C_2 + C_3 + C_4$$

$$= 6.4 + 6.4 + 6.4 + 6.4$$

$$= 4(6.4) = n \cdot C$$

2. If C is a constant (again say 6.4) then

$$\sum_{i=1}^{n} CX_i = C\sum_{i=1}^{n} X_i \tag{B.19}$$

EXAMPLE 2

$$\sum_{i=1}^{3} 6.4X_i = 6.4X_1 + 6.4X_2 + 6.4X_3$$

$$= 6.4(X_1 + X_2 + X_3) = 6.4\left(\sum_{i=1}^{3} X_i\right)$$

Note that this does *not* work for an expression of the type $\sum(2X_i + 1)^2$.

3.

$$\sum_{i=1}^{n} (X_i + Y_i) = \sum_{i=1}^{n} X_i + \sum_{i=1}^{n} Y_i \tag{B.20}$$

This is the only one of the three to use with extra caution.

Examine Equation B.17 and observe how you can drop all of the parentheses around the X and Y to obtain

$$\sum(X_i + Y_i) = X_1 + Y_1 + X_2 + Y_2 + X_3 + Y_3 + X_4 + Y_4 + X_5 + Y_5$$

Rearranging these, we could obtain

$$\sum_{i=1}^{5} (X_i + Y_i) = (X_1 + X_2 + X_3 + X_4 + X_5) + (Y_1 + Y_2 + Y_3 + Y_4 + Y_5)$$

$$= \sum_{i=1}^{5} X_i + \sum_{i=1}^{5} Y_i$$

PROBLEMS FOR APPENDIX B

For Problems B.1 to B.14 use the following values of X and Y.

$$X_1 = 1,\ X_2 = 3,\ X_3 = 4,\ X_4 = 2,\ X_5 = 3,\ X_6 = 1,\ X_7 = 4,\ X_8 = 3$$

$$Y_1 = 2,\ Y_2 = 4,\ Y_3 = 5,\ Y_4 = 3,\ Y_5 = 7,\ Y_6 = 2,\ Y_7 = 5,\ Y_8 = 5$$

Calculate the given summations.

B.1 $\sum_{i=1}^{3} X_i$

B.2 $\sum_{i=3}^{7} X_i$

B.3 $\sum_{i=1}^{4} X_i^2$

B.4 $\sum_{i=3}^{6} (X_i + 1)^2$

B.5 $\sum_{i=5}^{8} X_i + 2$

B.6 $\sum_{i=1}^{5} (X_i + Y_i)^2$

B.7 $\sum_{i=1}^{4} 2X_i$

B.8 $\sum_{i=1}^{8} 2(X_i + 2)$

B.9 $\sum_{i=1}^{4} \left(\frac{X_i}{Y_i}\right)$

B.10 $\sum_{i=5}^{8} X_i Y_i$

B.11 $\sum_{i=1}^{5} X_i Y_i^2$

B.12 $\sum_{i=1}^{4} X_i(X_i + Y_i)^2$

B.13 $\sum_{i=3}^{7} (2X_i + Y_i + 3X_i)$

B.14 $\sum_{i=1}^{5} (2X_i Y_i - 2X_i + X_i^2)$

BIBLIOGRAPHY

The accompanying list of references covers a broad range of general and specific texts in both elementary and advanced levels of statistics and probability. I have also provided a guide to these texts according to mathematical level and by topic. This bibliography is by no means exhaustive and merely serves as a point of departure for further work.

A. GENERAL BOOKS FOR THE LAYMAN. The following are essentially nonmathematical treatments of some areas of statistics and can be read for pleasure.

Huff (26)
Slonim (45)
Scott (43)
Claque (7)
Hauser (20)
Tanur et al. (49)
Spear (47)
Zeisel (54)

B. GENERAL STATISTICS BOOKS (basically nonprobability oriented). These should serve as good references for more examples on statistical analysis and other techniques.

Dixon and Massey (11)
Moroney (33)
Snedecor (46)
Weinberg and Schumaker (51)
Haber and Runyon (17)
Roscoe (41)
Walker and Lev (50)

C. ELEMENTARY STATISTICS BOOKS WITH A PROBABILITY ORIENTATION. These books generally first develop some of the probability theory and then begin the study of statistics.

Alder and Roessler (1)
Freund (14, 15)
Mosteller, et al. (34)
Mendenhall (30)
Wilks (52)

D. MORE ADVANCED STATISTICS BOOKS WITH EMPHASIS ON METHODOLOGY.

Anderson and Bancroft (2)
Brownlee (6)
Bennett and Franklin (3)
Miller and Freund (31)

E. MORE ADVANCED THEORETICAL OR MATHEMATICAL STATISTICS BOOKS.

Hoel (23)
Hogg and Craig (24)
Mood et al. (32)

F. PROBABILITY THEORY.

Feller (13)
Parzan (36)

G. TABLES.

Beyer (5)
Hald (18)
Owen (35)
Pearson and Hartley (37)
RAND (39)
Romig (40)

SOME SPECIALIZED STATISTICS BOOKS

H. NONPARAMETRIC STATISTICS.

Siegal (44)
Hollander and Wolfe (25)

I. QUALITY CONTROL.

Grant and Leavenworth (16)
Juran and Gryna (27)

J. DESIGN OF EXPERIMENTS.

Cochran and Cox (9)
Davies (10)
Hicks (22)

K. SAMPLING.

Cochran (8)
Sudman (48)
Mendenhall (29)

L. BUSINESS.

Hamberg (19)

M. REGRESSION.

Draper and Smith (12)

N. OVERVIEW AND READINGS.

Heerman & Brashan (21)
Barnett (3)
Kirk (28)

O. BAYESIAN STATISTICS.

Raiffa (38)
Schmitt (42)
Winkler (53)

1. Alder, H. L., and Roessler, E. B., *Introduction to Probability and Statistics*, 6th ed. W. H. Freeman, San Francisco, Calif., 1977.
2. Anderson, R. L., and Bancroft, T. A., *Statistical Theory in Research*. McGraw-Hill, New York, 1952.
3. Barnett, V., *Comparative Statistical Inference*. John Wiley, New York, 1973.
4. Bennett, C. A., and Franklin, N. L., *Statistical Analysis in Chemistry and the Chemical Industry*. John Wiley, New York, 1954.
5. Beyer, W. H., Ed., *Handbook of Tables for Probability and Statistics*, 2nd ed. Chemical Education Rubber Co., Easton, Pa., 1968.
6. Brownlee, K. A., *Statistical Theory and Methodology in Science and Engineering*, 2nd ed. John Wiley, New York, 1965.
7. Clague, E., *The Bureau of Labor Statistics*. Frederick A. Praeger, New York, 1968.
8. Cochran, W. G., *Sampling Techniques*, 2nd ed. John Wiley, New York, 1963.
9. Cochran, W. G., and Cox, G. M., *Experimental Designs*, 2nd ed. John Wiley, New York, 1957.
10. Davies, O. L., *Design and Analysis of Industrial Experiments*, 2nd ed. Hafner Publishing Co., New York, 1963.
11. Dixon, W. J., and Massey, F. J., Jr., *Introduction to Statistical Analysis*, 3rd ed. McGraw-Hill, New York, 1969.
12. Draper, N. R., and Smith, H., *Applied Regression Analysis*. John Wiley, New York, 1966.
13. Feller, W., *An Introduction to Probability Theory and Its Application*, Vol. 1, 3rd ed. John Wiley, New York, 1968.
14. Freund, J. E., *Modern Elementary Statistics*, 4th ed. Prentice-Hall, Englewood Cliffs, N.J., 1973.
15. Freund, J. E., *Statistics, A First Course*, 2nd ed. Prentice-Hall, Englewood Cliffs, N.J., 1976.
16. Grant, E. L., and Leavenworth, R. S., *Statistical Quality Control*, 4th ed. McGraw-Hill, New York, 1972.
17. Haber, A., and Runyon, R. P., *General Statistics*, 2nd ed. Addison Wesley, Reading, Mass., 1973.
18. Hald, A., *Statistical Tables and Formulas*. John Wiley, New York, 1952.
19. Hamberg, *Statistical Analysis for Decision Making*. Harcourt Brace Jovanovich, New York, 1970.
20. Hauser, P. M., *Social Statistics in Use*. Russell Sage Foundation, New York, 1975.
21. Heermann, E. F., Braskam, L. A., *Readings in Statistics for the Behavioral Sciences*. Prentice-Hall, Englewood Cliffs, N.J., 1970.
22. Hicks, C. R., *Fundamental Concepts in the Design of Experiments*. Holt, Rinehart, and Winston, New York, 1964.
23. Hoel, P. G., *Introduction to Mathematical Statistics*, 4th ed. John Wiley, New York, 1971.
24. Hogg, R. V., and Craig, A. T., *Introduction to Mathematical Statistics*, 3rd ed. MacMillan, New York, 1970.
25. Hollander, M., and Wolfe, D. A., *Nonparametric Statistical Methods*, John Wiley, New York, 1973.
26. Huff, D., *How to Lie with Statistics*. W. W. Norton, New York, 1954.
27. Juran, J. M., and Gryna, F. M., Jr., *Quality Planning and Analysis*. McGraw-Hill, New York, 1970.

28. Kirk, R. E., *Statistical Issues*. Brooks/Cole Pub. Co., Monterey, Calif., 1972.
29. Mendenhall, W., *Elementary Survey Sampling*. Duxbury Press, North Scitvate, Mass., 1971.
30. Mendenhall, W., *Introduction to Probability and Statistics*, 4th ed. Wadsworth Publishing Co., Belmont, Calif., 1975.
31. Miller, I., and Freund, J. E., *Probability and Statistics for Engineers*. Prentice-Hall, Englewood Cliffs, N.J., 1965.
32. Mood, A. M., et al., *Introduction to the Theory of Statistics*, 3rd ed. McGraw-Hill, New York, 1974.
33. Moroney, N. J., *Facts From Figures*, 3rd ed. Penguin Books, Baltimore, 1956.
34. Mosteller, F., Rourke, R. E. K., and Thomas, G. B., Jr., *Probability and Statistics*, 2nd ed. Addison Wesley, Reading, Mass., 1970.
35. Owen, D. B., *Handbook of Statistical Tables*. Addison Wesley, Reading, Mass., 1962.
36. Parzen, E., *Modern Probability Theory and Its Applications*. John Wiley, New York, 1960.
37. Pearson, E. S., and Hartley, H. O., *Biometrika Tables for Statisticians*, 3rd ed. Cambridge University Press, New York, 1966.
38. Raiffa, H., *Decision Analysis*. Addison Wesley, Reading, Mass., 1968.
39. RAND Corporation, *A Million Random Digits with 100,000 Normal Deviates*. The Free Press, Macmillan Corp., New York, 1955.
40. Romig, H. G., *50–100 Binomial Tables*. John Wiley, New York, 1953.
41. Roscoe, J. T., *Fundamental Research Statistics for the Behavioral Sciences*, 2nd ed. Holt, Rinehart, and Winston, New York, 1975.
42. Schmitt, S. A., *Measuring Uncertainty: An Elementary Introduction to Bayesian Statistics*. Addison Wesley, Reading, Mass., 1969.
43. Scott, A. H., *Census, U.S.A.* The Seabury Press, New York, 1968.
44. Siegel, S., *Nonparametric Statistics for the Behavioral Sciences*. McGraw-Hill, New York, 1956.
45. Slonim, M. J., *Sampling*. Sinom and Schuster, New York, 1960.
46. Snedecor, G. W., and Cochran, W. G., *Statistical Methods*, 6th ed. Iowa State University Press, Ames, Iowa, 1967.
47. Spear, M. E., *Practical Charting Techniques*. McGraw-Hill, New York, 1969.
48. Sudman, S., *Applied Sampling*. Academic Press, New York, 1976.
49. Tanur, et al., *Statistics: A Guide to the Unknown*. Holden-Day, Inc., San Francisco, 1972.
50. Walker, H. M., and Lev, J., *Elementary Statistical Methods*, 3rd ed. Holt, Rinehart, and Winston, New York, 1969.
51. Weinberg, G. M., and Schumaker, J. A., *Statistics: An Intuitive Approach*, 3rd ed., Wadsworth Publishing Co., Inc., Belmont, Calif., 1974.
52. Wilks, S. S., *Elementary Statistical Analysis*. Princeton University Press, Princeton, N.J., 1948.
53. Winkler, R. L., *Introduction to Bayesian Inference and Decision*. Holt, Rinehart and Winston, New York, 1972.
54. Zeisel, H., *Say it with Figures*, 5th ed. Harper & Row, New York, 1968.

ANSWERS TO MOST ODD PROBLEMS AND PORTIONS OF SOME EVEN PROBLEMS

1.1 For example: Has good flavor, aroma, sweetness; also, different grinds, different perking times, water temperature, amount of coffee. Perk coffee for different lengths of time using different amounts of coffee and have several people taste it.

1.3 a. Amount of reading (how many books), using a poll by librarian. Reading speed, using a test by a teacher.
b. Number, or proportion, of people wearing glasses, survey by an insurance company. Types of frames, survey by an eyeglass manufacturer.
c. Number of crimes reported, using state crime study. Impact of crime on community, using opinion survey.

1.5 For example: Long life, if bulbs are in out-of-the-way places; regular in other places. Wattage is associated with cost of operation. In some places the voltage is high and low voltage bulbs will burn out quickly. Amount of light (lumens) would be specific measure of the ability of the bulb to light up a particular area to a particular degree. Each requires a different instrument (clock, wattmeter, voltmeter, light meter).

1.7 Precision or fineness of the measuring instrument. Nonuniformity of the individual item being measured. Reading to reading variation in the instrument system. Sampling error.

1.11

a.

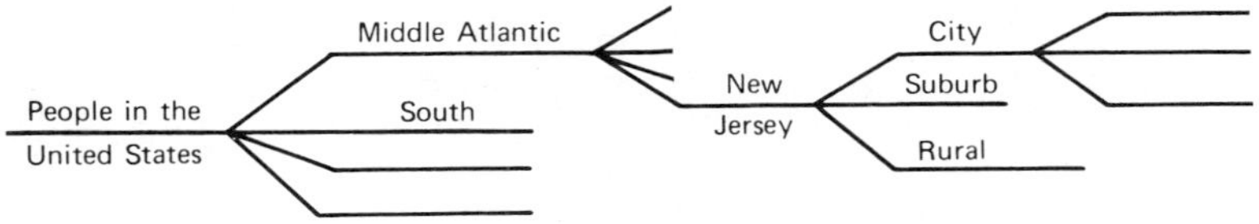

b.

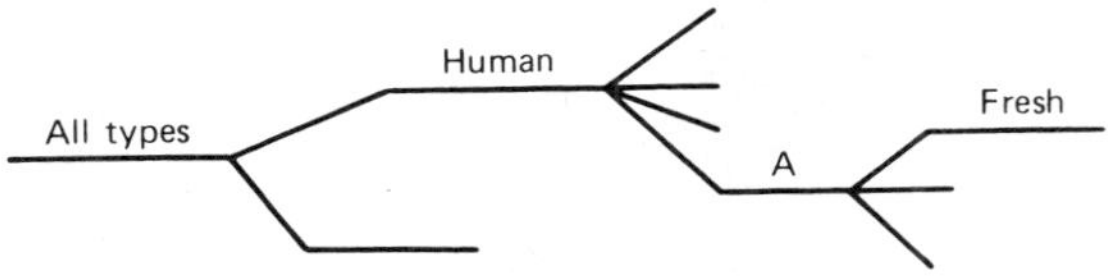

c.

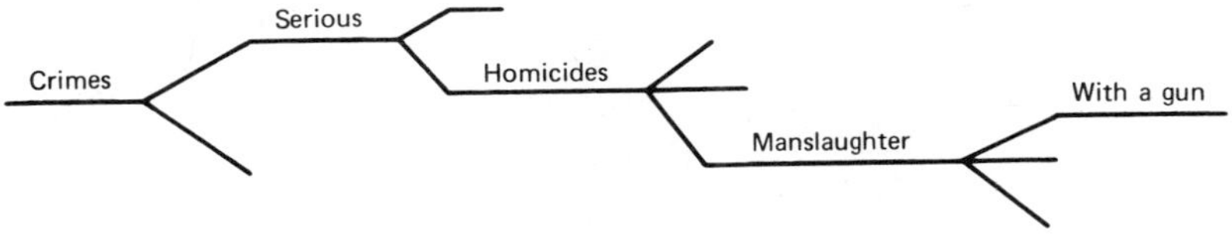

1.13 a. (1) a psychologist; (2) measure response time for a particular learning task; (3) all persons; (4) tiredness and illness not related to the experiment, different people might react very differently to the drugs; (5) might conclude that a particular drug helps learning when in fact it does not, and this might lead to a poor recommendation regarding its use; (6) if a person volunteered, determine why. Must make sure that there are control persons who are not using the drugs and the person using the drug should not be aware that he or she is definitely the one with the drug (must select the user randomly).

b. (1) a medical researcher and an insurance actuary; (2) noise level in a room (ordered), hearing (ordered); (3) all who attend rock performances; (4) different locations in the room and different people react differently; (5) could suggest no effect when damage will result and many people may loose some hearing; (6) cannot necessarily obtain random individuals (and randomize their seating arrangement)—would you go?

2.1

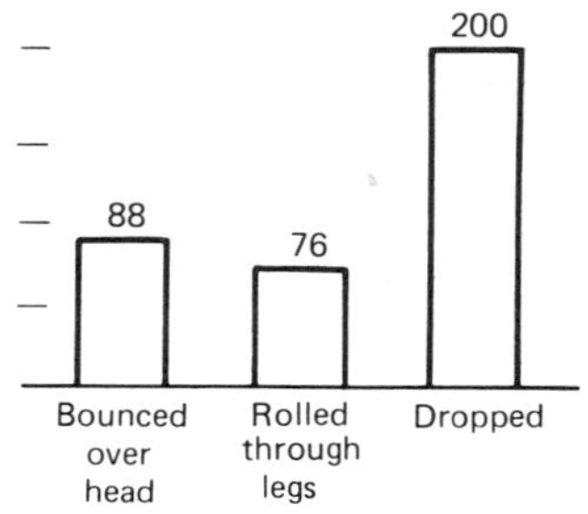

2.3

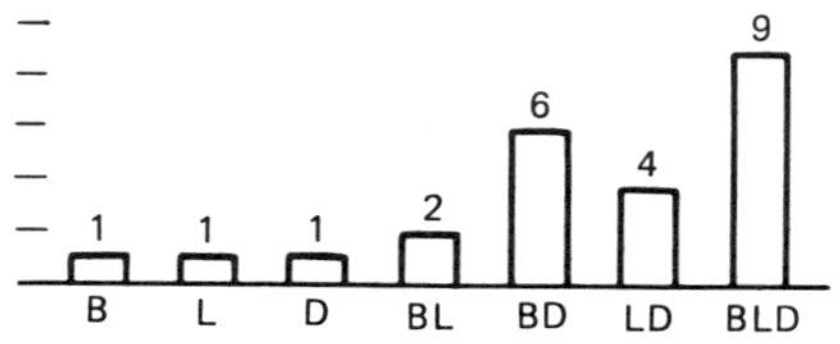

2.5

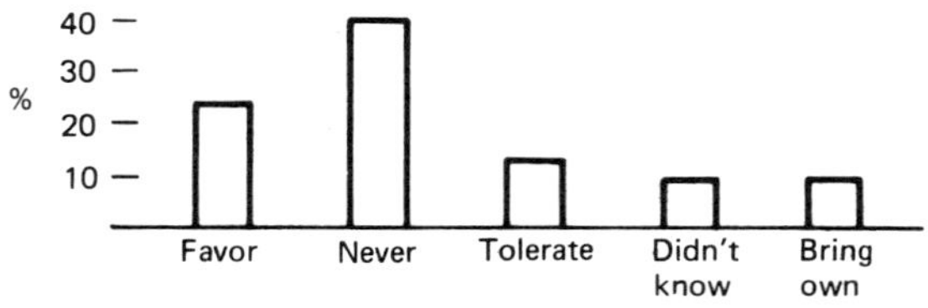

2.7

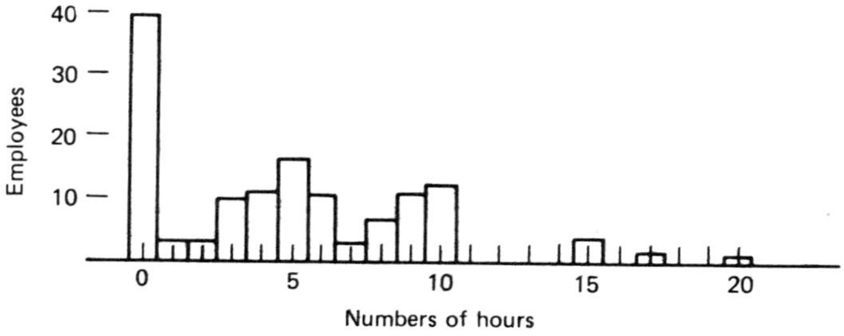

2.9 a. Generally very appropriate.

b. Probably only would be two categories—5 and 6—not enough.

c. Would probably be okay.
d. Everyone would be at a different location and we would probably need about 1500 places.

2.13

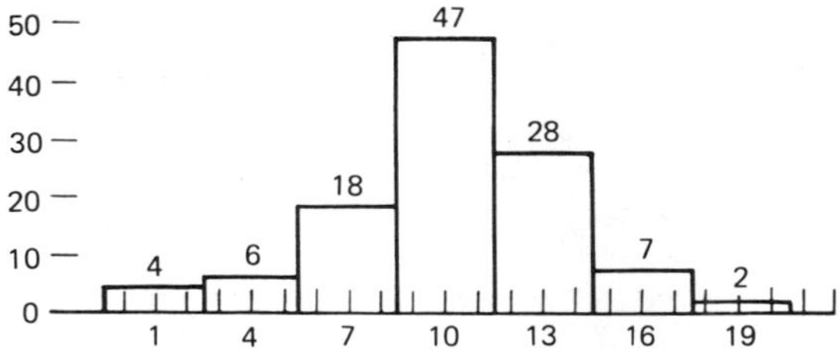

2.15 The histogram depends on the intervals selected. For example, I used the following

INTERVAL	MIDPOINT	FREQUENCY
19.955–64	19.9595 $\simeq$ 19.960	4
19.965–74	19.970	8
19.975–84	19.980	14
19.985–94	19.990	15
19.995–20.004	20.000	14
20.005–14	20.010	15
20.015–24	20.020	13
20.025–34	20.030	7
20.035–44	20.040	5
20.045–54	20.050	4
20.055–64	20.060	1

(Note that I have rounded up the midpoints in all cases.)

2.17

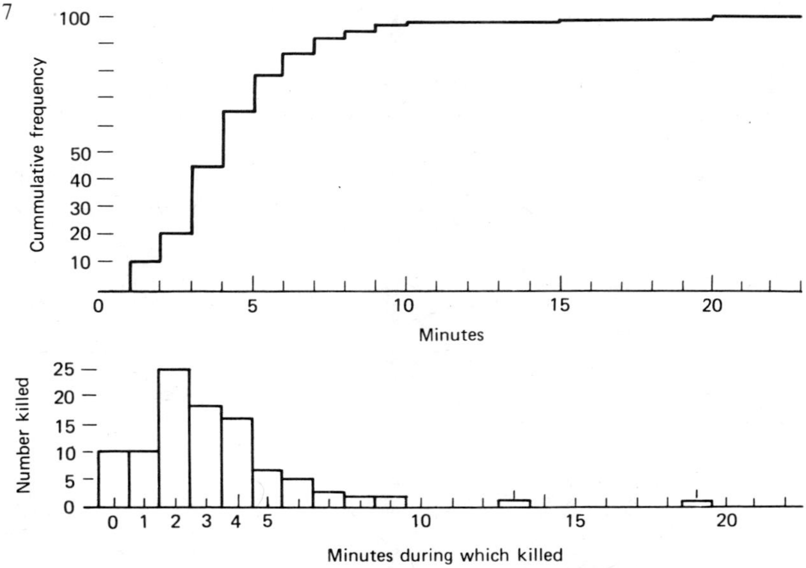

2.19

HOURS	CUMULATIVE PERCENT	HOURS	CUMULATIVE PERCENT
0	31.2	7	73.6
1	33.6	8	78.4
2	36.0	9	86.4
3	43.2	10	96.0
4	51.2	15	98.4
5	63.2	17	99.2
6	71.2	20	100.0

2.21 19.994, 20.007, 0.53

PERCENTILE	VALUE	PERCENTILE	VALUE
5	19.965	55	20.004
10	19.970	60	20.007
15	19.975	65	20.011
20	19.982	70	20.013
25	19.984	75	20.018
30	19.986	80	20.022
35	19.989	85	20.025
40	19.994	90	20.033
45	19.999	95	20.041
50	20.002	100	20.055

2.23 a. 25th.
b. 75th, Q_3.
c. 30th.
d. 60th.
e. P_{91} to P_{100}.
f. $Q_3 - Q_1$.

2.25 a. Skewed to left; most should do well.
b. Skewed to right; most will do very poorly.
c. Bimodal—2nd graders do poorly, freshmen do well.
d. Slightly skewed to right; 2nd graders do poorly and freshmen may do above poor (perhaps).

3.1

	MEDIAN	MODE	MIDRANGE	MEAN
a.	5	none	7.5	6.6
b.	6.5	8	7	6.33
c.	5.5	none	13.5	8
d.	3.5	none or 1 and 4	3	3
e.	3	3	7.5	4
f.	1	1	1.5	1.2

3.3 Median = 2, mode = 0, midrange = 4.5, mean = 2.82.
3.5 Median = 15, mean = 14.9.
3.7 Median = 49665, mode none, midrange = 49345, mean = 49478.
3.9 Mean = 1.28.
3.11 Must convert to numbers; mean = 2.25, median = mode = midrange = 2. All except the mean are about the same difficulty.
3.13 Can obtain a mode—never eat there. Data not even ordered; therefore no other "averages" can be found.
3.15 Mean = 4.6, median = 4, 3 modes—0, 5, and 12, midrange = 10. None really shows the center, since the distribution is unusually shaped.
3.17 Mean = 3.67, median = 3.5.
3.19 Depends on intervals, using 8 intervals of $500 each with midpoints starting at $47,500. Mean = 49536. Slight difference.
3.21 Using intervals of $0.05 starting at 0.85. $\overline{X} = 1.28$. No change.
3.23 Using the intervals in Problem 2.15, $\overline{X} = 20.0025$.
3.25 a. 11.
b. 10.3.
3.27 19.995 + 0.007 = 20.002.
3.29 3.25 + 0.39 = 3.64.
3.31 $2.875.
3.35

	RANGE	VARIANCE	STANDARD DEVIATION
a.	11	19.3	4.4
b.	10	13.9	3.72
c.	27	100	10
d.	6	4.57	2.14
e.	23	31	5.6
f.	1	0.18	0.4

3.37 Range = 9, $s^2 = 7.36$, $s = 2.7$.
3.39 Range = 11, $s^2 = 9.99$, $s = 3.2$.
3.41 $s = 917$, range = 2890, variance = 814141.
3.43 Range = 0.74, variance = 0.027, $s = 0.17$.
3.45 Range = 3.5, variance = 0.465, $s = 0.68$.
3.47 Range = 20, variance = 18.38, $s = 4.29$.
3.49 Variance = 0.000552, $s = 0.024$.
3.51 Variance = 902473, $s = 950$.
3.53 Variance = 0.0256, $s = 0.16$.
3.55 a. 3.52.
b. 3.
c. 6.83.
d. 1.75.
e. 3.7.
f. 0.32.
3.57 3.4, −2.4, 0, 157, 131.5, 148.5.
3.59 19.054, 20.050.
3.61 1 sigmas 0 versus 68%
3 sigmas 89 versus 99.74%

4.1 a. Subjective (possibly relative frequency).
b. Relative frequency.
c. Mathematical.
d. Subjective.
e. Relative frequency.

4.3 0 trues, 1 true, . . . up to 12 trues.

4.5 0.01, 0.001, 0.000001, 0.0000005, 0.0.

4.7 b. 0.380
c. 0.048
d. 0.143
e. 0.476
f. 0.476
g. 0.524
h. 0.238
i. 0.625
j. 0.286
k. 1.000

4.9 0.375, 0.35, 0.275.

Democrat	3	2	2	1	1	1	0	0	0	0
Republican	0	1	0	1	2	0	3	2	1	0
Not vote	0	0	1	1	0	2	0	1	2	3

4.11 a. 0.381
b. 0.952
c. 0.037
d. 0.0023
e. 0.286
f. 0.095
g. 0.190

4.13 a. 0.053.
b. 0.043.
c. 0.904.

4.15 0.9996.

4.17 52%.

4.19 a. 0.011.
b. 0.0579.
c. 0.254.

4.21 a.

Cost	21	22	23	24	25	26	27
Frequency	1	2	3	3	3	2	1

b. No, 20% of the cases will be $24.
c. Impossible, wrong book, or gypped.

4.23 For $24 the new population has about 12% chance and the original population has a 20% chance of $24.
For $28 about a 12% chance with the new population but impossible with the original population.

5.1 a. 0.0793.
c. 0.4292.
e. 0.6406.
g. 0.0778.
i. 0.9332.
k. 0.0918.

m. 0.6480.
o. 0.9582.
q. 0.8643.
s. 0.8665.
u. 0.7454.
w. 0.5000.

5.2 a. 0.25.
c. −0.319 to +0.319.
e. −0.93 to +0.93.
g. −0.674.
i. −0.842.

5.3 a. 0.1915.
c. 0.1586.
e. 0.2325.
g. 0.8664.
i. 0.0994.
k. 0.4400.

5.5 0.0918, $15.08.

5.7 a. 4.75%.
b. 19.71%.
c. 93.8.

5.9 0.0475, 0.4325, 0.3642.

5.11 a. 44%.
b. 24.1.
c. 12.4.

6.1 (a) For example, the two dies are the same or different; the sum is 7, or not 7. (b) For example, the party I call is there or is not there; I get the correct number or a wrong number.

6.3 Not replacing means the population is changed and p is not necessarily the same for the next sample.

6.5 Include the noncolors along with one of the other two categories (red or blacks). If you put 0 with red and 00 with black you have a 50–50 situation.

6.6 a. 0.3125.
c. 0.0413.
e. 0.7593
g. 0.1329, 0.0000, 0.2749
i. 0.0433.

6.7 a. 30%.
b. 0.1681.
c. 0.0282.
d.

None–0.0000	1–0.0001	2–0.0014
3–0.0090	4–0.0368	5–0.1029
6–0.2001	7–0.2668	8–0.2335
9–0.1211	10–0.0282	

6.9 a. 0.0625.
b. 0.0000.
c. 0.0207.
d. 0.9793.
e. 0.0000.

6.11 a. 0.2393.
b. 0.0026.
c. 0.0957
d. 0.9974.

6.13 5–0.0024 4–0.0284 3–0.1323
2–0.3087 1–0.3602 0–0.1681

6.15 a. 3.5.
b. 7, 1.45.
c. 7000.
d. Between 6908 and 7092.

6.17 10, 50, 5.

6.19 160, between 143 and 177.

6.21 0.1379, no.

6.23 0.0019.

7.3 $\bar{s} = 1.67$, $\bar{\sigma} = 1.45$.

7.5 $\mu = 3.5$, $\sigma^2 = 2.92$.

7.7 a. 3.54.
c. 2.5.
e. 10.
g. 0.03.

7.8 a. −7.08 to +7.08.
c. 995 to 1005.
e. 530 to 570.
g. 0.44 to 0.56.

7.9 a. 5, 15.54%
c. 2, 49.38%.
e. 0.04, 78.88%.
g. 71, 67.87%

7.11 0.0793.

7.13 51.3.

7.15 3.55 to 3.79.

7.17 98 to 102.

7.19 36.6 to 59.4

8.1 The person is *not* guilty.
Alternate: The person is guilty.
Errors: Decide an innocent person is guilty, decide a guilty person is innocent.
Consequence: Imprison an innocent person, let guilty person go free. Society generally considers the first as a more serious error.

8.3 You have cancer. You do not have cancer. The second is preferred.
Errors: Declare no cancer when you do have it (beta error).
Declare you have cancer when you do not (alpha error).
Consequence: May miss opportunity to treat it before it spreads.
May operate when unnecessary.

8.5 No change in yield. Alternate: There is an increase.
Not concerned with a decrease, a one-sided, one-directional test.

8.7 No. Hypothesis: Percent germinating is no different than 90%.
Alternate: Less than 90%, one direction.
85% of planted seeds germinate. The statistic would be the number of seeds germinating.

Errors: Can agree that 90% germinate but actually less germinate (beta error).
Can say that less than 90% germinate but actually 90% germinate (alpha error).

8.9 α would mean return good ammunition to store; β would mean keep bad ammunition.
If in a war the army has lives involved; a sportsman, perhaps at most may miss a duck or a deer.

8.11 Hypothesis: There is a very small proportion of bad nails (no difference than 1%).
Alternate: Proportion of bad nails is large.
Limit(s): 2
Test statistic: Number of bad nails in 30.
Decisions: Small proportion of bad nails.
Large proportion of bad nails.
Errors: α Get 2 or more bad in a good lot; don't buy good nails.
β Get 0 or 1 bad nail in a bad lot; buy bad nails.

8.13 Hypothesis: Texts cost no different than $13.
Alternate: Texts are different than $13.
Limits: 11.65 to 14.35.

8.15 Limits: 3.96 to 6.04, reject.

8.17 0.2119.

8.19 0.4602, 0.9916.

8.21 228.1 and 281.9; 0.1357; 0.0050.

9.1 Limit: $z = \pm 1.96$, $z = 4$, yes.

9.3 $z_{0.95} = 1.645$, $z = 1.22$, not significant, $-\$21,000$.

9.5 a. 62. b. 27.

9.7 87 days, 22 days.

9.9 a. $t = 1.93$, 3 *df* not significant at either.
c. $t = 2.26$, 5 *df*, significant at 0.10, not 0.01 or 0.05.
e. $t = 3.02$, 8 *df*, not significant at 0.01, significant at 0.05, significant at 0.02.

9.10 a. $t = 1.75$, 4 *df*, not significant at 0.10.
c. $t = 3.99$, 8 *df*, significant at 0.01.
e. $t = 2.07$, 9 *df*, not significant at 0.05, not significant at 0.01, significant at 0.10.

9.11 $t = 1.64$, not significant.

9.13 Limit: $t = \pm 2.262$, $t = -2.90$, significant, 44.7 to 49.3.

9.15 $t = 3.44$, very significant.

9.17 $z = 1.85$, one-sided test, limit $z = 1.645$.

9.19 $t = 3.87$, significant.

9.21 $t = 2.71$, significant.

9.23 $t = 4.66$, very significant.

9.25 $t = -0.94$, not significant.

9.27 $t = 1.87$, significant.

9.29 $t = 3.80$, highly significant.

9.31 a. 18.3.
b. 1.15.
c. 18.5.
d. 16.5 to 34.4.
e. Must go to another source—226.

9.33 $(n-1)s^2/\sigma^2 = 10.03$, $\chi^2_{0.05} = 14.1$, not significant.
9.35 Test statistic = 101.25, $\chi^2 = 30.6$, very significant.
9.37 Test statistic = 2.66. 0.10 limit = 10.9, significant.
9.39 a. 5.41.
b. 2.71.
c. 4.56.
d. 2.78.
e. 0.127.
f. 0.0002.
g. 0.070, 4.15.
9.41 $F = 10$, significant.
9.43 $F = 39.0$ significant.
9.45 $F = 1.78$, not significant. $t = 3.06$, significant.
10.1 a. Limit: $z_{0.99} = 2.33$; test value $z = 2.2$, significant at 0.025.
b. 18.0 to 36.6%.
c. 1080 to 2196.
10.3 Test limits = $z = \pm 1.96$, $z = 1.886$, not significant.
10.5 0.01, z-limit = -1.28, $z = -1$, not significant.
10.7 $z = 2.05$, yes; 95% confidence interval 11.8 to 17.6.
10.9 1024.3.
10.11 4160.
10.15 $z = 1.75$, not significant.
10.17 $z = 1.03$, not significant.
10.19 Two-sided test, $p = 0.1094$, not unusual.
10.21 7 Plus, 7 minus, 1 tie, not unusual.
10.23 At 0.01 level, expect from 13.9 to 31.1, significant change.
11.1 $\chi^2 = 2.68$, not significant.
11.3 $\chi^2 = 53.09$, very significant.
11.5 Using all categories, $\chi^2 = 46.45$, very significant.
Grouping $\chi^2 = 28.97$, still very significant.
11.7 $\chi^2 = 12.03$, very highly significant.
11.9 $\chi^2 = 22.50$, very significant.
11.11 $\chi^2 = 82.66$, very significant.
11.13 $\chi^2 = 1.97$, not significant.
11.15 $U = 14$, $U' = 34$, not significant.
11.17 $U = 33$, $U' = 44$, not significant.
11.19 $U = 169.5$, $z = 0.76$, not significant.
11.21 $r_{10} = 0.296$, not significant.
11.23 $r_{21} = 0.556$, not significant at 0.05.
11.25 Two sided $r_{21} = 0.333$ and 0.364, not significant.
11.27 $r_{22} = 0.319$, not significant.
12.1 MS between = 11.58; MS within = 1.17; $F = 9.93$. Significant at 0.01, $F_{0.99}(2,9) = 8.02$.
12.3 MS between = 6.67; MS within = 5.92; $F = 1.12$, not significant.
12.5 MS between = 0.84; MS within = 1.12; $F = 0.75$, not significant.
12.7 MS between = 175.9; MS within = 35.85; $F = 4.90$. Significant at 0.01, $F_{0.99}(4,20) = 4.43$.
12.11 MS between cars = 3.67; MS between brands = 6.67; MS residual = 6.67. Both F ratios are not significant.

12.13 MS years = 2.88; MS industry = 0.84; MS residual = 0.53. F for years = 5.40, not significant, F for industry = 1.58, not significant.

13.1 a. $Y_{\text{predicted}} = -0.69 + 1.27X$; for $X = 2$, $Y = 1.85$; for $X = 7$, $Y = 8.20$.
c. $Y_{\text{predicted}} = 1.41 + 0.54X$; 2.49; 5.19.
e. $Y_{\text{predicted}} = 7.70 - 0.63X$; 6.44; 3.29.

13.2 a. $X_{\text{predicted}} = 1.10 + 0.64Y$; for $Y = 2$, $X = 2.38$; for $X = 7$, $Y = 5.58$.
c. $X_{\text{prediced}} = 2.93 + 0.30Y$; 3.53; 5.03.
e. $X_{\text{predicted}} = 9.07 + 0.63X$; 7.81, 4.66.

13.3 Y = rating = $-0.55 + 1.01X$; 7.03; 4.5.

13.5 Y = staples = $27.6 - 1.3X$; 14.43; 1.22.

13.7 $Y = -2670.6 + 0.526X$, 2589.4, 23629.4, -1618.6.

13.9 a. $t = 4.84$, t for two tail 0.01 level = 4.032, very significant.
c. $t = 0.992$, not significant.
e. $t = -1.797$ two tail 10% level $t = \pm 2.015$, not significant.

13.10 a. $t = 4.57$, two tail 0.01 level $t = 4.032$, significant.
c. $t = 0.975$, not significant.
e. $t = -1.83$, not significant.

13.11 $t = 8.66$, two sided 0.01 t-value = 4.140, very significant.

13.13 $t = -13.23$, very highly significant.

13.15 $t = 21.61$, very significant.

13.17 a. 0.34 − 3.36; 4.37 − 12.03.
c. −1.47 − 6.45; −3.76 − 14.14.
e. 0.67 − 12.21; −6.91 − 13.49.

13.19 3.12 and 7.90.

13.21 a. 0.906.
c. 0.402.
e. −0.630.

13.22 See Problem 13.21.

13.23 $r = 0.918$.

13.25 $r = -0.978$.

13.27 $r = 0.989$.

13.29 a. Significant at 0.01, 82%.
c. Not significant, 16.2%.
e. Not significant, 39.7%.

13.31 Significant at 0.001, 84.3%

13.33 Significant at 0.01.

13.35 Significant at 0.001.

13.37 $r = 0.688$, not significant.

13.39 $Y_{11} = 10.58$. This is larger than the maximum possible reading. Rating = $3.32 + 0.76T$.

B.1 8.

B.3 30.

B.5 13.

B.7 20.

B.9 2.72.

B.11 317.

B.13 92.

INDEX

INDEX